AF411816

Springer Optimization and Its Applications

Aims and Scope
Optimization has been expanding in all directions at an astonishing rate during the last few decades. New algorithmic and theoretical techniques have been developed, the diffusion into other disciplines has proceeded at a rapid pace, and our knowledge of all aspects of the field has grown even more profound. At the same time, one of the most striking trends in optimization is the constantly increasing emphasis on the interdisciplinary nature of the field. Optimization has been a basic tool in all areas of applied mathematics, engineering, medicine, economics and other sciences.

The series *Springer Optimization and Its Applications* publishes undergraduate and graduate textbooks, monographs and state-of-the-art expository works that focus on algorithms for solving optimization problems and also study applications involving such problems. Some of the topics covered include nonlinear optimization (convex and nonconvex), network flow problems, stochastic optimization, optimal control, discrete optimization, multi-objective programming, description of software packages, approximation techniques and heuristic approaches.

More information about this series at http://www.springer.com/series/7393

Ramteen Sioshansi · Antonio J. Conejo

Optimization in Engineering

Models and Algorithms

 Springer

Ramteen Sioshansi
Department of Integrated Systems
  Engineering
The Ohio State University
Columbus, OH
USA

Antonio J. Conejo
Department of Integrated Systems
  Engineering and Department of Electrical
  and Computer Engineering
The Ohio State University
Columbus, OH
USA

ISSN 1931-6828   ISSN 1931-6836   (electronic)
Springer Optimization and Its Applications
ISBN 978-3-319-56767-9   ISBN 978-3-319-56769-3   (eBook)
DOI 10.1007/978-3-319-56769-3

Library of Congress Control Number: 2017937140

Mathematics Subject Classification (2010): 90-01, 90C05, 90C10, 90C11, 90C30, 90C39, 90C46, 90C90

Printed on acid-free paper

This Springer imprint is publshed by Springer Nature
The registered company is Springer International Publishing AG
The registered company address is: Gewerbestrasse 11, 6330 Cham, Switzerland

To my parents and Pedrom
To Núria

Preface

Optimization in Engineering: Models and Algorithms covers the fundamentals of optimization, including linear, mixed-integer linear, nonlinear, and dynamic optimization techniques, with a clear engineering focus.

This book carefully describes classical optimization models and algorithms using an engineering problem-solving perspective. Modeling issues are emphasized using many real-world examples, related to a variety of application areas.

This textbook is intended for advanced undergraduate and graduate-level teaching in industrial engineering and other engineering specialties. It is also of use to industry practitioners, due to the inclusion of real-world applications.

This book consists of five chapters and two appendices.

Chapter 1 introduces the book, providing an overview of the major classes of optimization problems that are relevant to the engineering profession.

Chapter 2 considers linear optimization problems. The practical significance of these problems is first shown through a number of examples. Precise formulations of these example problems are provided. The geometric and algebraic features of generic linear optimization problems are then analyzed. A well-known solution algorithm is next described, which is based on the algebraic features of linear optimization problems. Sensitivity analysis and the duality properties of linear optimization problems are then described.

Chapter 3 considers mixed-integer linear optimization problems. That is, problems that are linear in the decision variables but include variables that are restricted to take on integer values. The chapter provides a number of illustrative examples, showing the practical significance of mixed-integer linear optimization. Then, a general formulation of the mixed-integer linear optimization problem is provided. Next, the use of binary variables to model various types of nonlinearities and discontinuities, while maintaining a linear model structure, is described. This use of binary variables shows the true power of mixed-integer linear optimization models. Two solution techniques for mixed-integer linear optimization problems are then explained: branch-and-bound and cutting planes. The former is developed for general mixed-integer linear optimization problems, the latter for pure-integer linear optimization problems.

Chapters 4 and 5 consider nonlinear optimization problems. Chapter 4 begins with numerous application examples demonstrating the use of nonlinear optimization. This chapter then introduces the use of optimality conditions, which is an analytical approach to solving nonlinear optimization problems. Optimality conditions are properties that an optimal solution exhibits and can be used to find candidate optimal solutions to general nonlinear optimization problems.

Chapter 5 considers a generic iterative algorithm for solving unconstrained nonlinear optimization problems. Details on the two principal steps of this generic algorithm are then provided. Next, this chapter explains two ways in which the generic algorithm can be extended to solve constrained nonlinear optimization problems. One implicitly accounts for the constraints by incorporating them into the objective function through 'penalties', while the other explicitly accounts for problem constraints.

Chapter 6 covers dynamic optimization techniques. This chapter begins by first introducing a simple problem that can be formulated and solved using dynamic optimization techniques. The common elements that must be identified when formulating a dynamic optimization problem are then described. These elements 'encode' the structure and dynamics of the system being optimized. A standard algorithm used to solve dynamic optimization problems is then described.

Appendix A reviews Taylor approximations and definite matrices while Appendix B reviews convex sets and functions. The concepts covered in these appendices are vital for Chapters 4 and 5. The appendices are provided for readers needing to refresh these topics.

The material in this book can be arranged to address the needs of undergraduate teaching in two one-semester courses. Chapters 1–3 include the material for a linear and mixed-integer linear optimization course. On the other hand, Chapters 1, 4–6 and Appendices A and B include the material for a nonlinear and dynamic optimization course. Alternatively, Chapter 1 and Sections 2.1, 3.1, 3.3, and 4.1 of Chapters 2–4 could be used in a course that introduces optimization modeling *only*, without covering solution methods or theory.

The book provides an appropriate blend of practical applications and optimization theory. This feature makes the book useful to practitioners and to students in engineering, operations research, and business. It also provides the reader with a good sense of the power of optimization and the potential difficulties in applying optimization to modeling real-world systems. The algorithms and solution methods are developed intuitively and examples are used to illustrate their use. This text is written to avoid mathematical formality while still maintaining sufficient rigor for the reader to understand the subtleties of optimization. GAMS codes for many of the examples introduced in the text are also provided. These can be readily used and expanded upon by the reader. The aim is for the reader to be able to confidently apply the modeling and solution methodologies to problem domains that go beyond the examples used in the text. Overall, the reader benefits by understanding optimization, learning how to formulate problems, solve them, and interpret solution outputs. This is done using a problem-solving engineering approach.

This book opens the door to advanced courses on both modeling and algorithm development within the industrial engineering and operations research fields.

To conclude, we thank our colleagues and students at The Ohio State University for insightful observations, pertinent corrections, and helpful comments.

<table>
<tr><td>Columbus, Ohio, USA</td><td>Ramteen Sioshansi</td></tr>
<tr><td>February 2017</td><td>Antonio J. Conejo</td></tr>
</table>

Contents

Chapter 1
Optimization is Ubiquitous

This chapter introduces the relevance of **optimal thinking** and **optimization modeling** and describes the structure of the most common types of optimization problems that can be found in real-world practice. Importantly, optimization is ubiquitous in industry and life. Optimization problems formalize optimal decision making using mathematical models that describe how a system behaves based on decisions made by the decision maker. After this introduction to the concept of optimization, the chapter concludes by listing the types of optimization problems commonly encountered in the real world and outlines the types of problems discussed in the subsequent chapters of this text.

1.1 Industry (and Life!) is Optimization

Optimization is key to any industrial process. This is because optimization pursues the best way to achieve a particular **objective** (or multiple objectives) related to that process under a number of resource or other **constraints**. Optimization is key to life as well, as people (typically) strive to achieve the best for their lives under a number of personal and environmental constraints.

The field of optimization formalizes this concept of optimal thinking by making the process quantitative. This allows a decision maker to make well-informed decisions based on an objective numerical metric. This formalization uses what are known as mathematical **optimization problems** (which are also, for historical reasons, referred to as **mathematical programming problems**). As seen throughout this textbook, optimization problems can vary considerably in their complexity. Our use of the term 'complexity' here can refer to how complicated the structure of an optimization model is or the difficulty involved in solving a model to determine an optimal set of decisions.

© Springer International Publishing AG 2017
R. Sioshansi and A.J. Conejo, *Optimization in Engineering*,
Springer Optimization and Its Applications 120, DOI 10.1007/978-3-319-56769-3_1

In this chapter we briefly introduce the different classes of optimization problems. Subsequent chapters describe the structure of these problems and methods to solve them in more detail.

Most optimization problems involve three main elements, which are listed below.

1. The **decision variables** represent the decisions being optimized in the model. All optimization problems require at least one decision variable. Without decision variables there is nothing for the decision maker to decide, and thus no problem to solve.

   An important nuance in determining the decision variables of an optimization problem is that they should not be confused with problem data or parameters. Problem data represent exogenous information that the decision maker cannot choose or control. For instance, if a person is given $\$D$ to spend on either apples or oranges, we could denote x_A and x_O as two decision variables representing the amount spent on apples and oranges, respectively. Note, however, that D is not a decision variable. Rather, this is fixed problem data (otherwise the decision maker may choose to make D arbitrarily large to consume apples and oranges to his or her heart's desire).

2. The **objective function** is the numerical measure of how 'good' the decisions chosen are. Depending on the problem in question, the goal of the decision maker will be to either maximize or minimize this objective. For instance, one may model a problem in which a firm is making production decisions to maximize profit. As another example, one may model a problem in which a public health agency is allocating resources to minimize childhood mortality. Oftentimes, an optimization problem may be referred to as a 'maximization problem' or a 'minimization problem,' depending on the 'direction' in which the objective function is being optimized.

3. The **constraints** impose the physical, financial, or other limits on the decisions that can be made. Although some problems may not have any constraints, most problems often have implicit or explicit constraints. Going back to the example of the person having $\$D$ to spend on either apples or oranges, one explicit constraint is that no more than $\$D$ can be spent, or that:

$$x_A + x_O \leq D. \tag{1.1}$$

   Another pair of implicit constraints is that a non-negative amount of money must be spent on each:

$$x_A \geq 0,$$

   and:

$$x_O \geq 0.$$

   For most optimization problems, constraints can take one of three forms. We have just seen two of these forms—less-than-or-equal-to and greater-than-or-equal-to constraints—in the 'Apples and Oranges' example. Note that in almost all cases, these so-called inequality constraints must always been weak inequalities. Strict inequalities (*i.e.*, strictly less-than or strictly greater-than constraints) can introduce significant technical difficulties in solving an optimization problem. We discuss this point further in Section 2.1.1.

The third type of constraint is an equality constraint. As an example of this, suppose that the decision maker in the Apples and Oranges example must spend the entire D on either apples or oranges. If so, then constraint (1.1) would be changed to:

$$x_A + x_O = D.$$

The set of constraints in an optimization problem define the set of decision variables that can be feasibly chosen by the decision maker. The set of decision variables that are feasible define what is known as the **feasible region** of the problem. Section 2.1.1 discusses this point further and gives a graphical representation of the feasible region of a simple two-variable optimization problem.

Taking the three elements together, most optimization problems can be written in the very generic form:

$$\min_{x_1,\ldots,x_n} f(x_1, \ldots, x_n) \tag{1.2}$$

$$\text{s.t.} (x_1, \ldots, x_n) \in \Omega. \tag{1.3}$$

This generic problem has n decision variables, which are denoted $x_1, \ldots, x_n$. The objective function is $f(x_1, \ldots, x_n)$. Objective functions are always scalar-valued, meaning that they map the values of the n decision variables into a single scalar value that measures how good the outcome is in terms of the objective that the decision maker cares about (*i.e.*, we have that $f : \mathbb{R}^n \to \mathbb{R}$). This generic problem assumes that the objective function is being minimized. This is because of the min operator in objective function (1.2). If the problem was instead aiming to maximize the objective function, this would be denoted by replacing (1.2) with:

$$\max_{x_1,\ldots,x_n} f(x_1, \ldots, x_n).$$

It is customary to list the decision variables underneath the min or max operator. This way, there is no ambiguity regarding what are decision variables in the problem and what are not (*i.e.*, so as not to confuse decision variables with problem data or parameters).

The constraints are represented in the generic problem by (1.3). The abbreviation 's.t.' in (1.3) stands for 'subject to,' and means that the values of $x_1, \ldots, x_n$ chosen must satisfy the set of constraints listed. Here we let Ω denote the feasible region. Unless it is otherwise explicitly stated in a problem, we assume that the decision variables are continuous and can take on any real value that satisfies the constraints. We discuss classes of problems in which this assumption is relaxed and at least some of the decision variable must take on integer values in Section 1.3. We further discuss formulating and solving such problems in Chapters 3 and 6.

A vector of decision variables $(x_1, \ldots, x_n)$ that is in the feasible region, *i.e.*:

$$(x_1, \ldots, x_n) \in \Omega,$$

is said to be a **feasible solution**, whereas a vector that is not, *i.e.*:

$$(x_1, \ldots, x_n) \notin \Omega,$$

is said to be an **infeasible solution**.

Among the feasible solutions, the one (or ones) that optimize (*i.e.*, minimize or maximize) the objective function is said to be an **optimal solution**. We typically denote a set of optimal decisions by using asterisks, *i.e.*:

$$(x_1^*, \ldots, x_n^*).$$

We can finally note that it is customary to represent the generic optimization problem given by (1.2)–(1.3) in the more compact form:

$$\min_{x} f(x)$$

$$\text{s.t. } x \in \Omega,$$

where we define $x = (x_1, \ldots, x_n)$.

To give a simple starting example of an optimization problem, we can return to the Apples and Oranges example given above.

Example 1.1 A person has up to \$$D$ to spend on apples and oranges. Every dollar spent on apples brings her two units of happiness whereas each dollar spent on oranges brings her three units of happiness. Her goal is to maximize her happiness.

To write this optimization problem, we define two decision variables, x_A and x_O, which denote the amount of money spent on each of apples and oranges, respectively. The problem is then written as:

$$\max_{x_A, x_O} f(x_A, x_O) = 2x_A + 3x_O$$

$$\text{s.t. } x_A + x_O \leq D$$

$$x_A \geq 0$$

$$x_O \geq 0.$$

Because this is a problem in which the objective is being maximized, the max operator appears next to the objective function. Note that only x_A and x_O appear below the max operator, whereas D does not. This is because the \$$D$ that the decision maker has to spend is not a decision within her control. Instead, D is problem data.

The problem has three constraints. The first one is explicitly stated in the problem description (*i.e.*, that she only has at most \$$D$ to spend). The other two constraints are implicit, because we know that it is physically meaningless to spend a negative amount of money on apples or oranges. Furthermore, we work under the assumption that x_A and x_O can take on any continuous values (so long as they satisfy the constraints). Thus, for instance, if $D = 10$, then:

$$(x_A, x_O) = (1.00354, 5.23),$$

would be a feasible solution.

The constraints of the problem are explicitly written out, as opposed to being written implicitly via the feasible region. However, we could formulate the problem using the feasible region instead. To do this, we would first define the feasible region as:

$$\Omega = \{(x_A, x_O) | x_A + x_O \leq D, x_A \geq 0, x_O \geq 0\}.$$

The problem would then be formulated as:

$$\max_{x_A, x_O} f(x_A, x_O) = 2x_A + 3x_O$$

$$\text{s.t. } (x_A, x_O) \in \Omega.$$

$\square$

This chapter introduces the following classes of optimization problems, which are studied in subsequent chapters:

1. linear optimization problems,
2. mixed-integer linear optimization problems,
3. nonlinear optimization problems, and
4. dynamic optimization problems.

1.2 Linear Optimization Problems

A **linear optimization problem** or **linear programming problem** (LPP) has the following three important defining characteristics.

1. The decision variables are continuous (*i.e.*, they are not constrained to take on integer values) and thus we have:

$$(x_1, \ldots, x_n) \in \mathbb{R}^n.$$

2. The objective function is linear in the decision variables, and can thus be written as:

$$f(x_1, \ldots, x_n) = c_0 + c_1 x_1 + \cdots + c_n x_n = c_0 + \sum_{i=1}^{n} c_i x_i,$$

where $c_0, \ldots, c_n$ are constants.

3. The constraints are all equal-to, greater-than-or-equal-to, or less-than-or-equal-to constraints that are linear in the decision variables. Thus, the constraints can all be written as:

$$\sum_{i=1}^{n} A_{j,i}^{e} x_i = b_j^{e}, \qquad \forall \, j = 1, \ldots, m_e$$

$$\sum_{i=1}^{n} A_{j,i}^{g} x_i \geq b_j^{g}, \qquad \forall \, j = 1, \ldots, m_g$$

$$\sum_{i=1}^{n} A_{j,i}^{l} x_i \leq b_j^{l}, \qquad \forall \, j = 1, \ldots, m_l,$$

where m_e, m_g, and m_l are the numbers of equal-to, greater-than-or-equal-to, and less-than-or-equal-to constraints, respectively. Thus, $m = m_e + m_g + m_l$ is the total number of constraints. The coefficients, $A_{j,i}^{e}, \forall i = 1, \ldots, n, j = 1, \ldots, m_e$, $A_{j,i}^{g}, \forall i = 1, \ldots, n, j = 1, \ldots, m_g$, and $A_{j,i}^{l}, \forall i = 1, \ldots, n, j = 1, \ldots, m_l$, and the terms on the right-hand sides of the constraints, $b_j^{e}, \forall j = 1, \ldots, m_e, b_j^{g}, \forall j = 1, \ldots, m_g$, and $b_j^{l}, \forall j = 1, \ldots, m_l$, are all constants.

An LPP, thus, has the generic form:

$$\min_{x_1, \ldots, x_n} c_0 + \sum_{i=1}^{n} c_i x_i$$

$$\text{s.t.} \sum_{i=1}^{n} A_{j,i}^{e} x_i = b_j^{e}, \qquad \forall \, j = 1, \ldots, m_e$$

$$\sum_{i=1}^{n} A_{j,i}^{g} x_i \geq b_j^{g}, \qquad \forall \, j = 1, \ldots, m_g$$

$$\sum_{i=1}^{n} A_{j,i}^{l} x_i \leq b_j^{l}, \qquad \forall \, j = 1, \ldots, m_l.$$

An LPP does not need to include all of the different types of constraints (*i.e.*, equal-to, greater-than-or-equal-to, or less-than-or-equal-to) and may include only one or two types. For instance, the Apples and Oranges Example (*cf.* Example 1.1) has no equal-to constraints but does have the other two types.

The Apples and Oranges Example is one example of a linear optimization problem. Another example is:

$$\min_{x_1, x_2} x_1 + x_2$$

$$\text{s.t. } \frac{2}{3}x_1 + x_2 \leq 18$$

$$2x_1 + x_2 \geq 8$$

$$x_1 \leq 12$$

$$x_2 \leq 16$$

$$x_1, x_2 \geq 0.$$

The decision variables are x_1 and x_2, the linear objective function is $x_1 + x_2$, and the linear constraints are $(2/3)x_1 + x_2 \leq 18$, $2x_1 + x_2 \geq 8$, $x_1 \leq 12$, $x_2 \leq 16$, $x_1 \geq 0$ and $x_2 \geq 0$. As is common practice, the two non-negativity constraints are written together as:

$$x_1, x_2 \geq 0,$$

for sake of brevity. It is *very* important to stress that this constraint is not the same as:

$$x_1 + x_2 \geq 0.$$

The constraint, $x_1 + x_2 \geq 0$, would allow $x_1 = -1$ and $x_2 = 3$ as a feasible solution. However, these values for x_1 and x_2 do not satisfy the constraint, $x_1 \geq 0$.

Chapter 2 is devoted to the formulation and solution of linear optimization problems. A number of classic textbooks [5, 19, 24] also provide a more technically rigorous treatment of linear optimization.

1.3 Linear Optimization Problems with Integer Decisions

There is a special case of linear optimization problems in which some or all of the decision variables are restricted to take on integer values. Such problems in which some (but not necessarily all) of the variables are restricted to taking on integer values are called **mixed-integer linear optimization problems** or **mixed-integer linear programming problems** (MILPPs). A linear optimization problem in which all of the decision variables are restricted to take on integer values is sometimes called a pure-integer linear optimization problem. On occasion, some will also distinguish problems in which the integer decision variables can take on the values of 0 or 1 only. Such problems are referred to as mixed-binary or pure-binary linear optimization problems. We use the term MILPP throughout this text, however, because it is all-encompassing.

An MILPP is defined by the following three important characteristics.

1. Some set of the decision variables are restricted to take on integer values while others can take on real values. Thus we have:

$$x_i \in \mathbb{Z}, \text{ for some } i = 1, \ldots, n,$$

and:

$$x_i \in \mathbb{R}, \text{ for the remaining } i = 1, \ldots, n,$$

where $\mathbb{Z}$ denotes the set of integers.

2. The objective function is linear in the decision variables. Thus, we have that:

$$f(x_1, \ldots, x_n) = c_0 + \sum_{i=1}^{n} c_i x_i,$$

where $c_0, \ldots, c_n$ are constants.

3. The constraints are all equal-to, greater-than-or-equal-to, or less-than-or-equal-to constraints that are linear in the decision variables and can be written as:

$$\sum_{i=1}^{n} A_{j,i}^e x_i = b_j^e, \qquad \forall \, j = 1, \ldots, m_e$$

$$\sum_{i=1}^{n} A_{j,i}^g x_i \geq b_j^g, \qquad \forall \, j = 1, \ldots, m_g$$

$$\sum_{i=1}^{n} A_{j,i}^l x_i \leq b_j^l, \qquad \forall \, j = 1, \ldots, m_l,$$

where $m_e, m_g, m_l, A_{j,i}^e, A_{j,i}^g$, and $A_{j,i}^l, b_j^e, b_j^g$, and b_j^l have the same interpretations as in a linear optimization problem.

An MILPP, thus, has the generic form:

$$\min_{x_1, \ldots, x_n} c_0 + \sum_{i=1}^{n} c_i x_i$$
$$\text{s.t. } \sum_{i=1}^{n} A_{j,i}^e x_i = b_j^e, \qquad \forall \, j = 1, \ldots, m_e$$
$$\sum_{i=1}^{n} A_{j,i}^g x_i \geq b_j^g, \qquad \forall \, j = 1, \ldots, m_g$$
$$\sum_{i=1}^{n} A_{j,i}^l x_i \leq b_j^l, \qquad \forall \, j = 1, \ldots, m_l$$
$$x_i \in \mathbb{Z}, \qquad \text{for some } i = 1, \ldots, n$$
$$x_i \in \mathbb{R}, \qquad \text{for the remaining } i = 1, \ldots, n.$$

We normally do not explicitly write the last constraint:

$$x_i \in \mathbb{R}, \text{ for the remaining } i = 1, \ldots, n,$$

because of the implicit assumption that all of the decision variables can take on any real value (unless there is an explicit integrality constraint).

As mentioned above, in some cases we model problems in which variables are restricted to take on binary values (*i.e.*, 0 or 1 only). There are two ways that this can be done. To show these, let us suppose that x_i is the binary variable in question. One way to impose the binary restriction is to explicitly do so by replacing the integrality constraint:

$$x_i \in \mathbb{Z},$$

with the binary constraint:

$$x_i \in \{0, 1\}.$$

The other is to retain the integrality constraint:

$$x_i \in \mathbb{Z},$$

and to impose the two bound constraints:

$$x_i \geq 0,$$

and:

$$x_i \leq 1.$$

An example of mixed-integer linear optimization problem is:

$$\min_{p_1,p_2,p_3,x_1,x_2,x_3} (2p_1 + 5p_2 + 1p_3) + (40x_1 + 50x_2 + 35x_3)$$

$$\text{s.t.} \quad p_1 + p_2 + p_3 = 50$$
$$5x_1 \leq p_1 \leq 20x_1$$
$$6x_2 \leq p_2 \leq 40x_2$$
$$4x_3 \leq p_3 \leq 35x_3$$
$$p_1, p_2, p_3 \geq 0$$
$$x_1, x_2, x_3 \geq 0$$
$$x_1, x_2, x_3 \leq 1$$
$$x_1, x_2, x_3 \in \mathbb{Z}.$$

This problem has three decision variables that are real-valued—p_1, p_2, and p_3—and three integer-valued variables—x_1, x_2, and x_3. Indeed, because x_1, x_2, and x_3 are restricted to be between 0 and 1, these are binary variables. This problem also has examples of double-sided inequalities, such as:

$$5x_1 \leq p_1 \leq 20x_1.$$

This inequality represents the two constraints:

$$5x_1 \leq p_1,$$

and:

$$p_1 \leq 20x_1,$$

in a more compact form.

Chapter 3 is devoted to the formulation and solution of mixed-integer linear optimization problems. Although we discuss some two algorithms to solve linear optimization problems with integer variables, there are other more advanced techniques that must occasionally be employed to solve particularly complex problems. Interested readers are referred to more advances texts [6, 25] that discuss these solution techniques.

1.4 Nonlinear Optimization Problems

A **nonlinear optimization problem** or nonlinear programming problem (NLPP) has the following three defining characteristics.

1. The decision variables are continuous and we thus have that:

$$(x_1, \ldots, x_n) \in \mathbb{R}^n.$$

2. The objective function is a nonlinear real-valued function. Thus, we have that:

$$f : \mathbb{R}^n \to \mathbb{R}.$$

3. The constraints are nonlinear equality or less-than-or-equal-to constraints of the form:

$$h_i(x_1, \ldots, x_n) = 0, \forall i = 1, \ldots, m$$
$$g_j(x_1, \ldots, x_n) \leq 0, \forall j = 1, \ldots, r,$$

   where m is the number of equality constraints and r the number of inequality constraints, $h_i : \mathbb{R}^n \to \mathbb{R}, \forall i = 1, \ldots, m$, and $g_j : \mathbb{R}^n \to \mathbb{R}, \forall j = 1, \ldots, r$.

Therefore, an NLPP has the generic form:

$$\min_{x_1, \ldots, x_n} f(x)$$
$$\text{s.t. } h_i(x) = 0, \qquad \forall i = 1, \ldots, m$$
$$g_j(x) \leq 0, \qquad \forall j = 1, \ldots, r.$$

We assume that each of the constraints has a zero on its right-hand side and that all of the inequalities are less-than-or-equal-to constraints. It is straightforward to convert any generic NLPP to this form, as we show with the example problem:

$$\min_{x_1,x_2} \sqrt{x_1^2 + x_2^2}$$

$$\text{s.t. } x_2 - \frac{x_2}{x_1}w \geq h$$

$$x_1, x_2 \geq 0.$$

We can convert the inequalities into less-than-or-equal-to constraints with zeroes on their right-hand sides by subtracting terms on the greater-than-or-equal-to side of the inequalities to obtain:

$$\min_{x_1,x_2} \sqrt{x_1^2 + x_2^2}$$

$$\text{s.t. } h - x_2 + \frac{x_2}{x_1}w \leq 0$$

$$- x_1, -x_2 \leq 0.$$

Note that neither all of the constraints nor the objective function of an NLPP must be nonlinear. For instance, the constraints:

$$-x_1, -x_2 \leq 0,$$

in the example above are linear in the decision variables.

Chapters 4 and 5 are devoted to the formulation and solution of nonlinear optimization problems. Although we discuss a number of approaches to solving NLPPs, there are other textbooks [1, 4, 8, 19, 21] that introduce more advanced solution techniques. We should also note that we only study nonlinear optimization problems in which the variables take on real values. Although it is a straightforward extension of MILPPs and NLPPs to formulate mixed-integer nonlinear optimization problems, solving such problems can be extremely demanding. Indeed, very few textbooks on this topic exist and we must instead refer readers to research papers and monographs on the topic, as mixed-integer nonlinear optimization is still a burgeoning research topic. Nevertheless, Floudas [14] provides an excellent introduction to the formulation and solution of mixed-integer nonlinear programming problems.

1.5 Dynamic Optimization Problems

Dynamic optimization problems represent a vastly different way of modeling and solving optimization problems. The power of dynamic optimization is that it can be used to efficiently solve large problems by decomposing the problems into a sequence of stages. The solution technique works through the problem stages and determines what decision is optimal in each stage.

Dynamic optimization problems can be written using the same type of generic forms used to express LPPs, MILPPs, and NLPPs. However, this is not a particularly useful way of studying the structure and power of dynamic optimization. For this reason, we defer any further treatment of dynamic optimization problems to Chapter 6, where dynamic optimization is discussed in detail.

1.6 Technical Considerations

The four classes of problems introduced in the preceding sections cover many of the mainstream topics within the field of optimization. However, there are other classes of problems that go beyond those introduced here. Moreover, there are many technical issues related to optimization that go beyond the scope of an introductory textbook, such as this one. We introduce two technical considerations that may come up as one attempts to use optimization in the real world. Although these considerations are beyond the scope of this text, we provide references that can provide helpful methodologies for tackling such issues.

1.6.1 Large-Scale Optimization Problems

The optimization techniques introduced in this text can be fruitfully applied to relatively large problems. For instance, the algorithm introduced in Section 2.5 to solve linear optimization problems can efficiently handle problems with hundreds of thousands or even millions of variables. Nevertheless, one may encounter **large-scale optimization problems** with variables or constraints numbering in the tens of millions (or more). Moreover, some classes of problems (*e.g.*, MILPPs) may be much more difficult to solve than an LPP, even with only tens of thousands of variables.

Unfortunately, such problems are often unavoidable in practice and solving them 'directly' using the techniques introduced in this text may not yield a solution within a reasonable amount of time (or ever). Decomposition techniques (also sometimes called partitioning techniques) often help in solving such large-scale optimization problems. At a high level, these techniques work by breaking the large-scale problem into smaller subproblems, that can usually be solved directly. The important consideration in decomposing a large-scale problem is that one must ensure that the solutions given by the subproblems yield solutions that are optimal (or close to optimal) for the entire undecomposed problem.

While decomposition techniques are important, they are beyond the scope of this text. Interested readers are referred to other works, which provide introductions to decomposition techniques for different classes of optimization problems [2, 3, 12, 22].

1.6.2 Stochastic Optimization Problems

All of the models introduced in this section (and covered in this text) implicitly assume that the system being modeled is fully **deterministic**. This means that there is no uncertainty about the parameters of the problem. This assumption is clearly an abstraction of reality, as it is rare to have a system that is completely deterministic. Nevertheless, there are many settings in which the assumption of a deterministic system is suitable and the results given by a deterministic model are adequate.

There can, however, be settings in which there is sufficient uncertainty in a system (or in which the small amount of uncertainty is sufficiently important to have major impacts on making an optimal decision) that it is inappropriate to use a deterministic model. We introduce here a conceptual example of one way in which uncertainty can be introduced into an optimization problem. We assume that there are a set of uncertain parameters, which we denote $\xi_1, \ldots \xi_m$. We can then reformulate generic optimization problem (1.2)–(1.3) as the **stochastic optimization problem**:

$$\min_{x_1,\ldots,x_n} \mathbb{E}_{\xi_1,\ldots,\xi_m}[f(x_1, \ldots, x_n; \xi_1, \ldots, x_n)]$$

$$\text{s.t. } (x_1, \ldots, x_n) \in \Omega.$$

Here, we write the objective function as depending not only on the decisions chosen by the decision maker (*i.e.*, the x's) but also on the uncertain parameters (*i.e.*, the ξ's). The $\mathbb{E}_{\xi_1,\ldots,\xi_m}$ operator in the objective function is computing the expected value, with respect to the uncertain parameters, of the objective function. The expectation (or a similar) operator is needed here because the uncertain parameters make the objective function uncertain. In essence, the expectation operator is needed for the stochastic optimization problem to 'make sense.'

Stochastic optimization problems can very easily become large-scale because many **scenarios** (alternate values of the uncertain parameters) need to be considered. It should be noted that this formulation only captures uncertainty in the objective-function value. There could be cases in which the constraints are uncertain. Decomposition techniques [12] are generally helpful in solving stochastic optimization problems. Stochastic optimization problems are beyond the scope of this text. Interested readers are referred to the classical monograph on the topic [7] and to a more application-oriented text [11].

1.7 Optimization Environments and Solvers

Solving most classes of optimization problems requires two pieces of software. The first is a **mathematical programming language,** which allows the user to formulate the problem (*i.e.*, specify problem data, decision variables, the objective function, and constraints) in a human-readable format.

Most mathematical programming languages also include facilities for reading and writing data from and to standard formats (*e.g.*, Microsoft Excel workbooks, csv files, or databases) and have scripting features. These scripting features can be especially useful if modeling a system requires multiple interrelated optimization problems to be run in succession, as such a process can be automated.

Some mathematical programming languages also include built-in presolving features, which can reduce the complexity of a problem through simple arithmetic substitutions and operations before having the problem solved. The pioneering mathematical programming language is GAMS [9] and for this reason all of the sample codes given in this book use GAMS. Other mathematical programming languages include AMPL [15], AIMMS [23], and JuliaOpt [18].

The mathematical programming language translates or compiles the human-readable problem into a machine-readable format that is used by a **solver**. The solver does the actual work of solving the optimization problem (beyond the presolving feature available in many mathematical programming languages). Most solvers are tailored to solve a specific class of optimization problems. For instance, CPLEX [17] and GUROBI [16] are two state-of-the-art solvers for LPPs and MILPPs. State-of-the-art NLPP solvers include KNITRO [10], MINOS [20], and CONOPT [13].

Dynamic optimization problems typically require problem-specific codes. This is because solving a dynamic optimization problem requires exploiting the structure of the problem. For this reason, general solvers for dynamic optimization problems are not available.

1.8   Scope of the Book

This book considers the following classes of optimization problems in further detail in the following chapters:

1. linear optimization problems (Chapter 2),
2. mixed-integer linear optimization problems (Chapter 3),
3. nonlinear optimization problems (Chapters 4 and 5), and
4. dynamic optimization problems (Chapter 6).

1.9   Final Remarks

We conclude this chapter by noting that our approach in writing this book is reader friendly. This approach relies largely on illustrative and insightful examples, avoiding formal mathematical proof except when absolutely needed or relatively simple. The aim is for the text to (hopefully) provide the student with a pain-free introduction to optimization.

Students and readers desiring a more rigorous treatment of optimization theory or solution algorithms are referred to a number of more advanced textbooks [1–5, 19, 21, 22, 24, 25]. These texts provide technical details and mathematical formalism that we exclude.

This textbook provides GAMS codes for many of the illustrative examples used in the subsequent chapters. However, we do not formally introduce the use of GAMS or other mathematical programming software. Interested readers are referred to a number of texts that introduce the use of mathematical programming languages. For instance, Brook *et al.* [9] provide an encyclopedic tome on the use of GAMS. However, Chapter 2 of this user guide provides an introduction to the major features of the software package. Fourer *et al.* [15] similarly provide an introduction to the use of the AMPL programming language. Similar resources exist for AIMMS [23], JuliaOpt [18], and other mathematical programming languages.

References

1. Bazaraa MS, Sherali HD, Shetty CM (2006) Nonlinear programming: theory and algorithms, 3rd edn. Wiley-Interscience, Hoboken
2. Bertsekas DP (2012) Dynamic programming and optimal control, vol 1, 4th edn. Athena Scientific, Belmont
3. Bertsekas DP (2012) Dynamic programming and optimal control, vol 2, 4th edn. Athena Scientific, Belmont
4. Bertsekas D (2016) Nonlinear programming, 3rd edn. Athena Scientific, Belmont
5. Bertsimas D, Tsitsiklis J (1997) Introduction to linear optimization. Athena Scientific, Belmont
6. Bertsimas D, Weismantel R (2005) Optimization over integers. Dynamic Ideas, Belmont
7. Birge JR, Louveaux F (1997) Introduction to stochastic programming, corrected edn. Springer, New York
8. Boyd S, Vandenberghe L (2004) Convex optimization. Cambridge University Press, Cambridge
9. Brook A, Kendrick D, Meeraus, A (1988) GAMS – a user's guide. ACM, New York. www.gams.com
10. Byrd RH, Nocedal J, Waltz RA (2006) KNITRO: an integrated package for nonlinear optimization. In: di Pillo G, Roma M (eds) Large-scale nonlinear optimization, pp 35–59. www.artelys.com/en/optimization-tools/knitro
11. Conejo AJ, Carrión M, Morales JM (2010) Decision making under uncertainty in electricity markets. International series in operations research and management science, vol 153. Springer, Berlin
12. Conejo AJ, Castillo E, Minguez R, Garcia-Bertrand R (2006) Decomposition techniques in mathematical programming: engineering and science applications. Springer, Berlin
13. Drud AS (1994) CONOPT – a large-scale GRG code. ORSA J Comput 6(2):207–216. www.conopt.com
14. Floudas CA (1995) Nonlinear and mixed-integer optimization: fundamentals and applications. Oxford University Press, Oxford
15. Fourer R, Gay DM, Kernighan BW (2002) AMPL: a modeling language for mathematical programming, 2nd edn. Cengage Learning, Boston. www.ampl.com
16. Gurobi Optimization, Inc. (2010) Gurobi optimizer reference manual, version 3.0. Houston, Texas. www.gurobi.com
17. IBM ILOG CPLEX Optimization Studio (2016) CPLEX user's manual, version 12 release 7. IBM Corp. www.cplex.com

18. Lubin M, Dunning I (2015) Computing in operations research using Julia. INFORMS J Comput 27(2):238–248. www.juliaopt.org
19. Luenberger DG, Ye Y (2016) Linear and nonlinear programming, 4th edn. Springer, New York
20. Murtagh BA, Saunders MA (1995) MINOS 5.4 user's guide. Technical report SOL 83-20R, Systems Optimization Laboratory, Department of Operations Research, Stanford University, Stanford, California. www.sbsi-sol-optimize.com/asp/sol_product_minos.htm
21. Nocedal J, Wright S (2006) Numerical optimization, 2nd edn. Springer, New York
22. Powell WB (2007) Approximate dynamic programming: solving the curses of dimensionality. Wiley-Interscience, Hoboken
23. Roelofs M, Bisschop J (2016) AIMMS the language reference, AIMMS B.V., Haarlem, The Netherlands. www.aimms.com
24. Vanderbei RJ (2014) Linear programming: foundations and extensions, 4th edn. Springer, New York
25. Wolsey LA, Nemhauser GL (1999) Integer and combinatorial optimization. Wiley-Interscience, New York

Chapter 2
Linear Optimization

In this chapter we consider linear programming problems (LPPs). We show the practical significance of LPPs through a number of energy-related examples, providing a precise formulation for such problems. We then analyze the geometric and algebraic features of generic LPPs, using some of the energy-related examples as specific cases. We then describe a well known solution algorithm, which is based on the algebraic features of LPPs, show how to perform a sensitivity analysis, and provide and discuss the dual form of an LPP. We finally conclude with a number of practical observations and end-of-chapter exercises.

2.1 Motivating Examples

This introductory section provides a number of energy-related motivating examples for the use of linear optimization. It illustrates that optimization is an everyday endeavor.

2.1.1 Electricity-Production Problem

An electricity producer operates two production facilities that have capacities of 12 and 16 units per hour, respectively. This producer sells the electricity produced at $1 per unit per hour. The two production facilities share a cooling system that restricts their operation, from above and below. More specifically, the sum of the hourly output from facility 2 and twice the hourly output of facility 1 must be at least 8 units. Moreover, the sum of the hourly output from facility 2 and two-thirds of the hourly output from facility 1 must be no more than 18 units. The producer wishes to

© Springer International Publishing AG 2017
R. Sioshansi and A.J. Conejo, *Optimization in Engineering*,
Springer Optimization and Its Applications 120, DOI 10.1007/978-3-319-56769-3_2

determine hourly production from the two facilities to maximize total revenues from energy sales.

To formulate this and any optimization problem, there are three basic problem elements that must be identified. The first is the **decision variables**, which represent the decisions being made in the problem. In essence, the decision variables represent the elements of the system being modeled that are under the decision maker's control, in the sense that their values can be changed. This should be contrasted with problem data or parameters, which are fixed and cannot be changed by the decision maker. In the context of our electricity-production problem, the decisions being made are how many units to produce from each production facility in each hour. We denote these decisions by the two variables, x_1 and x_2, being cognizant of what units these production decisions are being measured in.

The second element of an optimization problem is the **objective function**. The objective function is the metric upon which the decision variables are chosen. Depending on the problem context, the objective is either being minimized or maximized. An optimization problem will often be referred to as either a minimization or maximization problem, depending on what 'direction' the objective function is being optimized in. An important property of an LPP is that the objective function must be *linear* in the decision variables. In the electric-production problem, we are told that the objective is to maximize total revenues. Because the two production facilities sell their outputs, which are represented by x_1 and x_2, at a unit price of \$1 per unit per hour, the objective function can be written as:

$$\max_{x_1, x_2} \ 1x_1 + 1x_2.$$

We typically write the decision variables underneath the 'min' or 'max' operator in the objective function, to make it easy for anyone to know what the problem variables are.

The final problem element is any **constraint**. The constraints indicate what, if any, restrictions there are on the decision variables. Most constraints are given in a problem's description. For instance, we are told that the two facilities have production limits of 12 and 16 units per hour. We can express these restrictions as the two constraints:

$$x_1 \leq 12,$$

and:

$$x_2 \leq 16.$$

We are also told that there are upper and lower limits imposed by the shared cooling system. These restrictions can be expressed mathematically as:

$$\frac{2}{3}x_1 + x_2 \leq 18,$$

and:

$$2x_1 + x_2 \geq 8.$$

In addition to these four explicit constraints, we know that it is physically impossible for either facility to produce a negative amount of electricity. Although this is not given as an explicit restriction in the problem description, we must include the following two non-negativity constraints:

$$x_1, x_2 \geq 0,$$

to complete the problem formulation.

Taking all of these together, the problem formulation can be written compactly as:

$$\max_{x_1,x_2} z = x_1 + x_2 \tag{2.1}$$

$$\text{s.t.} \quad \frac{2}{3}x_1 + x_2 \leq 18 \tag{2.2}$$

$$2x_1 + x_2 \geq 8 \tag{2.3}$$

$$x_1 \leq 12 \tag{2.4}$$

$$x_2 \leq 16 \tag{2.5}$$

$$x_1, x_2 \geq 0. \tag{2.6}$$

The z in (2.1) represents the objective function value. The abbreviation 's.t.' stands for 'subject to,' and denotes that the following lines have problem constraints. The total set of constraints (2.2)–(2.6) define the problem's **feasible region** or feasible set. The feasible region is the set of values that the decision variables can take and satisfy the problem constraints. Importantly, it should be clear that an optimal solution of the problem must satisfy the constraints, meaning it must belong to the feasible region.

All of the constraints in an LPP must be one of three types: (i) less-than-or-equal-to inequalities, (ii) greater-than-or-equal-to inequalities, or (iii) equalities. Moreover, all of the constraints of an LPP must be *linear* in the decision variables. No other constraint types can be used in an LPP. Constraints (2.2), (2.4), and (2.5) are examples of less-than-or-equal-to inequalities while constraints (2.3) and (2.6) are examples of greater-than-or-equal-to constraints. This problem does not include any equality constraints, however, the Natural Gas-Transportation Problem, which is introduced in Section 2.1.2, does.

It is also important to stress that LPPs cannot include any *strict* inequality constraints. That is to say, each inequality constraint must either have a '$\leq$' or '$\geq$' in it and cannot have a '$<$' or '$>$.' The reason for this is that a strict inequality can give us an LPP that does not have a well defined optimal solution. To see a very simple example of this, consider the following optimization problem with a single variable, which we denote as x:

$$\min_{x} \; z = x$$

$$\text{s.t.} \;\; x > 0.$$

Examining this problem should reveal that it is impossible to find an optimal value for x. This is because for any strictly positive value of x, regardless of how close to 0 it is, $x/2$ will also be feasible and give a slightly smaller objective-function value. This simple example illustrates why we do not allow strict inequality constraints in LPPs.

If a problem does require a strict inequality, we can usually approximate it using a weak inequality based on physical realities of the system being modeled. For instance, suppose that in the Electricity-Production Problem we are told that facility 1 must produce a strictly positive amount of electricity. This constraint would take the form:

$$x > 0, \tag{2.7}$$

which cannot be included in an LPP. Suppose that the control system on the facility cannot realistically allow a production level less than 0.0001 units. Then, we can instead substitute strict inequality (2.7) with:

$$x \geq 0.0001,$$

which is a weak inequality that can be included in an LPP.

Figure 2.1 shows a geometrical representation of the Electricity-Production Problem. The two axes represent different values for the two decision variables. The boundary defined by each constraint is given by a blue line and the two small arrows at the end of each line indicate which side of the line satisfies the constraint. For instance, the blue line that goes through the points $(x_1, x_2) = (0, 8)$ and $(x_1, x_2) = (4, 0)$ defines the boundary of constraint (2.3) and the arrows indicate that points above and to the right of this line satisfy constraint (2.3).

For a problem that has more than two dimensions (meaning that it has more than two decision variables) the boundary of each constraint defines a **hyperplane**. A hyperplane is simply a higher-dimensional analogue to a line. The set of points that satisfies an inequality constraint in more than two dimensions is called a **halfspace**.

The feasible region of the Electricity-Production Problem is the interior of a **polygon**. For problems that have more than two dimensions the feasible region becomes a **polytope**. A polytope is simply a higher-dimensional analogue to a polygon. An important property of LPPs is that because the problem constraints are all linear in the decision variables, the feasible region is always a polytope. It is important to stress, however, that the polytope is not necessarily **bounded**, as the one for the Electricity-Production Problem is. That is to say, for some problems there may be a direction in which the decision variables can keep increasing or decreasing without any limit. The implications of having a feasible region that is not bounded are discussed later in Section 2.3.1. The feasible region of the Electricity-Production Problem has six

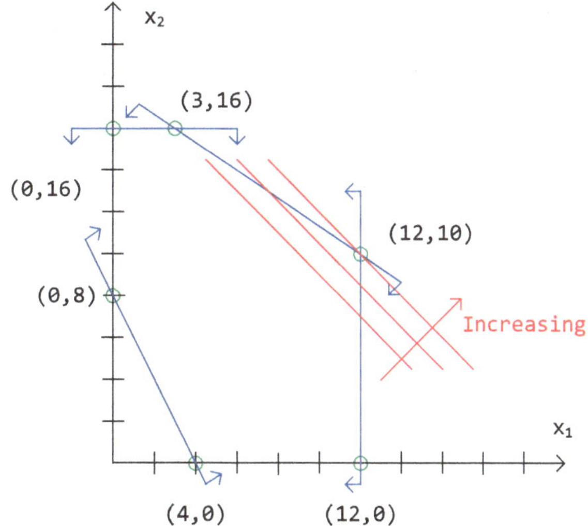

Fig. 2.1 Geometrical representation of the Electricity-Production Problem

corners, which are $(4, 0)$, $(12, 0)$, $(12, 10)$, $(3, 16)$, $(0, 16)$, and $(0, 8)$. The corners of a polytope are often called **vertices** or **extreme points**.

The objective function is represented by its **contour plot**. The contour plot represents sets of values for the decision variables that give the same objective function value. The contour plot for the Electricity-Production Problem is represented by the parallel red lines in Figure 2.1. The red arrow that runs perpendicular to the red lines indicates the direction in which the objective function is increasing. An important property of LPPs is that because the objective function is linear in the decision variables, the contour plot is always parallel lines (for problems that have more than two dimensions, the contour plot is parallel hyperplanes) and the objective is always increasing/decreasing in the same direction that runs perpendicular to the contour plot.

The solution of this problem is easily obtained by inspecting Figure 2.1: the point in the feasible region that corresponds with the contour line with the highest value is vertex $(12, 10)$. Thus, vertex $(12, 10)$ is the optimal solution with an objective-function value of 22. We typically identify an optimal solution using stars, meaning that we write $(x_1^*, x_2^*) = (12, 10)$ and $z^* = 22$.

It is finally worth noting that the Electricity-Production Problem is a simplified example of what is known as a production-scheduling problem.

2.1.2 Natural Gas-Transportation Problem

A natural gas producer owns two gas fields and serves two markets. Table 2.1 summarizes the capacity of each field and Table 2.2 gives the demand of each market, which must be satisfied exactly. Finally, Table 2.3 summarizes the per-unit cost of transporting gas from each field to each market. The company would like to determine how to transport natural gas from the two fields to the two markets to minimize its total transportation cost.

There are four decision variables in this problem, which are:

- $x_{1,1}$: units of natural gas transported from field 1 to market 1;
- $x_{1,2}$: units of natural gas transported from field 1 to market 2;
- $x_{2,1}$: units of natural gas transported from field 2 to market 1; and
- $x_{2,2}$: units of natural gas transported from field 2 to market 2.

Table 2.1 Capacity of each gas field in the Natural Gas-Transportation Problem

Field	Capacity [units]
1	7
2	12

Table 2.2 Demand of each market in the Natural Gas-Transportation Problem

Market	Demand [units]
1	10
2	8

Table 2.3 Transportation cost between each gas field and market [\$/unit] in the Natural Gas-Transportation Problem

	Market 1	Market 2
Field 1	5	4
Field 2	3	6

For many problems, it can be cumbersome to list all of the problem variables explicitly. To shorten the variable definition, we can introduce **index sets**, over which the variables are defined. In this problem, the variables are indexed by two sets. The first is the field from which the gas is being transported and the second index set is the market to which it is being shipped. If we let i denote the index for the field and j the index for the market, we can more compactly define our decision variables as $x_{i,j}$, which represents the units of natural gas transported from field i to market j. Of course when defining the decision variables this way we know that $i = 1, 2$ (because there are two fields) and $j = 1, 2$ (because there are two markets). However, typically the two (or more) index sets over which a variable is defined do not necessarily have the same number of elements, as we have in this example.

The objective of this problem is to minimize total transportation cost, which is given by:

$$\min_{x} \; 5x_{1,1} + 4x_{1,2} + 3x_{2,1} + 6x_{2,2},$$

where we have listed the decision variables compactly as x, and have that $x = (x_{1,1}, x_{1,2}, x_{2,1}, x_{2,2})$.

This problem has three types of constraints. The first are capacity limits on how much can be produced by each field:

$$x_{1,1} + x_{1,2} \leq 7,$$

and:

$$x_{2,1} + x_{2,2} \leq 12.$$

The second are constraints that ensure that the demand in each market is exactly satisfied:

$$x_{1,1} + x_{2,1} = 10,$$

and:

$$x_{1,2} + x_{2,2} = 8.$$

Note that these demand conditions are equality constraints. We finally need non-negativity constraints:

$$x_{1,1}, x_{1,2}, x_{2,1}, x_{2,2} \geq 0.$$

The non-negativity constraints can be written more compactly either as:

$$x_{i,j} \geq 0, \forall\, i = 1, 2;\ j = 1, 2;$$

or as:

$$x \geq 0.$$

Taking all of these elements together, the entire LPP can be written as:

$$
\begin{aligned}
\min_{x}\ z = {}& 5x_{1,1} + 4x_{1,2} + 3x_{2,1} + 6x_{2,2} \\
\text{s.t. }\ & x_{1,1} + x_{1,2} \leq 7 \\
& x_{2,1} + x_{2,2} \leq 12 \\
& x_{1,1} + x_{2,1} = 10 \\
& x_{1,2} + x_{2,2} = 8 \\
& x_{i,j} \geq 0, \forall\, i = 1, 2;\ j = 1, 2.
\end{aligned}
\tag{2.8}\tag{2.9}
$$

The Natural Gas-Transportation Problem is a simplified instance of a transportation problem.

2.1.3 Gasoline-Mixture Problem

A gasoline refiner needs to produce a cost-minimizing blend of ethanol and traditional gasoline. The blend needs to have at least 65% burning efficiency and a pollution level no greater than 85%. The burning efficiency, pollution level, and per-ton cost of ethanol and traditional gasoline are given in Table 2.4.

Table 2.4 Burning efficiency, pollution level, and per-ton cost of ethanol and traditional gasoline in the Gasoline-Mixture Problem

Product	Efficiency [%]	Pollution [%]	Cost [\$/ton]
Gasoline	70	90	200
Ethanol	60	80	220

To formulate this problem, we model the refiner as determining a least-cost mixture of gasoline and ethanol to produce one ton of blend. The resulting optimal mixture can then be scaled up or down by the refiner depending on how much blend it actually wants to produce. There are two decision variables in this problem—x_1 and x_2 denote how many tons of gasoline and ethanol are used in the blend, respectively.

The objective is to minimize the cost of the blend:

$$\min_{x}\ 200x_1 + 220x_2.$$

There are three sets of problem constraints. The first ensures that the blend meets the minimum burning efficiency level:

$$0.7x_1 + 0.6x_2 \geq 0.65,$$

and the maximum pollution level:

$$0.9x_1 + 0.8x_2 \leq 0.85.$$

Next we must ensure that we produce one ton of the blend:

$$x_1 + x_2 = 1.$$

Finally, the decision variables must be non-negative, because it is physically impossible to have negative tons of ethanol or gasoline in the blend:

$$x_1, x_2 \geq 0.$$

Putting all of the problem elements together, the LPP can be written as:

$$\min_{x} z = 200x_1 + 220x_2$$

$$\text{s.t. } 0.7x_1 + 0.6x_2 \geq 0.65$$
$$0.9x_1 + 0.8x_2 \leq 0.85$$
$$x_1 + x_2 = 1$$
$$x_1, x_2 \geq 0.$$

The Gasoline-Mixture Problem is a simplified example of a commodity-mixing problem.

2.1.4 *Electricity-Dispatch Problem*

The electric power network in Figure 2.2 includes two production plants, at nodes 1 and 2, and demand at node 3. The production plants at nodes 1 and 2 have production capacities of 6 and 8 units, respectively, and their per-unit production costs are \$1 and \$2, respectively. There is demand for 10 units of energy at node 3.

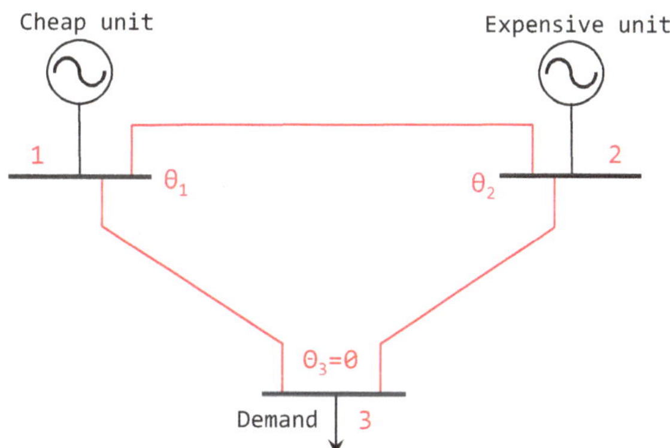

Fig. 2.2 Electric power network in the Electricity-Dispatch Problem

The operation of the network is governed by differences in the electrical heights of the three nodes. More specifically, the flow of electricity through any line is proportional to the difference of electrical heights of the initial and final nodes of the line. This means that the amount of energy produced at node 1 is equal to the difference between the electrical heights of nodes 1 and 2 plus the difference between the electrical heights of nodes 1 and 3. Electricity produced at node 2 is similarly equal to the difference between the electrical heights of nodes 2 and 1 plus the difference between the electrical heights of nodes 2 and 3. Finally, the electricity consumed at node 3 is defined as the difference between the electrical-heights of nodes 1 and 3 plus the difference between the electrical heights of nodes 2 and 3 (the electrical-height differences are opposite to those for nodes 1 and 2 because energy is consumed at node 3 as opposed to being produced).

The network operator seeks to produce electricity at the plants and operate the network in such a way to serve the demand at node 3 at minimum cost.

This problem has five decision variables. We let x_1 and x_2 denote the units of electricity produced at nodes 1 and 2, respectively. We also let θ_1, θ_2, and θ_3 denote the electrical heights of the three nodes.

The objective is to minimize total production cost:

$$\min_{x,\theta}\ 1x_1 + 2x_2.$$

There are three sets of problem constraints. The first defines the amount produced and consumed at each node in terms of the electrical-height differences. The node 1 constraint is:

$$x_1 = (\theta_1 - \theta_2) + (\theta_1 - \theta_3),$$

the node 2 constraint is:

$$x_2 = (\theta_2 - \theta_1) + (\theta_2 - \theta_3),$$

and the node 3 constraint is:

$$10 = (\theta_1 - \theta_3) + (\theta_2 - \theta_3).$$

As noted above, the electrical-height differences defining consumption at node 3 is opposite to the height differences defining production at nodes 1 and 2. The second set of constraints imposes the production limits at the two nodes:

$$x_1 \leq 6,$$

and:

$$x_2 \leq 8.$$

We finally need non-negativity for the production variables *only*:

$$x_1, x_2 \geq 0.$$

The θ's can take negative values, because they are used to define the relative electrical heights of the three nodes.

This LPP can be slightly simplified. This is because the production and consumption levels at the three nodes are defined in terms of electrical height differences. As such, we can arbitrarily fix one of the three θ's and only keep the other two as variables. If we fix $\theta_3 = 0$, the LPP is further simplified to:

$$\min_{x,\theta} z = x_1 + 2x_2$$

$$\text{s.t. } x_1 = 2\theta_1 - \theta_2$$

$$x_2 = 2\theta_2 - \theta_1$$

$$10 = \theta_1 + \theta_2$$

$$x_1 \leq 6$$

$$x_2 \leq 8$$

$$x_1, x_2 \geq 0.$$

The Electricity-Dispatch Problem is a simplified instance of a scheduling problem with network constraints.

2.2 Forms of Linear Optimization Problems

As noted before, linear optimization problems vary in terms of a number of attributes. The objective can either be a minimization or maximization. Moreover, it can include a mixture of less-than-or-equal-to, greater-than-or-equal-to, or equality constraints. In this section we first give the general form of a linear optimization problem. We then discuss two important special forms that any linear optimization problem can be converted to: standard and canonical forms. These forms are used later to solve and study algebraic features of linear optimization problems. In addition to introducing these special forms, we discuss how to convert any generic linear optimization problem into these two forms.

2.2.1 General Form of Linear Optimization Problems

As noted in Section 2.1.1, linear optimization problems have a very important defining feature. This feature is that the objective function and all of the constraints are linear in all of the decision variables. Because LPPs have this special feature, we can write them generically as:

$$\min_{x_1,\ldots,x_n} \sum_{i=1}^{n} c_i x_i \tag{2.10}$$

$$\text{s.t. } \sum_{i=1}^{n} A_{j,i}^e x_i = b_j^e, \qquad \forall\, j = 1, \ldots, m_e \tag{2.11}$$

$$\sum_{i=1}^{n} A_{j,i}^g x_i \geq b_j^g, \qquad \forall\, j = 1, \ldots, m_g \tag{2.12}$$

$$\sum_{i=1}^{n} A_{j,i}^{l} x_i \le b_j^{l}, \qquad\qquad \forall\, j = 1, \ldots, m_l. \qquad (2.13)$$

This generic problem has n decision variables: $x_1, \ldots, x_n$. The c_i's in the objective function and the $A_{j,i}^{e}$'s, $A_{j,i}^{g}$'s, $A_{j,i}^{l}$'s, b_j^{e}'s, b_j^{g}'s, and b_j^{l}'s in the constraints are constants. Thus the objective function and the constraints are linear in the decision variables, because each decision variable is multiplied by a constant coefficient in the objective function and in each constraint and those products are summed together.

Although this generic LPP is written as a minimization problem, we could have just as easily written it as a maximization problem. This generic problem has m_e equality, m_g greater-than-or-equal-to, and m_l less-than-or-equal-to constraints. This means that there are $m = m_e + m_g + m_l$ constraints in total. An LPP does not have to include all of the three types of constraints. For instance, the Electricity-Production Problem, which is introduced in Section 2.1.1, does not have any equality constraints, meaning that $m_e = 0$ for that particular problem.

As noted in Section 2.1.1, LPPs can only include weak inequalities. Strict inequalities cannot be used, because they typically raise technical issues. If a problem calls for the use of a strict inequality, for instance of the form:

$$\sum_{i=1}^{n} A_{j,i} y_i > b_j,$$

this can be approximated by introducing a sufficiently small positive constant, ε_j, and replacing the strict inequality with a weak inequality of the form:

$$\sum_{i=1}^{n} A_{j,i} y_i \ge b_j + \varepsilon_j.$$

Oftentimes, the physical properties of the system being modeled may allow for such a value of ε_j to be chosen (we discuss one example of such a physical property in Section 2.1.1). If not, then one must simply choose a very small value for ε_j to ensure that the final problem solution is not drastically affected by it. A strictly less-than inequality of the form:

$$\sum_{i=1}^{n} A_{j,i} y_i < b_j,$$

can be similarly approximated by replacing it with a weak inequality of the form:

$$\sum_{i=1}^{n} A_{j,i} y_i \le b_j - \varepsilon_j,$$

where ε_j is again a sufficiently small positive constant.

2.2.2 Standard and Canonical Forms of Linear Optimization Problems

Although any LPP can be written in the generic form just introduced, we occasionally want to write a problem in one of two more tailored forms. These forms—the so-called standard and canonical forms—are used because they make solving or analyzing a linear optimization problem more straightforward. We introduce each of these forms in turn and then discuss how to convert any generic LPP into them.

2.2.2.1 Standard Form of Linear Optimization Problems

The **standard form** of a linear optimization problem has three defining features. First, the objective function is a minimization. The other two properties are that the problem has two types of constraints. The first types are non-negativity constraints, which require *all* of the decision variables to be non-negative. The other types are **structural constraints**, which are any constraints other than non-negativity constraints. All of the structural constraints of a linear optimization problem in standard form must be equality constraints.

Converting a generic linear optimization problem to standard form requires several steps. We begin with the decision variables. Any decision variables that are non-negative in the original generic problem are already in the correct form for the standard form of the problem. If a decision variable has a non-positivity restriction of the form:

$$y \leq 0,$$

it can be replaced throughout the LPP with a new variable, $\tilde{y}$, which is defined as:

$$\tilde{y} = -y.$$

Clearly, the new variable would have a non-negativity restriction, because the original non-positivity constraint:

$$y \leq 0,$$

could be rewritten as:

$$-y \geq 0,$$

by multiplying the constraint through by -1. We would then substitute $\tilde{y}$ for $-y$ in the left-hand side of this constraint, which gives:

$$\tilde{y} \geq 0.$$

If a variable is unrestricted in sign, a similar type of substitution can be done. More specifically, suppose a variable, y, in a generic LPP is unrestricted in sign. We can then introduce two new non-negative variables, y^- and y^+, and define them as:

$$y^+ - y^- = y.$$

We then substitute y with $y^+ - y^-$ throughout the LPP and also add two non-negativity constraints:

$$y^-, y^+ \geq 0.$$

Note that because y is defined as the difference between two non-negative variables, y can be made positive or negative depending on which of y^- or y^+ is bigger (or, if we want $y = 0$, we would have $y^- = y^+$).

After all of the variables have been made non-negative, we next turn our attention to the structural constraints. If a structural constraint in a generic LPP is an equality, then it is already in the correct format for the standard form LPP and no further work is needed. If, however, we have a less-than-or-equal-to constraint of the form:

$$\sum_{i=1}^{n} A_{j,i}^{l} x_i \leq b_j^l,$$

we can convert this to an equality constraint by introducing a non-negative **slack variable**, which we will denote as s_j. With this slack variable, we can replace the less-than-or-equal-to constraint with the equivalent equality constraint:

$$\sum_{i=1}^{n} A_{j,i}^{l} x_i + s_j = b_j^l,$$

and also add the non-negativity constraint:

$$s_j \geq 0.$$

A greater-than-or-equal-to constraint of the form:

$$\sum_{i=1}^{n} A_{j,i}^{g} x_i \geq b_j^g,$$

can be similarly converted to an equality constraint by introducing a non-negative **surplus variable**, which we denote r_j. With this surplus variable, we can replace the greater-than-or-equal-to constraint with the equivalent equality constraint:

$$\sum_{i=1}^{n} A_{j,i}^{g} x_i - r_j = b_j^g.$$

We must also add the non-negativity constraint:

$$r_j \geq 0.$$

The slack and surplus variables introduced to convert inequalities into equalities can be interpreted as measuring the difference between the left- and right-hand sides of the original inequality constraints.

The final step to convert a generic LPP to standard form is to ensure that the objective function is a minimization. If the objective of the generic problem is a minimization, then no further work is needed. Otherwise, if the objective is maximization, it can be converted by multiplying the objective through by -1.

We demonstrate the use of these steps to convert a generic LPP into standard form with the following example.

Example 2.1 Consider the following LPP:

$$\max_{x} \; 3x_1 + 5x_2 - 3x_3 + 1.3x_4 - x_5$$
$$\text{s.t.} \quad x_1 + x_2 - 4x_4 \leq 10$$
$$x_2 - 0.5x_3 + x_5 = -1$$
$$x_3 \geq 5$$
$$x_1, x_2 \geq 0$$
$$x_4 \leq 0.$$

To convert this generic LPP into standard form, we begin by first noting that both x_1 and x_2 are non-negative, thus no substitutions have to be made for these variables. The variable x_4 is non-positive, thus we define a new variable, $\tilde{x}_4 = -x_4$. Substituting $\tilde{x}_4$ for x_4 in the LPP gives:

$$\max_{x} \; 3x_1 + 5x_2 - 3x_3 - 1.3\tilde{x}_4 - x_5$$
$$\text{s.t.} \quad x_1 + x_2 + 4\tilde{x}_4 \leq 10$$
$$x_2 - 0.5x_3 + x_5 = -1$$
$$x_3 \geq 5$$
$$x_1, x_2, \tilde{x}_4 \geq 0.$$

The signs of the coefficients in the objective function and first constraint on $\tilde{x}_4$ have been changed, because we have defined $\tilde{x}_4$ as being equal to x_4. Next, we note that because x_3 and x_5 are unrestricted in sign, we must introduce four new non-negative variables, x_3^-, x_3^+, x_5^-, and x_3^+, and define them as:

$$x_3^+ - x_3^- = x_3,$$

and:

$$x_5^+ - x_5^- = x_5.$$

We make these substitutions for x_3 and x_5 and add the non-negativity constraints, which gives:

$$\max_{x} \; 3x_1 + 5x_2 - 3(x_3^+ - x_3^-) - 1.3\tilde{x}_4 - (x_5^+ - x_5^-)$$

$$\text{s.t.} \; x_1 + x_2 + 4\tilde{x}_4 \leq 10$$
$$x_2 - 0.5(x_3^+ - x_3^-) + (x_5^+ - x_5^-) = -1$$
$$(x_3^+ - x_3^-) \geq 5$$
$$x_1, x_2, x_3^-, x_3^+, \tilde{x}_4, x_5^-, x_5^+ \geq 0.$$

Next, we must add a non-negative slack and subtract a non-negative surplus variable, which we call s_1 and r_1, to and from structural constraints 1 and 3, respectively. This gives:

$$\max_{x,s,r} \; 3x_1 + 5x_2 - 3x_3^+ + 3x_3^- - 1.3\tilde{x}_4 - x_5^+ + x_5^-$$

$$\text{s.t.} \; x_1 + x_2 + 4\tilde{x}_4 + s_1 = 10$$
$$x_2 - 0.5x_3^+ + 0.5x_3^- + x_5^+ - x_5^- = -1$$
$$x_3^+ - x_3^- - r_1 = 5$$
$$x_1, x_2, x_3^-, x_3^+, \tilde{x}_4, x_5^-, x_5^+, s_1, r_1 \geq 0.$$

Finally, we convert the objective function to a minimization, by multiplying it through by -1, giving:

$$\min_{x,s,r} \; -3x_1 - 5x_2 + 3x_3^+ - 3x_3^- + 1.3\tilde{x}_4 + x_5^+ - x_5^-$$

$$\text{s.t.} \; x_1 + x_2 + 4\tilde{x}_4 + s_1 = 10$$
$$x_2 - 0.5x_3^+ + 0.5x_3^- + x_5^+ - x_5^- = -1$$
$$x_3^+ - x_3^- - r_1 = 5$$
$$x_1, x_2, x_3^-, x_3^+, \tilde{x}_4, x_5^-, x_5^+, s_1, r_1 \geq 0,$$

which is the standard form of our starting LPP.

$\square$

Example 2.2 Consider the Gasoline-Mixture Problem, which is introduced in Section 2.1.3. This problem is formulated generically as:

$$\min_{x} 200x_1 + 220x_2$$

$$\text{s.t.} \; 0.7x_1 + 0.6x_2 \geq 0.65$$
$$0.9x_1 + 0.8x_2 \leq 0.85$$
$$x_1 + x_2 = 1$$
$$x_1, x_2 \geq 0.$$

To convert this to standard form, we simply need to introduce one non-negative surplus variable, r_1, and a non-negative slack variable, s_1. The standard form of the LPP would then be:

$$\min_{x,r,s} 200x_1 + 220x_2$$

$$\text{s.t. } 0.7x_1 + 0.6x_2 - r_1 = 0.65$$
$$0.9x_1 + 0.8x_2 + s_1 = 0.85$$
$$x_1 + x_2 = 1$$
$$x_1, x_2, r_1, s_1 \geq 0. \qquad \square$$

An LPP in standard form can be generically written as:

$$\min_{x_1,\ldots,x_n} \sum_{i=1}^{n} c_i x_i$$

$$\text{s.t. } \sum_{i=1}^{n} A_{j,i} x_i = b_j, \qquad \forall\, j = 1, \ldots, m$$
$$x_i \geq 0, \qquad \forall\, i = 1, \ldots, n.$$

We can write the generic standard form even more compactly. This is done by first defining a vector of objective-function coefficients:

$$c = \begin{pmatrix} c_1 \\ c_2 \\ \vdots \\ c_n \end{pmatrix},$$

a vector of decision variables:

$$x = \begin{pmatrix} x_1 \\ x_2 \\ \vdots \\ x_n \end{pmatrix},$$

a matrix of constraint coefficients:

$$A = \begin{bmatrix} a_{1,1} & a_{1,2} & \cdots & a_{1,n} \\ a_{2,1} & a_{2,2} & \cdots & a_{2,n} \\ \vdots & \vdots & \ddots & \vdots \\ a_{m,1} & a_{m,2} & \cdots & a_{m,n} \end{bmatrix},$$

and a vector of constraint right-hand-side constants:

$$b = \begin{pmatrix} b_1 \\ b_2 \\ \vdots \\ b_m \end{pmatrix}.$$

The standard-form LPP is then compactly written as:

$$\min_{x} \; c^\top x \tag{2.14}$$

$$\text{s.t.} \; Ax = b \tag{2.15}$$

$$x \geq 0. \tag{2.16}$$

2.2.2.2   Canonical Form of Linear Optimization Problems

The **canonical form** of a linear optimization problem has: (i) an objective function that is a minimization, (ii) greater-than-or-equal-to structural constraints, and (iii) non-negative decision variables. Converting a generic linear optimization problem to standard form requires several steps. Ensuring that the decision variables are non-negative and that the objective is a minimization are handled in the same manner as they are in converting an LPP to standard form.

As for the structural constraints, any constraints that are greater-than-or-equal-to require no further work. A less-than-or-equal-to constraint of the form:

$$\sum_{i=1}^{n} A^l_{j,i} x_i \leq b^l_j,$$

can be converted to a greater-than-or-equal-to constraint by multiplying both sides by -1. This converts the constraint to:

$$-\sum_{i=1}^{n} A^l_{j,i} x_i \geq -b^l_j.$$

Finally, an equality constraint of the form:

$$\sum_{i=1}^{n} A^e_{j,i} x_i = b^e_j,$$

can be replaced by two inequalities of the form:

$$\sum_{i=1}^{n} A^e_{j,i} x_i \leq b^e_j,$$

and:

$$\sum_{i=1}^{n} A_{j,i}^{e} x_i \geq b_j^{e}.$$

We can then convert the first inequality constraint into a greater-than-or-equal-to constraint by multiplying both sides by -1. Thus, the equality constraint is replaced with:

$$-\sum_{i=1}^{n} A_{j,i}^{e} x_i \geq -b_j^{e},$$

and:

$$\sum_{i=1}^{n} A_{j,i}^{e} x_i \geq b_j^{e}.$$

It should be noted that this transformation of an equality constraint when converting to canonical form can create numerical issues when solving the LPP. The reason for this is that the two inequalities 'fight' one another to bring the solution to its corresponding 'side' of b_j^{e}. Because a feasible solution is right in the middle, not in either of these two sides, this 'fight' may result in a sluggish back-and-forth progression to the middle, where a feasible solution lies.

Example 2.3 Recall the generic LPP, which is introduced in Example 2.1:

$$\max_{x} \ 3x_1 + 5x_2 - 3x_3 + 1.3x_4 - x_5$$

$$\text{s.t.} \ \ x_1 + x_2 - 4x_4 \leq 10$$

$$x_2 - 0.5x_3 + x_5 = -1$$

$$x_3 \geq 5$$

$$x_1, x_2 \geq 0$$

$$x_4 \leq 0.$$

To convert this LPP to canonical form, we first undertake the same steps to have all of the decision variables non-negative, which gives:

$$\max_{x} \ 3x_1 + 5x_2 - 3x_3^{+} + 3x_3^{-} - 1.3\tilde{x}_4 - x_5^{+} + x_5^{-}$$

$$\text{s.t.} \ \ x_1 + x_2 + 4\tilde{x}_4 \leq 10$$

$$x_2 - 0.5x_3^{+} + 0.5x_3^{-} + x_5^{+} - x_5^{-} = -1$$

$$x_3^{+} - x_3^{-} \geq 5$$

$$x_1, x_2, x_3^{-}, x_3^{+}, \tilde{x}_4, x_5^{-}, x_5^{+} \geq 0.$$

We next convert the first structural inequality into a greater-than-or-equal-to by multiplying both sides by -1, which gives:

$$\max_{x} \ 3x_1 + 5x_2 - 3x_3^+ + 3x_3^- - 1.3\tilde{x}_4 - x_5^+ + x_5^-$$

$$\text{s.t.} \quad -x_1 - x_2 - 4\tilde{x}_4 \geq -10$$
$$x_2 - 0.5x_3^+ + 0.5x_3^- + x_5^+ - x_5^- = -1$$
$$x_3^+ - x_3^- \geq 5$$
$$x_1, x_2, x_3^-, x_3^+, \tilde{x}_4, x_5^-, x_5^+ \geq 0.$$

We then replace the second structural constraint, which is an equality, with two inequalities, giving:

$$\max_{x} \ 3x_1 + 5x_2 - 3x_3^+ + 3x_3^- - 1.3\tilde{x}_4 - x_5^+ + x_5^-$$

$$\text{s.t.} \quad -x_1 - x_2 - 4\tilde{x}_4 \geq -10$$
$$x_2 - 0.5x_3^+ + 0.5x_3^- + x_5^+ - x_5^- \leq -1$$
$$x_2 - 0.5x_3^+ + 0.5x_3^- + x_5^+ - x_5^- \geq -1$$
$$x_3^+ - x_3^- \geq 5$$
$$x_1, x_2, x_3^-, x_3^+, \tilde{x}_4, x_5^-, x_5^+ \geq 0.$$

We convert the first of these two into a greater-than-or-equal-to by multiplying it through by -1, giving:

$$\max_{x} \ 3x_1 + 5x_2 - 3x_3^+ + 3x_3^- - 1.3\tilde{x}_4 - x_5^+ + x_5^-$$

$$\text{s.t.} \quad -x_1 - x_2 - 4\tilde{x}_4 \geq -10$$
$$-x_2 + 0.5x_3^+ - 0.5x_3^- - x_5^+ + x_5^- \geq 1$$
$$x_2 - 0.5x_3^+ + 0.5x_3^- + x_5^+ - x_5^- \geq -1$$
$$x_3^+ - x_3^- \geq 5$$
$$x_1, x_2, x_3^-, x_3^+, \tilde{x}_4, x_5^-, x_5^+ \geq 0.$$

We finally convert the objective function into a minimization by multiplying through by -1, which gives:

$$\min_{x} \ -3x_1 - 5x_2 + 3x_3^+ - 3x_3^- + 1.3\tilde{x}_4 + x_5^+ - x_5^-$$

$$\text{s.t.} \quad -x_1 - x_2 - 4\tilde{x}_4 \geq -10$$
$$-x_2 + 0.5x_3^+ - 0.5x_3^- - x_5^+ + x_5^- \geq 1$$
$$x_2 - 0.5x_3^+ + 0.5x_3^- + x_5^+ - x_5^- \geq -1$$
$$x_3^+ - x_3^- \geq 5$$
$$x_1, x_2, x_3^-, x_3^+, \tilde{x}_4, x_5^-, x_5^+ \geq 0,$$

which is the canonical form of the starting LPP.

$\square$

Example 2.4 Consider the Electricity-Production Problem, which is introduced in Section 2.1.1. The generic formulation of this problem is:

$$\max_{x_1,x_2} \; x_1 + x_2$$

$$\text{s.t.} \quad \frac{2}{3}x_1 + x_2 \le 18$$

$$2x_1 + x_2 \ge 8$$

$$x_1 \le 12$$

$$x_2 \le 16$$

$$x_1, x_2 \ge 0.$$

To convert this problem to canonical form, we must multiply the objective function and the first, third, and fourth structural constraints through by -1. This gives:

$$\min_{x_1,x_2} \; -x_1 - x_2$$

$$\text{s.t.} \quad -\frac{2}{3}x_1 - x_2 \ge -18$$

$$2x_1 + x_2 \ge 8$$

$$-x_1 \ge -12$$

$$-x_2 \ge -16$$

$$x_1, x_2 \ge 0,$$

as the canonical form.

$\square$

Example 2.5 Consider the Natural Gas-Transportation Problem, which is introduced in Section 2.1.2. This problem is generically formulated as:

$$\min_{x} \; z = 5x_{1,1} + 4x_{1,2} + 3x_{2,1} + 6x_{2,2}$$

$$\text{s.t.} \; x_{1,1} + x_{1,2} \le 7$$

$$x_{2,1} + x_{2,2} \le 12$$

$$x_{1,1} + x_{2,1} = 10$$

$$x_{1,2} + x_{2,2} = 8$$

$$x_{i,j} \ge 0, \forall \, i = 1, 2; \, j = 1, 2.$$

To convert this to canonical form, both sides of the first two inequalities must be multiplied by -1. Moreover, the two equality constraints must be replaced with two inequalities, one of each of which is multiplied by -1 to convert all of the structural constraints into greater-than-or-equal-to constraints. This gives:

$$\min_{x} z = 5x_{1,1} + 4x_{1,2} + 3x_{2,1} + 6x_{2,2}$$

$$\text{s.t. } -x_{1,1} - x_{1,2} \geq -7$$
$$-x_{2,1} - x_{2,2} \geq -12$$
$$-x_{1,1} - x_{2,1} \geq -10$$
$$x_{1,1} + x_{2,1} \geq 10$$
$$-x_{1,2} - x_{2,2} \geq -8$$
$$x_{1,2} + x_{2,2} \geq 8$$
$$x_{i,j} \geq 0, \forall i = 1, 2; j = 1, 2,$$

as the canonical form of this LPP.

$\square$

The canonical form of an LPP can be written generically as:

$$\min_{x_1,\ldots,x_n} \sum_{i=1}^{n} c_i x_i$$

$$\text{s.t. } \sum_{i=1}^{n} A_{j,i} x_i \geq b_j, \qquad \forall j = 1, \ldots, m$$

$$x_i \geq 0, \qquad \forall i = 1, \ldots, n.$$

This can also be written more compactly as:

$$\min_{x} c^{\top} x \tag{2.17}$$

$$\text{s.t. } Ax \geq b \tag{2.18}$$

$$x \geq 0, \tag{2.19}$$

where c, x, A, and b maintain the same definitions as in the compact standard-form LPP, given by (2.14)–(2.16).

2.3   Basic Feasible Solutions and Optimality

This section provides both geometric and algebraic analyses of the feasible region and objective function of linear optimization problems. Based on the geometric analysis, we draw some conclusions regarding the geometrical properties of an optimal solution of a linear optimization problem. We then use an algebraic analysis of the constraints of a linear optimization problem to determine a way to characterize points that may be optimal solutions of an LPP. This algebraic analysis is the backbone of the algorithm used to solve LPPs, which is later introduced in Section 2.5.

2.3.1 Geometric View of Linear Optimization Problems

Recall the Electricity-Production Problem, which is introduced in Section 2.1.1. This problem is formulated as:

$$\max_{x_1,x_2} \; z = x_1 + x_2$$

$$\text{s.t.} \quad \frac{2}{3}x_1 + x_2 \leq 18$$

$$2x_1 + x_2 \geq 8$$

$$x_1 \leq 12$$

$$x_2 \leq 16$$

$$x_1, x_2 \geq 0.$$

Figure 2.1 shows the feasible region of this problem and the contour plot of its objective function.

From the discussion in Section 2.1.1, we know that linear optimization problems have two important geometric properties, which are due to the linearity of their constraints and objective function. The first is that the feasible region of a linear optimization problem is always a polytope, which is the multidimensional analogue of a polygon. We also see from Figure 2.1 that the contour plot of the objective function of an LPP is always a set of parallel hyperplanes, which are the multidimensional analogue of lines. Moreover, the objective function is always increasing or decreasing in the same direction, which is perpendicular to the contours.

This latter geometric property of LPPs, that the contours are parallel and always increasing or decreasing in the same direction, implies that we find an optimal solution by moving as far as possible within the feasible region until hitting a boundary. Put another way, we always find an optimal solution to a linear optimization problem on the boundary of its feasible region. The first geometric property of LPPs, that the feasible region is a polytope, allows us to make an even stronger statement about their optimal solutions. Because the feasible set of an LPP is a polytope, we can always find a vertex or extreme point of the feasible region that is optimal. This is, indeed, one way of stating the fundamental theorem of linear optimization. Figure 2.3 shows the feasible region of the Electricity-Production Problem and identifies its extreme points. We know from the discussion in Section 2.1.1 that $(x_1^*, x_2^*) = (12, 10)$ is the optimal extreme point of this problem.

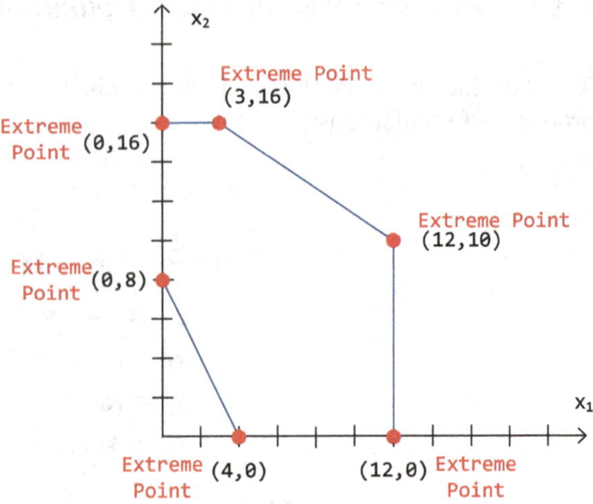

Fig. 2.3 Feasible region of Electricity-Production Problem and its extreme points

There are also some 'pathological' cases that may arise with an LPP. The first is that all of the points along a line or hyperplane defining a boundary of the feasible are optimal solutions. This occurs if the contour lines of the objective function are parallel to that side of the polytope. We call this a case of **multiple optimal solutions**. To see how this happens, suppose that the objective function of the Electricity-Production Problem is changed to:

$$\max_{x_1,x_2} \; z = \frac{2}{3}x_1 + x_2.$$

Figure 2.4 shows the contour plot of the objective function in this case. Note that all of the points highlighted in purple are now optimal solutions of the LPP.

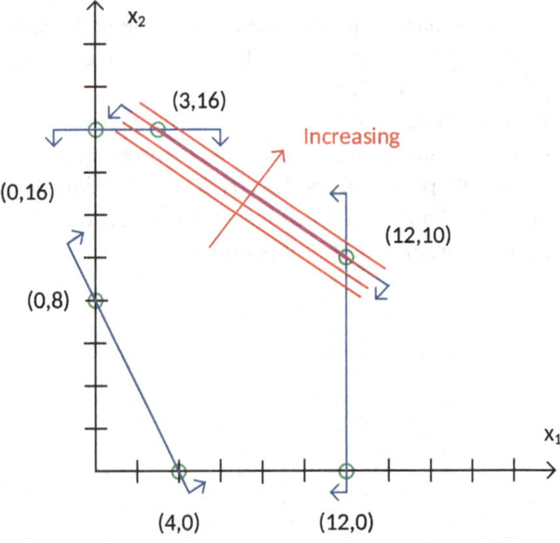

Fig. 2.4 Geometrical representation of the Electricity-Production Problem with $\frac{2}{3}x_1 + x_2$ as its objective function

Another issue arises if the feasible region is not bounded. Recall from the discussion in Section 2.1.1 that the feasible region of an optimization problem does not necessarily have to be bounded. The feasible region of the Electricity-Production Problem is bounded, as illustrated in Figure 2.1. A linear optimization problem with a bounded feasible region is guaranteed to have an optimal solution. Otherwise, if the feasible region is **unbounded**, the problem may have an optimal solution or it may be possible to have the objective increase or decrease without limit.

To understand how an LPP with an unbounded feasible region may have an optimal solution, suppose that we remove constraints (2.3) and (2.6) from the Electricity-Production Problem. The LPP would then be:

$$\max_{x_1, x_2} z = x_1 + x_2$$

$$\text{s.t.} \quad \frac{2}{3}x_1 + x_2 \leq 18$$

$$x_1 \leq 12$$

$$x_2 \leq 16.$$

Figure 2.5 shows the feasible region of the new LPP, which is indeed unbounded (we can make both x_1 and x_2 go to $-\infty$ without violating any of the constraints). Note, however, that the same point, $(x_1^*, x_2^*) = (12, 10)$, that is optimal in the original LPP is optimal in the new problem as well. This is because the side of the polytope that is unbounded is not the side in which the objective improves.

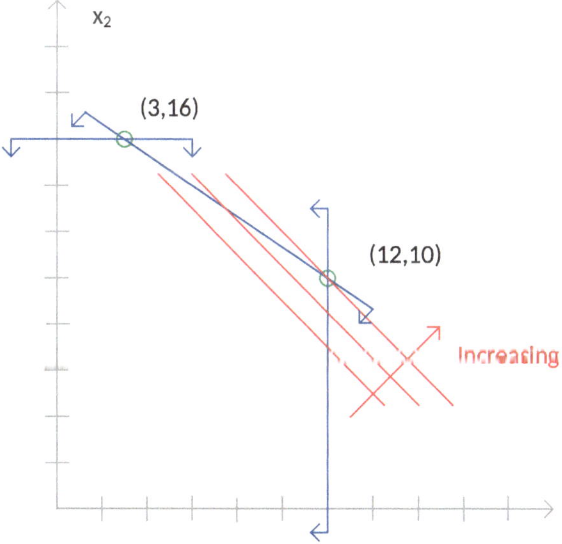

Fig. 2.5 Geometrical representation of the Electricity-Production Problem with constraints (2.3) and (2.6) removed

Consider, as an opposite example, if constraints (2.2) and (2.4) are removed from the Electricity-Production Problem. Our LPP would then be:

$$\max_{x_1,x_2} z = x_1 + x_2$$

$$\text{s.t.} \quad 2x_1 + x_2 \geq 8$$

$$x_2 \leq 16$$

$$x_1, x_2 \geq 0.$$

Figure 2.6 shows the feasible region of this LPP, which is also unbounded. Note that there is an important distinction between this LPP and that shown in Figure 2.5. The new LPP no longer has an optimal solution, because the objective function can be made arbitrarily large without violating any of the constraints (we can make x_1 go to $+\infty$ without violating any constraints, and doing so makes the objective function go to $+\infty$). This LPP is said to be **unbounded**.

An unbounded optimization problem is said to have an optimal objective function value of either $-\infty$ or $+\infty$ (depending on whether the problem is a minimization or maximization). Unbounded optimization problems are uncommon in practice, because the physical and economic worlds are bounded. Thus, we do not study unbounded problems in much detail, although we do discuss in Section 2.5.6 how to determine analytically (as opposed to graphically) if an LPP is unbounded.

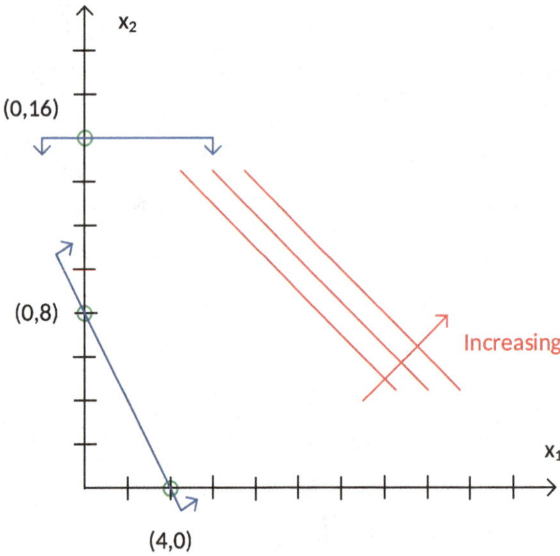

Fig. 2.6 Geometrical representation of the Electricity-Production Problem with constraints (2.2) and (2.4) removed

Finally, we should stress that there are two notions of boundedness and unboundedness in the context of optimization. One is whether the feasible region of an LPP is bounded or unbounded. This is a property of the constraints. The second is whether the problem is bounded or unbounded. An LPP must have an unbounded feasible region for it to be unbounded. Moreover, the objective function must improve (either increase or decrease, depending on whether we are considering a maximization or minimization problem) in the direction that the feasible region is unbounded.

The final pathological case is one in which the feasible region is empty. In such a case there are no feasible solutions that satisfy all of the constraints and we say that such a problem is **infeasible**. To illustrate how a problem can be infeasible, suppose that the Electricity-Production Problem is changed to:

$$\max_{x_1,x_2} \ z = x_1 + x_2$$

$$\text{s.t.} \ \ \frac{2}{3}x_1 + x_2 \geq 18$$

$$2x_1 + x_2 \leq 8$$

$$x_1 \leq 12$$

$$x_2 \leq 16$$

$$x_1, x_2 \geq 0.$$

Figure 2.7 shows the feasible region of this LPP, which is indeed empty. It is empty because there are no points that simultaneously satisfy the $\frac{2}{3}x_1 + x_2 \geq 18$, $2x_1 + x_2 \leq 8$, and $x_1 \geq 0$ constraints. These three constraints conflict with one another.

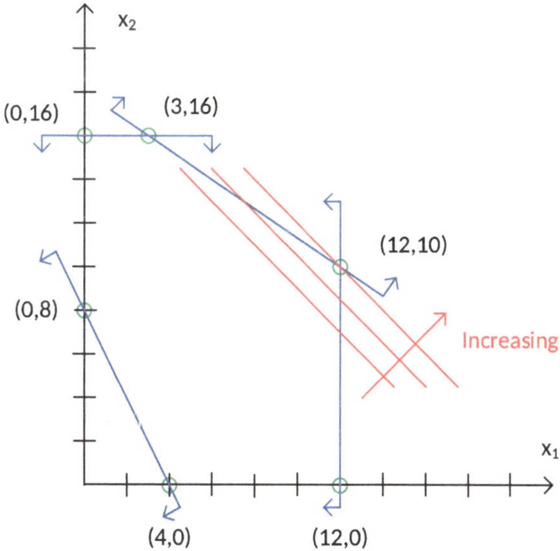

Fig. 2.7 Geometrical representation of the infeasible variant of the Electricity-Production Problem

In practice, an infeasible LPP may indicate that there are problem constraints that are not properly specified. This is because the physical world is (normally) feasible. For this reason, we do not pay particular attention to infeasible problems.

An exception to this is if a hypothetical system is being modeled. For instance, suppose that a system is being designed to meet certain criteria. If the resulting model is infeasible, this may be an indication that the system cannot feasibly meet the design criteria specified in the model constraints.

It should, finally, be noted that infeasibility is not caused by a single constraint. Rather, it is caused by two or more constraints that conflict with each other. Thus, when diagnosing the cause of infeasibility of a model, one must identify two or more conflicting constraints.

2.3.2 *Algebraic View of Linear Optimization Problems*

We now focus our attention on bounded and feasible linear optimization problems. Based on our discussion in Section 2.3.1, we note that every bounded and feasible linear optimization problem has an extreme point that is an optimal solution. Thus, we now work on determining if we can characterize extreme points algebraically, by analyzing the constraints of the LPP.

To gain this insight, we transform the Electricity-Production Problem, which is introduced in Section 2.1.1, into standard form, which is:

$$\min_{x} \ z = -x_1 - x_2$$

$$\text{s.t.} \ \ \frac{2}{3}x_1 + x_2 + x_3 = 18$$

$$2x_1 + x_2 - x_4 = 8$$

$$x_1 + x_5 = 12$$

$$x_2 + x_6 = 16$$

$$x_1, x_2, x_3, x_4, x_5, x_6 \geq 0.$$

The standard form-version of this problem can be written more compactly in matrix form as:

$$\min_{x} \ z = \begin{pmatrix} -1 & -1 & 0 & 0 & 0 & 0 \end{pmatrix} \begin{pmatrix} x_1 \\ x_2 \\ x_3 \\ x_4 \\ x_5 \\ x_6 \end{pmatrix} \tag{2.20}$$

$$\text{s.t.} \quad \begin{bmatrix} 2/3 & 1 & 1 & 0 & 0 & 0 \\ 2 & 1 & 0 & -1 & 0 & 0 \\ 1 & 0 & 0 & 0 & 1 & 0 \\ 0 & 1 & 0 & 0 & 0 & 1 \end{bmatrix} \begin{pmatrix} x_1 \\ x_2 \\ x_3 \\ x_4 \\ x_5 \\ x_6 \end{pmatrix} = \begin{pmatrix} 18 \\ 8 \\ 12 \\ 16 \end{pmatrix} \qquad (2.21)$$

$$\begin{pmatrix} x_1 \\ x_2 \\ x_3 \\ x_4 \\ x_5 \\ x_6 \end{pmatrix} \geq \begin{pmatrix} 0 \\ 0 \\ 0 \\ 0 \\ 0 \\ 0 \end{pmatrix}. \qquad (2.22)$$

One way to generate candidate points that may be feasible optimal solutions of the LPP is to focus on the algebraic properties of structural equality constraints (2.21). More specifically, we see that this is a system of four equations with six variables that we are solving for. This means that if we fix the values of $6 - 4 = 2$ variables, we can solve for the remaining variables using the structural equality constraints. Once we have solved the structural equality constraints, we then verify whether the resulting values for x are all non-negative, which is the other constraint in the standard-form LPP.

To make the algebra (*i.e.*, the solution of the structural equalities) easier, we fix the $6 - 4 = 2$ variables equal to zero. Solutions that have this structure (*i.e.*, setting a subset of variables equal to zero and solving for the remaining variables using the equality constraints) are called **basic solutions**. A solution that has this structure and also satisfies the non-negativity constraint is called a **basic feasible solution**.

To illustrate how we find basic solutions, let us take the case in which we set x_1 and x_2 equal to zero and solve for the remaining variables using the structural equality constraints. In this case, constraint (2.21) becomes:

$$\begin{bmatrix} 2/3 & 1 & 1 & 0 & 0 & 0 \\ 2 & 1 & 0 & -1 & 0 & 0 \\ 1 & 0 & 0 & 0 & 1 & 0 \\ 0 & 1 & 0 & 0 & 0 & 1 \end{bmatrix} \begin{pmatrix} 0 \\ 0 \\ x_3 \\ x_4 \\ x_5 \\ x_6 \end{pmatrix} = \begin{pmatrix} 18 \\ 8 \\ 12 \\ 16 \end{pmatrix}. \qquad (2.23)$$

Note, however, that because x_1 and x_2 are set equal to zero in equation (2.23), we can actually ignore the first two columns of the matrix on the left-hand-side of the equality. This is because all of the entries in those columns are multiplied by zero. Thus, equation (2.23) can be further simplified to:

$$\begin{bmatrix} 1 & 0 & 0 & 0 \\ 0 & -1 & 0 & 0 \\ 0 & 0 & 1 & 0 \\ 0 & 0 & 0 & 1 \end{bmatrix} \begin{pmatrix} x_3 \\ x_4 \\ x_5 \\ x_6 \end{pmatrix} = \begin{pmatrix} 18 \\ 8 \\ 12 \\ 16 \end{pmatrix},$$

which has the solution:

$$(x_3, x_4, x_5, x_6) = (18, -8, 12, 16).$$

This means that we have found a solution to the structural equality constraints, which is:

$$(x_1, x_2, x_3, x_4, x_5, x_6) = (0, 0, 18, -8, 12, 16).$$

Because we found the values for x by first setting a subset of them equal to zero and solving for the remainder in the equality constraints, this is a basic solution. Note that because $x_4 = -8$ is not non-negative, this is not a basic feasible solution but rather a **basic infeasible solution**.

It is worth noting that whenever we find a basic solution, we solve a square system of equations given by the structural equality constraints. This is because we can neglect the columns of the coefficient matrix that correspond to the variables fixed equal to zero. For instance, if we fix x_5 and x_6 equal to zero in the standard-form version of the Electricity-Production Problem, structural equality constraint (2.21) becomes:

$$\begin{bmatrix} 2/3 & 1 & 1 & 0 \\ 2 & 1 & 0 & -1 \\ 1 & 0 & 0 & 0 \\ 0 & 1 & 0 & 0 \end{bmatrix} \begin{pmatrix} x_1 \\ x_2 \\ x_3 \\ x_4 \end{pmatrix} = \begin{pmatrix} 18 \\ 8 \\ 12 \\ 16 \end{pmatrix}.$$

As a matter of terminology, the variables that are fixed equal to zero when solving for a basic solution are called **non-basic variables**. The other variables, which are solved for using the structural equality constraints, are called **basic variables**.

The number of basic variables that an LPP has is determined by the number of structural equality constraints and the variables that its standard form has. This is because some subset of the variables is set equal to zero to find a basic solution. The standard form of the Electricity-Production Problem has six variables and two of them must be chosen to be set equal to zero. This means that the Electricity-Production Problem has:

$$\binom{6}{2} = \frac{6!}{2!(6-2)!} = 15,$$

basic solutions.

Table 2.5 lists the 15 basic solutions of the Electricity-Production Problem. Each solution is characterized (in the second column of the table) by which variables are basic. The third column of the table gives the values for the basic variables, which are found by setting the non-basic variables equal to zero and solving the structural equality constraints. Two of the basic solutions, numbers 7 and 12, are listed as singular. What this means is that when the non-basic variables are fixed equal to zero, the structural equality constraints do not have a solution. Put another way, when we select the subset of columns of the coefficient matrix that defines the structural constraints, that submatix is singular. We discuss the geometric interpretation of these types of basic solutions later.

Table 2.5 Basic solutions of the Electricity-Production Problem

Solution #	Basic Variables	Basic-Variable Values	Objective-Function Value	x_1	x_2
1	1, 2, 3, 4	12, 16, −6, 32	basic infeasible solution		
2	1, 2, 3, 5	−4, 16, 14/3, 16	basic infeasible solution		
3	1, 2, 3, 6	12, −16, 26, 32	basic infeasible solution		
4	1, 2, 4, 5	3, 16, 14, 9	−19	3	16
5	1, 2, 4, 6	12, 10, 26, 6	−22	12	10
6	1, 2, 5, 6	−15/2, 23, 39/2, −7	basic infeasible solution		
7	1, 3, 4, 5	singular			
8	1, 3, 4, 6	12, 10, 16, 16	−12	12	0
9	1, 3, 5, 6	4, 46/3, 8, 16	−4	4	0
10	1, 4, 5, 6	27, 46, −15, 16	basic infeasible solution		
11	2, 3, 4, 5	16, 2, 8, 12	−16	0	16
12	2, 3, 4, 6	singular			
13	2, 3, 5, 6	8, 10, 12, 8	−8	0	8
14	2, 4, 5, 6	18, 10, 12, −2	basic infeasible solution		
15	3, 4, 5, 6	18, −8, 12, 16	basic infeasible solution		

Table 2.5 also shows that some of the basic solutions, specifically solutions 1 through 3, 6, 10, 14, and 15, are basic infeasible solutions. This is because at least one of the basic variables turns out to have a negative value when the structural equality constraints are solved. For the remaining six basic feasible solutions (*i.e.*, those that are neither singular nor a basic infeasible solution), the last three columns of the table provides the objective-function value and the values of the two variables in the original generic formulation of the problem, x_1 and x_2.

Note that because the standard-form LPP is a minimization problem, basic feasible solution number 5 is the best one from among the six basic feasible solutions found. Moreover, this solution corresponds to the optimal solution that is found graphically in Section 2.1.1. It gives the same objective-function value (when we take into account the fact that the objective function is multiplied by −1 to convert it into a minimization) and the values for the decision variables are the same as well.

Inspecting the six basic feasible solutions in Table 2.5 reveals that they correspond to the extreme points of the feasible region in Figure 2.1. Figure 2.8 shows the feasible region of the Electricity-Production Problem only (*i.e.*, without the contour plot of the objective function) and the six basic feasible solutions that are listed in Table 2.5.

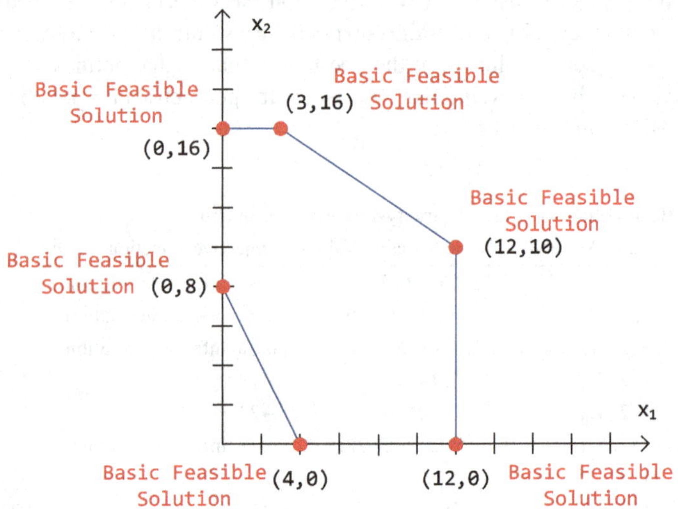

Fig. 2.8 Basic feasible solutions of Electricity-Production Problem

This is a fundamental property of linear optimization: each extreme point of the polytope is a basic feasible solution and each basic feasible solution is an extreme point of the polytope. Proving this property is beyond the scope of this book, and more advanced texts [1] provide the formal proof. The important takeaway from this observation is that if an LPP has an optimal solution, there must be a basic feasible solution that is optimal. This is because the shape of the feasible region and contour plot of the objective imply that there must be an optimal extreme point. Thus, a possible approach to solving an LPP is to enumerate all of the basic feasible solutions and select the one that provides the best objective-function value. This can be quite cumbersome, however, because the number of basic feasible solutions grows exponentially in the problem size (*i.e.*, number of constraints and variables). A more efficient way to solve an LPP is to find a starting basic feasible solution. From this starting point, we then look to see if there is a basic feasible solution next to it that improves the objective function. If not, the basic feasible solution we are currently at is optimal. If so, we move to that basic feasible solution and repeat the process (*i.e.*, determine if there is another basic feasible solution next to the new one that improves the objective). This process is done iteratively until we arrive at a basic feasible solution where the objective cannot be improved. This algorithm, known as the Simplex method, is a standard technique for solving LPPs and is fully detailed in Section 2.5.

Table 2.5 also lists seven basic infeasible solutions, which are shown in Figure 2.9. Basic infeasible solutions are found in the same way as basic feasible solutions— they are 'corners' of the feasible region that are defined by the intersection of the boundaries of two of the linear constraints. However, the intersection point found is infeasible, because it violates at least one other constraint. For instance, the point $(x_1, x_2) = (0, 0)$ is found by intersecting the boundaries of $x_1 \geq 0$ and $x_2 \geq 0$ constraints. However, this point violates the $2x_1 + x_2 \geq 8$ constraint and is, thus, infeasible.

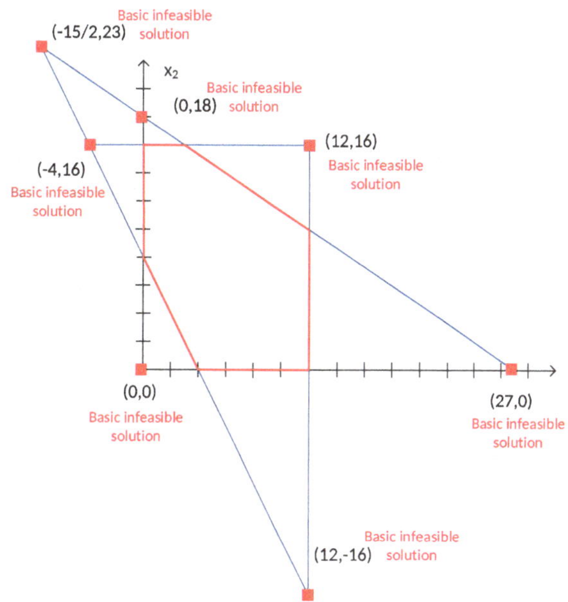

Fig. 2.9 Basic infeasible solutions of Electricity-Production Problem

We can also 'visualize' the two basic solutions in Table 2.5 that are labeled as 'singular.' These basic solutions are defined by trying to intersect constraint boundaries that do not actually intersect. For instance, solution number 7 corresponds to the intersection between the boundaries of the $x_2 \geq 0$ and $x_2 \leq 16$ constraints. However, the boundaries of these two constraints are parallel to one another, meaning that there is no basic solution at their intersection. The other singular basic solution corresponds to intersecting the boundaries of the $x_1 \geq 0$ and $x_1 \leq 12$ constraints.

2.4   A Clever Partition

This section establishes three important foundations of the Simplex method. First, we derive expressions that allow us to find the objective-function and basic-variable values of an LPP in terms of the values of the non-basic variables. These expressions are useful when we want to determine in the Simplex method whether a given basic solution is optimal or not. If it is not, these expressions are useful for finding a new basic solution. We then introduce a tabular format to efficiently organize all of the calculations associated with a given basic solution. Finally, we discuss the algebraic steps used to move from one basic solution to another in the Simplex method.

2.4.1 The Partition

Recall that a basic solution is found by setting one set of variables (the non-basic variables) equal to zero and solving for the others (the basic variables) using the structural equality constraints. For this reason, it is often useful to partition the variables, the coefficients from the left-hand side of the equality constraints, and the objective-function coefficients between the basic and non-basic variables. Moreover, it is useful for developing the Simplex method to derive expressions that give us the values of the objective function and the basic variables in terms of the values of the non-basic variables.

We begin by first examining the structural equality constraints of a standard-form LPP, which can be written as:

$$Ax = b,$$

where A is an $m \times n$ coefficient matrix and b is an m-dimensional vector of constraint right-hand-side constants (*cf.* Equation (2.15) in Section 2.2.2.1). The order that the variables are listed in the x vector is arbitrary. Thus, we can write the x vector as:

$$x = \begin{pmatrix} x_B \\ x_N \end{pmatrix},$$

where x_B are the basic variables and x_N the non-basic variables. If we reorder the x's in this way, then the columns of the A matrix must also be reordered. We do this by writing A as:

$$A = \begin{bmatrix} B & N \end{bmatrix},$$

where B is a submatrix with the constraint coefficients on the basic variables and N is a submatrix with the constraint coefficients on the non-basic variables. We know that x_N is an $(n - m)$-dimensional vector (if there are m structural equality constraints and n variables in the standard-form LPP, we must set $(n - m)$ non-basic variables equal to zero). Thus, B is an $m \times m$ matrix and N is an $m \times (n - m)$ matrix.

Recall from the discussion in Section 2.3.2 and the derivation of the basic solutions for the Electricity-Production Problem in particular, that basic solutions are found by setting by non-basic variables equal to zero. When we do this, we can ignore the columns of the A matrix that are associated with the non-basic variables and solve for the basic variables. This means that we find the basic variables by solving $Bx_B = b$, which gives $x_B = B^{-1}b$. Thus, whenever we find a basic solution, the B matrix must be full-rank.

Based on these observations, we now know that the structural equality constraints (2.15) can be written as:

$$\begin{bmatrix} B & N \end{bmatrix} \begin{pmatrix} x_B \\ x_N \end{pmatrix} = b,$$

or as:

$$Bx_B + Nx_N = b.$$

Because B is full-rank, we can solve for the basic variables in terms of the non-basic variables, giving:

$$\begin{aligned} x_B &= B^{-1}b - B^{-1}Nx_N \\ &= \tilde{b} + \tilde{N}x_N, \end{aligned} \tag{2.24}$$

where $\tilde{b} = B^{-1}b$ and $\tilde{N} = -B^{-1}N$. As a matter of terminology, when solving for a basic solution in this manner the submatrix B is called the **basis**. It is also common to say that basic variables are in the basis, while non-basic variables are said to not be in the basis.

We can also express the objective function in terms of the non-basic variables. To do this, we first note that the standard-form objective function:

$$z = c^\top x,$$

can be written as:

$$z = \begin{bmatrix} c_B & c_N \end{bmatrix} \begin{pmatrix} x_B \\ x_N \end{pmatrix},$$

where c_B and c_N are vectors with the objective-function coefficients in c reordered in the same way that the x vector is reordered into x_B and x_N. Using Equation (2.24) we can write this as:

$$\begin{aligned} z &= \begin{bmatrix} c_B & c_N \end{bmatrix} \begin{pmatrix} x_B \\ x_N \end{pmatrix} \\ &= c_B^\top x_B + c_N^\top x_N \\ &= c_B^\top \cdot (\tilde{b} + \tilde{N}x_N) + c_N^\top x_N \\ &= c_B^\top \tilde{b} + (c_B^\top \tilde{N} + c_N^\top)x_N \\ &= \tilde{c}_0 + \tilde{c}^\top x_N, \end{aligned} \tag{2.25}$$

where $\tilde{c}_0 = c_B^\top \tilde{b}$ and $\tilde{c}^\top = c_B^\top \tilde{N} + c_N^\top$.

Finally, we can arrange the objective function and structural equality constraints of a standard-form LPP in matrix form, which gives:

$$\begin{pmatrix} z \\ x_B \end{pmatrix} = \begin{bmatrix} \tilde{c}_0 & \tilde{c}^\top \\ \tilde{b} & \tilde{N} \end{bmatrix} \begin{pmatrix} 1 \\ x_N \end{pmatrix}.$$

This can be written more explicitly as:

$$\begin{pmatrix} z \\ x_{B,1} \\ \vdots \\ x_{B,r} \\ \vdots \\ x_{B,m} \end{pmatrix} = \begin{bmatrix} \tilde{c}_0 & \tilde{c}_1 & \cdots & \tilde{c}_s & \cdots & \tilde{c}_{n-m} \\ \tilde{b}_1 & \tilde{N}_{1,1} & \cdots & \tilde{N}_{1,s} & \cdots & \tilde{N}_{1,n-m} \\ \vdots & \vdots & \ddots & \vdots & \ddots & \vdots \\ \tilde{b}_r & \tilde{N}_{r,1} & \cdots & \tilde{N}_{r,s} & \cdots & \tilde{N}_{r,n-m} \\ \vdots & \vdots & \ddots & \vdots & \ddots & \vdots \\ \tilde{b}_m & \tilde{N}_{m,1} & \cdots & \tilde{N}_{m,s} & \cdots & \tilde{N}_{m,n-m} \end{bmatrix} \begin{pmatrix} 1 \\ x_{N,1} \\ \vdots \\ x_{N,s} \\ \vdots \\ x_{N,n-m} \end{pmatrix}. \tag{2.26}$$

These matrix expressions, which give the objective-function and basic-variable values in terms of the non-basic-variable values, form the algebraic backbone of the Simplex method, which is developed in Section 2.5.

Example 2.6 Consider the Electricity-Production Problem, which is introduced in Section 2.1.1. When converted to standard form, we can write the objective-function coefficients as:

$$c^\top = \begin{pmatrix} -1 & -1 & 0 & 0 & 0 & 0 \end{pmatrix},$$

the constraint coefficients as:

$$A = \begin{bmatrix} 2/3 & 1 & 1 & 0 & 0 & 0 \\ 2 & 1 & 0 & -1 & 0 & 0 \\ 1 & 0 & 0 & 0 & 1 & 0 \\ 0 & 1 & 0 & 0 & 0 & 1 \end{bmatrix},$$

and the constraint right-hand-side constants as:

$$b^\top = \begin{pmatrix} 18 & 8 & 12 & 16 \end{pmatrix}.$$

If we let $x_B = (x_1, x_2, x_4, x_5)$ and $x_N = (x_3, x_6)$, then we would have:

$$c_B^\top = \begin{pmatrix} -1 & -1 & 0 & 0 \end{pmatrix},$$

$$c_N^\top = \begin{pmatrix} 0 & 0 \end{pmatrix},$$

$$B = \begin{bmatrix} 2/3 & 1 & 0 & 0 \\ 2 & 1 & -1 & 0 \\ 1 & 0 & 0 & 1 \\ 0 & 1 & 0 & 0 \end{bmatrix},$$

and:

$$N = \begin{bmatrix} 1 & 0 \\ 0 & 0 \\ 0 & 0 \\ 0 & 1 \end{bmatrix}.$$

Using the definitions above, and Equations (2.24) and (2.25) in particular, we have:

$$\tilde{b} = B^{-1}b = \begin{bmatrix} 2/3 & 1 & 0 & 0 \\ 2 & 1 & -1 & 0 \\ 1 & 0 & 0 & 1 \\ 0 & 1 & 0 & 0 \end{bmatrix}^{-1} \begin{pmatrix} 18 \\ 8 \\ 12 \\ 16 \end{pmatrix} = \begin{pmatrix} 3 \\ 16 \\ 14 \\ 9 \end{pmatrix},$$

$$\tilde{N} = -B^{-1}N = \begin{bmatrix} 2/3 & 1 & 0 & 0 \\ 2 & 1 & -1 & 0 \\ 1 & 0 & 0 & 1 \\ 0 & 1 & 0 & 0 \end{bmatrix}^{-1} \begin{bmatrix} 1 & 0 \\ 0 & 0 \\ 0 & 0 \\ 0 & 1 \end{bmatrix} = \begin{bmatrix} -3/2 & 3/2 \\ 0 & -1 \\ -3 & 2 \\ 3/2 & -3/2 \end{bmatrix},$$

$$\tilde{c}_0 = c_B^\top \tilde{b} = \begin{pmatrix} -1 & -1 & 0 & 0 \end{pmatrix} \begin{pmatrix} 3 \\ 16 \\ 14 \\ 9 \end{pmatrix} = -19,$$

and:

$$\tilde{c}^\top = c_B^\top \tilde{N} + c_N^\top = \begin{pmatrix} -1 & -1 & 0 & 0 \end{pmatrix} \begin{bmatrix} -3/2 & 3/2 \\ 0 & -1 \\ -3 & 2 \\ 3/2 & -3/2 \end{bmatrix} + \begin{pmatrix} 0 & 0 \end{pmatrix} = \begin{pmatrix} 3/2 & -1/2 \end{pmatrix}.$$

These simple matrix operations can be effortlessly carried out using the public-domain Octave software package [6] or the MATLAB commercial software package [10].

$\square$

2.4.2 The Tableau

Matrix expression (2.26) forms the backbone of the Simplex method. This expression can be even more compactly arranged in what is called a **tableau**, which takes the general form shown in Table 2.6.

Table 2.6 General form of tableau

	1	$x_{N,1}$	$\cdots$	$x_{N,s}$	$\cdots$	$x_{N,n-m}$	$\Longleftarrow$ non-basic-variable block
z	$\tilde{c}_0$	$\tilde{c}_1$	$\cdots$	$\tilde{c}_s$	$\cdots$	$\tilde{c}_{n-m}$	$\Longleftarrow$ objective-function block
$x_{B,1}$	$\tilde{b}_1$	$\tilde{N}_{1,1}$	$\cdots$	$\tilde{N}_{1,s}$	$\cdots$	$\tilde{N}_{1,n-m}$	
$\vdots$	$\vdots$	$\vdots$	$\ddots$	$\vdots$	$\ddots$	$\vdots$	
$x_{B,r}$	$\tilde{b}_r$	$\tilde{N}_{r,1}$	$\cdots$	$\tilde{N}_{r,s}$	$\cdots$	$\tilde{N}_{r,n-m}$	$\Longleftarrow$ basic-variable block
$\vdots$	$\vdots$	$\vdots$	$\ddots$	$\vdots$	$\ddots$	$\vdots$	
$x_{B,m}$	$\tilde{b}_m$	$\tilde{N}_{m,1}$	$\cdots$	$\tilde{N}_{m,s}$	$\cdots$	$\tilde{N}_{m,n-m}$	

Table 2.6 identifies three blocks of rows in the tableau. Comparing these three blocks to Equation (2.26) provides some insight into the structure of the tableau. First, the bottom 'basic-variable block' of the tableau is associated with all but the first row of Equation (2.26). These rows of Equation (2.26) each have one basic variable on their left-hand sides, which are in the first column of the basic-variable block of the tableau. Moreover, the right-hand sides of these rows of Equation (2.26) have $\tilde{b}$ and $\tilde{N}$ terms, all of which are in the second and remaining columns of the basic-variable block of the tableau.

Next, the 'objective-function block' of the tableau corresponds to the first row of Equation (2.26). It contains z, which is on the left-hand side of the first row of Equation (2.26), in its first column and the $\tilde{c}$ terms in the remaining columns. Finally, inspecting the 'non-basic-variable block' of the tableau reveals that each column in the tableau is associated with a non-basic variable. These non-basic variables appear in the vector on the right-hand side of Equation (2.26).

It is important to stress that each basic solution has a tableau associated with it. That is because once the variables are partitioned into basic and non-basic variables, that partition determines the entries that label each row and column of the tableau. Moreover, the values of $\tilde{b}$, $\tilde{c}$, and $\tilde{N}$ that go in the tableau are also determined by which columns of the A matrix are put into the B and N submatrices.

The tableau also allows us to easily find a basic variable by setting all of the non-basic variables equal to zero (as noted in Sections 2.3.2 and 2.4.1). If we do this, we obtain basic-variable values:

$$x_B = \tilde{b},$$

and the objective-function value:

$$z = \tilde{c}_0.$$

These values for the basic variables and objective function follow immediately from the discussions and algebraic manipulations of the LPP carried out in Sections 2.3.2 and 2.4.1.

Example 2.7   Recall Example 2.6. If we let $x_B = (x_1, x_2, x_4, x_5)$ and $x_N = (x_3, x_6)$, then we would have:

$$\tilde{b} = B^{-1}b = \begin{pmatrix} 3 \\ 16 \\ 14 \\ 9 \end{pmatrix},$$

$$\tilde{N} = -B^{-1}N = \begin{bmatrix} -3/2 & 3/2 \\ 0 & -1 \\ -3 & 2 \\ 3/2 & -3/2 \end{bmatrix},$$

$$\tilde{c}_0 = c_B^\top \tilde{b} = -19,$$

and:

$$\tilde{c}^\top = c_B^\top \tilde{N} + c_N^\top = \left(3/2 \ -1/2\right).$$

The tableau associated with this basic solution is shown in Table 2.7. The 'B' and 'N' subscripts on the basic and non-basic variables have been omitted, because it is clear from the way that the tableau is arranged that x_1, x_2, x_4, and x_5 are basic variables while x_3 and x_6 are non-basic variables.

Table 2.7 Tableau for Example 2.7

	1	x_3	x_6
z	-19	$3/2$	$-1/2$
x_1	3	$-3/2$	$3/2$
x_2	16	0	-1
x_4	14	-3	2
x_5	9	$3/2$	$-3/2$

We can also determine the basic-variable and objective-function values when the non-basic variables are fixed equal to zero. The basic variables take on the values given in the second column of the tableau that are next to each basic variable (*i.e.*, $(x_1, x_2, x_4, x_5) = (3, 16, 14, 9)$ when the non-basic variables are set equal to zero) and the objective function takes on the value next to it in the second column of the tableau (*i.e.*, $z = -19$). □

2.4.3 *Pivoting*

One way of characterizing a basic solution is by which variables are basic variables (we also say that such variables are in the basis) and which variables are non-basic variables (we say that such variables are not in the basis). One of the features of the Simplex method is that the basic solutions found in each successive iteration differ by one variable. That is to say, if we compare the basic solutions found in two successive iterations, one of the basic variables in the first basic solution will be a non-basic variable in the second solution and one of the non-basic variables in the first solution will be a basic variable in the second solution.

Section 2.4.1 shows how the basic-variable and objective-function values can be expressed in terms of the values of the non-basic variables. Section 2.4.2 further shows how this information can be compactly written in tableau form. As successive iterations of the Simplex method are carried out, the tableau must be updated to reflect the fact that we move from one basic solution to another. Of course, the tableau could be updated in each iteration simply by redefining B, N, c_B, and c_N and applying Equations (2.24) and (2.25) to compute $\tilde{b}$, $\tilde{N}$, and $\tilde{c}$. This can be very computationally expensive, however, especially if the Simplex method is being applied by hand.

There is, however, a shortcut to update the tableau, which is called **pivoting**. Pivoting relies on the property of successive basic solutions found in the Simplex method that they differ only in that one basic variable leaves that basis and one non-basic variable enters the basis. To demonstrate the pivoting operation, suppose that we are currently at a basic solution and would like to move to a new basic solution. In the new basic solution there is a basic variable, $x_{B,r}$, that leaves the basis and a non-basic variable, $x_{N,s}$, that enters the basis. Table 2.8 shows the initial tableau before the pivoting operation. For notational convenience, we omit the 'B,' and 'N' subscripts on the variables exiting and entering the basis. Thus, these two variables are labeled x_r and x_s in the tableau.

Table 2.8 Initial tableau before pivoting

	1	$x_{N,1}$	$\cdots$	x_s	$\cdots$	$x_{N,n-m}$
z	$\tilde{c}_0$	$\tilde{c}_1$	$\cdots$	$\tilde{c}_s$	$\cdots$	$\tilde{c}_{n-m}$
$x_{B,1}$	$\tilde{b}_1$	$\tilde{N}_{1,1}$	$\cdots$	$\tilde{N}_{1,s}$	$\cdots$	$\tilde{N}_{1,n-m}$
$\vdots$	$\vdots$	$\vdots$		$\vdots$		$\vdots$
x_r	$\tilde{b}_r$	$\tilde{N}_{r,1}$	$\cdots$	$\tilde{N}_{r,s}$	$\cdots$	$\tilde{N}_{r,n-m}$
$\vdots$	$\vdots$	$\vdots$		$\vdots$		$\vdots$
$x_{B,m}$	$\tilde{b}_m$	$\tilde{N}_{m,1}$	$\cdots$	$\tilde{N}_{m,s}$	$\cdots$	$\tilde{N}_{m,n-m}$

To derive the pivoting operation, we first write the row associated with x_r in Table 2.8 explicitly as:

$$x_r = \tilde{b}_r + \tilde{N}_{r,1}x_{N,1} + \cdots + \tilde{N}_{r,s}x_s + \cdots + \tilde{N}_{r,n-m}x_{N,n-m}. \qquad (2.27)$$

We then manipulate Equation (2.24) to express x_s as a function of x_r and the remaining non-basic variables:

$$x_s = -\frac{\tilde{b}_r}{\tilde{N}_{r,s}} - \frac{\tilde{N}_{r,1}}{\tilde{N}_{r,s}}x_{N1} - \cdots + \frac{1}{\tilde{N}_{r,s}}x_r - \cdots - \frac{\tilde{N}_{r,n-m}}{\tilde{N}_{r,s}}x_{N,n-m}. \tag{2.28}$$

Next, we write the objective-function row of the tableau in Table 2.8 explicitly as:

$$z = \tilde{c}_0 + \tilde{c}_1 x_{N,1} + \cdots + \tilde{c}_s x_s + \cdots + \tilde{c}_{n-m} x_{N,n-m}. \tag{2.29}$$

We then use Equation (2.28) to substitute for x_s in Equation (2.29), which gives:

$$z = \left(\tilde{c}_0 - \frac{\tilde{c}_s}{\tilde{N}_{r,s}}\tilde{b}_r\right) + \left(\tilde{c}_1 - \tilde{c}_s\frac{\tilde{N}_{r,1}}{\tilde{N}_{r,s}}\right)x_{N,1} + \cdots + \frac{\tilde{c}_s}{\tilde{N}_{r,s}}x_r \tag{2.30}$$
$$+ \cdots + \left(\tilde{c}_{n-m} - \tilde{c}_s\frac{\tilde{N}_{r,n-m}}{\tilde{N}_{r,s}}\right)x_{N,n-m}.$$

Next we write the row associated with $x_{B,1}$ in Table 2.8 as:

$$x_{B,1} = \tilde{b}_1 + \tilde{N}_{1,1}x_{N,1} + \cdots + \tilde{N}_{1,s}x_s + \cdots + \tilde{N}_{1,n-m}x_{N,n-m}.$$

If we use Equation (2.28) to substitute for x_s in this expression we have:

$$x_{B,1} = \left(\tilde{b}_1 - \frac{\tilde{N}_{1,s}}{\tilde{N}_{r,s}}\tilde{b}_r\right) + \left(\tilde{N}_{1,1} - \tilde{N}_{1,s}\frac{\tilde{N}_{r,1}}{\tilde{N}_{r,s}}\right)x_{N,1} + \cdots + \frac{\tilde{N}_{1,s}}{\tilde{N}_{r,s}}x_r \tag{2.31}$$
$$+ \cdots + \left(\tilde{N}_{1,n-m} - \tilde{N}_{1,s}\frac{\tilde{N}_{r,n-m}}{\tilde{N}_{r,s}}\right)x_{N,n-m}.$$

A similar manipulation of the row associated with $x_{B,m}$ in Table 2.8 yields:

$$x_{Bm} = \left(\tilde{b}_m - \frac{\tilde{N}_{m,s}}{\tilde{N}_{r,s}}\tilde{b}_r\right) + \left(\tilde{N}_{m,1} - \tilde{N}_{m,s}\frac{\tilde{N}_{r,1}}{\tilde{N}_{r,s}}\right)x_{N,1} + \cdots + \frac{\tilde{N}_{m,s}}{\tilde{N}_{r,s}}x_r \tag{2.32}$$
$$+ \cdots + \left(\tilde{N}_{m,n-m} - \tilde{N}_{m,s}\frac{\tilde{N}_{r,n-m}}{\tilde{N}_{r,s}}\right)x_{N,n-m}. \tag{2.33}$$

Using Equations (2.28), (2.30), (2.31), and (2.32), which express x_s, z, $x_{B,1}$, and $x_{B,m}$ in terms of x_r, we can update the tableau to that given in Table 2.9.

Table 2.9 Updated tableau after pivoting

	1	$x_{N,1}$	$\cdots$	x_r	$\cdots$	$x_{N,n-m}$
z	$\tilde{c}_0 - \dfrac{\tilde{c}_s}{\tilde{N}_{r,s}}\tilde{b}_r$	$\tilde{c}_1 - \tilde{c}_s\dfrac{\tilde{N}_{r,1}}{\tilde{N}_{r,s}}$	$\cdots$	$\dfrac{\tilde{c}_s}{\tilde{N}_{r,s}}$	$\cdots$	$\tilde{c}_{n-m} - \tilde{c}_s\dfrac{\tilde{N}_{r,n-m}}{\tilde{N}_{r,s}}$
$x_{B,1}$	$\tilde{b}_1 - \dfrac{\tilde{N}_{1,s}}{\tilde{N}_{r,s}}\tilde{b}_r$	$\tilde{N}_{1,1} - \tilde{N}_{1,s}\dfrac{\tilde{N}_{r,1}}{\tilde{N}_{r,s}}$	$\cdots$	$\dfrac{\tilde{N}_{1,s}}{\tilde{N}_{r,s}}$	$\cdots$	$\tilde{N}_{1,n-m} - \tilde{N}_{1,s}\dfrac{\tilde{N}_{r,n-m}}{\tilde{N}_{r,s}}$
$\vdots$	$\vdots$	$\vdots$	$\ddots$	$\vdots$	$\ddots$	$\vdots$
x_s	$-\dfrac{\tilde{b}_r}{\tilde{N}_{r,s}}$	$-\dfrac{\tilde{N}_{r,1}}{\tilde{N}_{r,s}}$	$\cdots$	$\dfrac{1}{\tilde{N}_{r,s}}$	$\cdots$	$-\dfrac{\tilde{N}_{r,n-m}}{\tilde{N}_{r,s}}$
$\vdots$	$\vdots$	$\vdots$	$\ddots$	$\vdots$	$\ddots$	$\vdots$
$x_{B,m}$	$\tilde{b}_m - \dfrac{\tilde{N}_{m,s}}{\tilde{N}_{r,s}}\tilde{b}_r$	$\tilde{N}_{m,1} - \tilde{N}_{m,s}\dfrac{\tilde{N}_{r,1}}{\tilde{N}_{r,s}}$	$\cdots$	$\dfrac{\tilde{N}_{m,s}}{\tilde{N}_{r,s}}$	$\cdots$	$\tilde{N}_{m,n-m} - \tilde{N}_{m,s}\dfrac{\tilde{N}_{r,n-m}}{\tilde{N}_{r,s}}$

Example 2.8 Recall Example 2.7 and that if we let $x_B = (x_1, x_2, x_4, x_5)$ and $x_N = (x_3, x_6)$ then the tableau is given by Table 2.7. Let us conduct one pivot operation in which x_5 leaves the basis and x_6 enters it. From the x_5 row of Table 2.7 we have:

$$x_5 = 9 + \frac{3}{2}x_3 - \frac{3}{2}x_6.$$

This can be rewritten as:

$$x_6 = 6 + x_3 - \frac{1}{2}x_5. \tag{2.34}$$

Substituting Equation (2.34) into the objective-function row of Table 2.7:

$$z = -19 + \frac{3}{2}x_3 - \frac{1}{2}x_6,$$

gives:

$$z = -22 + x_3 + \frac{1}{3}x_5. \tag{2.35}$$

We next consider the rows of Table 2.7, which are:

$$x_1 = 3 - \frac{3}{2}x_3 + \frac{3}{2}x_6,$$

$$x_2 = 16 - x_6,$$

and:

$$x_4 = 14 - 3x_3 + 2x_6.$$

Substituting Equation (2.34) into these three equations gives:

$$x_1 = 12 - x_5, \tag{2.36}$$

$$x_2 = 10 - x_3 + \frac{2}{3}x_5, \tag{2.37}$$

and:

$$x_4 = 26 - x_3 - \frac{4}{3}x_5. \tag{2.38}$$

Substituting Equations (2.34) through (2.38) into Table 2.7 gives an updated tableau after the pivot operation, which is given in Table 2.10.

Table 2.10 Updated tableau after conducting pivot operation in Example 2.8

	1	x_3	x_5
z	-22	1	1/3
x_1	12	0	-1
x_2	10	-1	2/3
x_4	26	-1	$-4/3$
x_6	6	1	$-1/2$

It is finally worth noting that we have performed a step-by-step pivot operation using the expressions derived above. One could also obtain the new tableau directly by using the expressions in Table 2.9. Doing so yields the same tableau.

$\square$

2.5   The Simplex Method

The Simplex method is the most commonly used approach to solving LPPs. At its heart, the Simplex method relies on two important properties of linear optimization problems. First, as noted in Sections 2.1.1 and 2.3.1, if an LPP has an optimal solution, then there must be at least one extreme point of the feasible set that is optimal. Secondly, as discussed in Section 2.3.2, there is a one-to-one correspondence between extreme points of the feasible set and basic feasible solutions. That is to say, each basic feasible solution corresponds to an extreme point of the feasible set and each extreme point of the feasible set corresponds to a basic feasible solution.

Building off of these two properties, the Simplex method solves an LPP by following two major steps. First, it works to find a starting basic feasible solution. Once it has found a basic feasible solution, the Simplex method iteratively determines if there is another corner of the feasible region next to the corner that the current basic feasible solution corresponds to that gives a better objective-function value. If there is, the Simplex method moves to this new basic feasible solution. Otherwise, the algorithm terminates because the basic feasible solution it is currently at is optimal.

In the following sections we describe each of these steps of the Simplex method in turn. We then provide an overview of the entire algorithm. We finally discuss some technical issues, such as guaranteeing that the Simplex method terminates and detecting if an LPP is unbounded, infeasible, or has multiple optimal solutions.

2.5.1 Finding an Initial Basic Feasible Solution

We know from the discussions in Sections 2.3.2 and 2.4.1 that finding a basic solution is relatively straightforward. All we must do is partition the variables into basic and non-basic variables, x_B and x_N, respectively. Once we have done this we determine B, N, c_B, and c_N and then compute:

$$\tilde{b} = B^{-1}b,$$

$$\tilde{N} = -B^{-1}N,$$

$$\tilde{c}_0 = c_B^\top \tilde{b},$$

and:

$$\tilde{c}^\top = c_B^\top \tilde{N} + c_N^\top.$$

To the extent possible, one can choose x_B in a way such that the B matrix is relatively easy to invert.

After these calculations are done, we can put them into a tableau, such as the one given in Table 2.11. We know from the discussion in Section 2.4.2 that for the chosen basic solution, the values of the basic variables can be easily read from the tableau as the value of $\tilde{b}$. We further know that if $\tilde{b} \geq 0$, then the basic solution that we have found is a basic feasible solution and no further work must be done (*i.e.*, we can proceed to the next step of the Simplex method, which is discussed in Section 2.5.2). Otherwise, we must conduct what is called a regularization step.

Table 2.11 The tableau for a starting basic solution

	1	$x_{N,1}$	$\cdots$	$x_{N,s}$	$\cdots$	$x_{N,n-m}$
z	$\tilde{c}_0$	$\tilde{c}_1$	$\cdots$	$\tilde{c}_s$	$\cdots$	$\tilde{c}_{n-m}$
$x_{B,1}$	$\tilde{b}_1$	$\tilde{N}_{1,1}$	$\cdots$	$\tilde{N}_{1,s}$	$\cdots$	$\tilde{N}_{1,n-m}$
$\vdots$	$\vdots$	$\vdots$	$\ddots$	$\vdots$	$\ddots$	$\vdots$
$x_{B,r}$	$\tilde{b}_r$	$\tilde{N}_{r,1}$	$\cdots$	$\tilde{N}_{r,s}$	$\cdots$	$\tilde{N}_{r,n-m}$
$\vdots$	$\vdots$	$\vdots$	$\ddots$	$\vdots$	$\ddots$	$\vdots$
$x_{B,m}$	$\tilde{b}_m$	$\tilde{N}_{m,1}$	$\cdots$	$\tilde{N}_{m,s}$	$\cdots$	$\tilde{N}_{m,n-m}$

In the regularization step we add one new column to the tableau, which is highlighted in boldface in Table 2.12. This added column has a new non-basic variable, which we call $x_{N,n-m+1}$, all ones in the basic-variable rows, and a value of K in the objective-function row. The value of K is chosen to be larger than any of the other existing values in the objective-function row of the tableau. We next conduct one pivot operation in which the added variable, $x_{N,n-m+1}$, becomes a basic variable. The basic variable that exits the basis is the one that has the smallest or most negative $\tilde{b}$ value. That is, the variable that exits the basis corresponds to:

$$\min \left\{ \tilde{b}_1, \ldots, \tilde{b}_r, \ldots, \tilde{b}_m \right\}.$$

Table 2.12 The tableau after the regularization step

	1	$x_{N,1}$	$\cdots$	$x_{N,s}$	$\cdots$	$x_{N,n-m}$	$x_{N,n-m+1}$
z	$\tilde{c}_0$	$\tilde{c}_1$	$\cdots$	$\tilde{c}_s$	$\cdots$	$\tilde{c}_{n-m}$	K
$x_{B,1}$	$\tilde{b}_1$	$\tilde{N}_{1,1}$	$\cdots$	$\tilde{N}_{1,s}$	$\cdots$	$\tilde{N}_{1,n-m}$	1
$\vdots$	$\vdots$	$\vdots$	$\ddots$	$\vdots$	$\ddots$	$\vdots$	$\vdots$
$x_{B,r}$	$\tilde{b}_r$	$\tilde{N}_{r,1}$	$\cdots$	$\tilde{N}_{r,s}$	$\cdots$	$\tilde{N}_{r,n-m}$	1
$\vdots$	$\vdots$	$\vdots$	$\ddots$	$\vdots$	$\ddots$	$\vdots$	$\vdots$
$x_{B,m}$	$\tilde{b}_m$	$\tilde{N}_{m,1}$	$\cdots$	$\tilde{N}_{m,s}$	$\cdots$	$\tilde{N}_{m,n-m}$	1

After this pivoting operation is conducted, we can guarantee that we have a basic feasible solution (*i.e.*, all of the $\tilde{b}$'s are non-negative after the tableau is updated). We now show this formally.

> **Regularization Property:** After conducting the regularization step and a pivot operation, the updated $\tilde{b}$ will be non-negative.

To show this, suppose that we denote the basic variable that will be exiting the basis as $x_{B,r}$. The basic-variable rows of the tableau in Table 2.12 can be written as:

$$x_{B,1} = \tilde{b}_1 + \tilde{N}_{1,1}x_{N,1} + \cdots + \tilde{N}_{1,s}x_{N,s} + \cdots \tilde{N}_{1,n-m}x_{N,n-m} + x_{N,n-m+1},$$

$$\vdots$$

$$x_{B,r} = \tilde{b}_r + \tilde{N}_{r,1}x_{N,1} + \cdots + \tilde{N}_{r,s}x_{N,s} + \cdots \tilde{N}_{r,n-m}x_{N,n-m} + x_{N,n-m+1},$$

$$\vdots$$

$$x_{B,m} = \tilde{b}_m + \tilde{N}_{m,1}x_{N,1} + \cdots + \tilde{N}_{m,s}x_{N,s} + \cdots \tilde{N}_{m,n-m}x_{N,n-m} + x_{N,n-m+1}.$$

After the pivot operation $x_{N,1}, \ldots, x_{N,n-m}$ will remain non-basic variables that are fixed equal to zero, thus these equations can be simplified to:

$$x_{B,1} = \tilde{b}_1 + x_{N,n-m+1}, \tag{2.39}$$

$$\vdots$$

$$x_{B,r} = \tilde{b}_r + x_{N,n-m+1}, \tag{2.40}$$

$$\vdots$$

$$x_{B,m} = \tilde{b}_m + x_{N,n-m+1}. \tag{2.41}$$

From Equation (2.40) we have the value that $x_{N,n-m+1}$ takes when it becomes a basic variable as:

$$x_{N,n-m+1} = x_{B,r} - \tilde{b}_r = -\tilde{b}_r,$$

where the second equality follows because we know $x_{B,r}$ is becoming a non-basic variable after the pivot operation is completed. Because r is chosen such that $\tilde{b}_r < 0$, we know that $x_{N,n-m+1} > 0$ after this pivot operation is completed.

Moreover, if we substitute this value of $x_{N,n-m+1}$ into the remainder of Equations (2.39) through (2.41) then we have:

$$x_{B,1} = \tilde{b}_1 - \tilde{b}_r,$$

$$\vdots$$

$$x_{B,m} = \tilde{b}_m - \tilde{b}_r.$$

Note, however, that because r is chosen such that it gives the most negative value of $\tilde{b}$, the right-hand sides of all of these equations are non-negative. Thus, our new basic solution is guaranteed to be feasible.

This Regularization Property implies that for any LPP we must do at most one regularization step to find a starting basic feasible solution. The idea of the regularization step is that we add a new **artificial variable** to the LPP and set its value in a way that all of the variables take on non-negative values. Of course, adding this new variable is 'cheating' in the sense that it is not a variable of the original LPP. Thus, adding the artificial variable changes the problem's feasible region.

The value of K in the new tableau is intended to take care of this. As we see in Section 2.5.2, the Simplex method determines whether the current basic feasible solution is optimal by examining the values in objective-function row of the tableau.

The high value of K works to add a penalty to the objective function for allowing the artificial variable to take on a value greater than zero (it should be easy to convince yourself that if the artificial variable is equal to zero in a basic feasible solution, then that solution is feasible in the original LPP without the artificial variable added). The high value for K will have the Simplex method try to reduce the value of the artificial variable to zero. Once the simplex method drives the value of the artificial variable to zero, then we have found a basic feasible solution that is feasible in the original LPP without the artificial variable added. Once the value of the artificial variable has been driven to zero, it can be removed from the tableau and the Simplex method can be further applied without that variable in the problem.

It should be further noted that in some circumstances, an artificial variable may not need to be added to conduct the regularization step. This would be the case if the starting tableau already has a non-basic variable with a column of ones in the basic-variable rows. If so, one can conduct a pivot operation in which this non-basic variable enters the basis and the basic variable with the most negative $\tilde{b}$ value exits to obtain a starting basic feasible solution.

2.5.2 Moving Between Basic Feasible Solutions

The main optimization step of the Simplex method checks to see whether the current basic feasible solution is optimal or not. If it is optimal, then the method terminates. Otherwise, the Simplex method moves to a new basic feasible solution. To determine whether the current basic feasible solution is optimal or not, we examine the objective-function row of the tableau.

Recall from Equation (2.25) that the objective-function row of the tableau can be written as:

$$z = \tilde{c}_0 + \tilde{c}^\top x_N,$$

which expresses the objective function value of the LPP in terms of the values of the non-basic variables. If any of the elements of $\tilde{c}$ are negative, this implies that increasing the value of the corresponding non-basic variable from zero to some positive value improves (decreases) the objective function. Thus, the Simplex method determines whether the current basic feasible solution is optimal or not by checking the signs of the $\tilde{c}$ values in the tableau. If they are all non-negative, then the current solution is optimal and the algorithm terminates. Otherwise, if at least one of the values is negative, then the current solution is not optimal.

In this latter case that one or more of the $\tilde{c}$'s is negative, one of the non-basic variables with a negative $\tilde{c}$ is chosen to enter the basis. Any non-basic variable with a negative $\tilde{c}$ can be chosen to enter the basis. However, in practice it is common to choose the non-basic variable with the most negative $\tilde{c}$. This is because each unit increase in the value of the non-basic variable with the most negative $\tilde{c}$ gives the greatest objective-function decrease. We let s denote the index of the non-basic variable that enters the basis.

The next step of the Simplex method is to determine which basic variable exits the basis when $x_{N,s}$ enters it. To determine this, we note that after we swap $x_{N,s}$ for whatever basic variable exits the basis, we want to ensure that we are still at a basic feasible solution. This means that we want to ensure that the basic variables all have non-negative values after the variables are swapped. To ensure this, we examine the problem tableau, which is shown in Table 2.13.

Table 2.13 The tableau for the current basic feasible solution

		1	$x_{N,1}$	$\cdots$	$x_{N,s}$	$\cdots$	$x_{N,n-m}$
z		$\tilde{c}_0$	$\tilde{c}_1$	$\cdots$	$\tilde{c}_s$	$\cdots$	$\tilde{c}_{n-m}$
$x_{B,1}$		$\tilde{b}_1$	$\tilde{N}_{1,1}$	$\cdots$	$\tilde{N}_{1,s}$	$\cdots$	$\tilde{N}_{1,n-m}$
$\vdots$		$\vdots$	$\vdots$	$\ddots$	$\vdots$	$\ddots$	$\vdots$
$x_{B,r}$		$\tilde{b}_r$	$\tilde{N}_{r,1}$	$\cdots$	$\tilde{N}_{r,s}$	$\cdots$	$\tilde{N}_{r,n-m}$
$\vdots$		$\vdots$	$\vdots$	$\ddots$	$\vdots$	$\ddots$	$\vdots$
$x_{B,m}$		$\tilde{b}_m$	$\tilde{N}_{m,1}$	$\cdots$	$\tilde{N}_{m,s}$	$\cdots$	$\tilde{N}_{m,n-m}$

The basic-variable rows of the tableau can be expanded as:

$$x_{B,1} = \tilde{b}_1 + \tilde{N}_{1,1}x_{N,1} + \cdots + \tilde{N}_{1,s}x_{N,s} + \cdots \tilde{N}_{1,n-m}x_{N,n-m},$$

$$\vdots$$

$$x_{B,r} = \tilde{b}_r + \tilde{N}_{r,1}x_{N,1} + \cdots + \tilde{N}_{r,s}x_{N,s} + \cdots \tilde{N}_{r,n-m}x_{N,n-m},$$

$$\vdots$$

$$x_{B,m} = \tilde{b}_m + \tilde{N}_{m,1}x_{N,1} + \cdots + \tilde{N}_{m,s}x_{N,s} + \cdots \tilde{N}_{m,n-m}x_{N,n-m}.$$

These equations define the value of the basic variables in terms of the values of the non-basic variables. These equations simplify to:

$$x_{B,1} = \tilde{b}_1 + \tilde{N}_{1,s}x_{N,s}, \tag{2.42}$$

$$\vdots$$

$$x_{B,r} = \tilde{b}_r + \tilde{N}_{r,s}x_{N,s}, \tag{2.43}$$

$$\vdots$$

$$x_{B,m} = \tilde{b}_m + \tilde{N}_{m,s}x_{N,s}, \tag{2.44}$$

because after $x_{N,s}$ enters the basis the other variables that are non-basic at the current basic feasible solution remain non-basic. Inspecting Equations (2.42)–(2.44), we see that increasing the value of $x_{N,s}$ has two possible effects. A basic variable that has a negative $\tilde{N}$ coefficient on $x_{N,s}$ in the equation defining its value decreases as $x_{N,s}$ increases. On the other hand, a basic variable that has a zero or positive $\tilde{N}$ coefficient on $x_{N,s}$ remains the same or increases as $x_{N,s}$ increases. Thus, to ensure that our new basic solution is feasible, we only need to concern ourselves with basic variables that have a negative $\tilde{N}$ coefficient in the tableau (because we do not want any of the basic variables to become negative at our new basic solution).

This means that we can restrict attention to the subset of Equations (2.42)–(2.44) that have a negative $\tilde{N}$ coefficient on $x_{N,s}$. We write these equations as:

$$x_{B,1} = \tilde{b}_1 + \tilde{N}_{1,s} x_{N,s},$$

$$\vdots$$

$$x_{B,r} = \tilde{b}_r + \tilde{N}_{r,s} x_{N,s},$$

$$\vdots$$

$$x_{B,m'} = \tilde{b}_{m'} + \tilde{N}_{m',s} x_{N,s},$$

where we let $x_{B,1}, \ldots, x_{B,m'}$ be the subset of basic variables that have negative $\tilde{N}$ coefficients on $x_{N,s}$. We want all of our basic variables to be non-negative when we increase the value of $x_{N,s}$, which we can write as:

$$x_{B,1} = \tilde{b}_1 + \tilde{N}_{1,s} x_{N,s} \geq 0,$$

$$\vdots$$

$$x_{B,r} = \tilde{b}_r + \tilde{N}_{t,s} x_{N,s} \geq 0,$$

$$\vdots$$

$$x_{B,m'} = \tilde{b}_{m'} + \tilde{N}_{m',s} x_{N,s} \geq 0.$$

Subtracting $\tilde{b}$ from both sides of each inequality and dividing both sides of each by $\tilde{N}$ gives:

$$x_{N,s} \leq -\frac{\tilde{b}_1}{\tilde{N}_{1,s}},$$

$$\vdots$$

$$x_{N,s} \leq -\frac{\tilde{b}_r}{\tilde{N}_{r,s}},$$

$$\vdots$$

$$x_{N,s} \leq -\frac{\tilde{b}_{m'}}{\tilde{N}_{m',s}},$$

where the directions of the inequalities change because we are focusing on basic variables that have a negative $\tilde{N}$ coefficient on $x_{N,s}$. Also note that because we are only examining basic variables that have a negative $\tilde{N}$ coefficient on $x_{N,s}$, all of the ratios, $-\tilde{b}_1/\tilde{N}_{1,s}, \ldots, -\tilde{b}_{m'}/\tilde{N}_{m',s}$, are positive.

Taken together, these inequalities imply that the largest $x_{N,s}$ can be made without causing any of the basic variables to become negative is:

$$x_{N,s} = \min\left\{-\frac{\tilde{b}_1}{\tilde{N}_{1,s}}, \ldots, -\frac{\tilde{b}_r}{\tilde{N}_{r,s}}, \ldots, -\frac{\tilde{b}_{m'}}{\tilde{N}_{m',s}}\right\}.$$

Because we restricted our attention to basic variables that have a negative $\tilde{N}$ coefficient on $x_{N,s}$ in the tableau, we can also write this maximum value that $x_{N,s}$ can take as:

$$x_{N,s} = \min_{i=1,\ldots,m:\tilde{N}_{i,s}<0}\left\{-\frac{\tilde{b}_i}{\tilde{N}_{i,s}}\right\}. \tag{2.45}$$

If we define r as the index of the basic variable that satisfies condition (2.45) then we know that when we increase $x_{N,s}$ to:

$$x_{N,s} = -\frac{\tilde{b}_r}{\tilde{N}_{r,s}},$$

$x_{B,r}$ becomes equal to zero. This means that $x_{B,r}$ becomes the new non-basic variable when $x_{N,s}$ becomes a basic variable.

Once the basic variable that enters the basis, $x_{N,s}$, and the non-basic variable that exits the basis, $x_{B,r}$, are identified, a pivot operation (*cf.* Section 2.4.3) is conducted and the tableau is updated. The process outlined in the current section to determine if the new basic feasible solution found after the pivoting operation is optimal or not is then applied to the updated tableau. If the updated tableau (specifically, the values in the objective-function row) indicates that the new basic feasible solution is optimal, then the Simplex method terminates. Otherwise, a non-basic variable is chosen to enter the basis and the ratio test shown in Equation (2.45) is conducted to determine which basic variable exits the basis. This process is repeated iteratively until the Simplex method terminates.

2.5.3 Simplex Method Algorithm

We now provide a more general outline of how to apply the Simplex method to solve any linear optimization problem. We assume that the problem has been converted to standard form and that we have chosen a starting set of basic and non-basic variables. Note, however, that the basic and non-basic variables chosen do not necessarily have to give us a basic feasible solution. If they do not, we conduct the regularization step to make the basic-variable values all non-negative. Otherwise, we skip the regularization step and proceed to conducting Simplex iterations to move between basic feasible solutions while improving the objective function.

The following algorithm outlines the major steps of the Simplex method. We begin in Step 2 by computing the starting tableau, based on the chosen partition of the variables into basic and non-basic variables. In Step 3 we determine if the regularization step is needed. Recall that if $\tilde{b} \geq 0$, then our starting basic solution is also a basic feasible solution and regularization is not needed. Otherwise, if at least one component of $\tilde{b}$ is negative, regularization must be conducted. Regularization consists of first adding an artificial variable to the tableau in Step 4. We then select which basic variable exits the basis in Step 5 and conduct a pivot operation in Step 6. Recall from the discussion in Section 2.5.1 that after this one regularization step, the new $\tilde{b}$ vector is guaranteed to be non-negative and no further regularization steps are needed.

Simplex Method Algorithm

1: **procedure** SIMPLEX METHOD
2:     Compute $\tilde{b} \leftarrow B^{-1}b$, $\tilde{N} \leftarrow -B^{-1}N$, $\tilde{c}_0 \leftarrow c_B^\top \tilde{b}$, $\tilde{c}^\top \leftarrow c_B^\top \tilde{N} + c_N^\top$
3:     **if** $\tilde{b} \ngeq 0$ **then**
4:         Add non-basic variable, $x_{N,n-m+1}$, with ones in basic-variable rows and K larger than all other $\tilde{c}$'s in objective-function row of tableau
5:         $r \leftarrow \arg\min_i\{\tilde{b}_i\}$
6:         Conduct a pivot in which $x_{N,n-m+1}$ enters the basis and $x_{B,r}$ exits
7:     **end if**
8:     **while** $c \ngeq 0$ **do**
9:         Select a non-basic variable, $_{N,s}$, with $\tilde{c}_s < 0$ to enter the basis
10:       Select a basic variable, $x_{B,r}$, with $r = \arg\min_{i:\tilde{N}_{i,s}<0} -\tilde{b}_i/\tilde{N}_{i,s}$ to exit the basis
11:       Conduct a pivot in which $x_{N,s}$ enters the basis and $x_{B,r}$ exits
12:     **end while**
13: **end procedure**

The main Simplex iteration takes place in Steps 8 through 12. We first determine in Step 8 whether we are currently at an optimal basic feasible solution. If the $\tilde{c}$ vector is non-negative, this means that we cannot improve the objective function by increasing the values of any of the non-basic variables. Thus, the current basic feasible solution is optimal. This means that Step 8 constitutes the termination criterion of the Simplex method—we conduct iterations until $\tilde{c} \geq 0$.

If at least one component of $\tilde{c}$ is negative, then the objective-function value can be improved by increasing the value of the corresponding non-basic variable. This means that the current basic feasible solution is not optimal. In this case, one of the non-basic variables with a negative $\tilde{c}$ coefficient is chosen to enter the basis (Step 9). In Step 10 the ratio test outlined in Equation (2.45) is conducted to determine which basic variable exits the basis. A pivot operation is then conducted to update the tableau in Step 11. After the tableau is updated we return to Step 8 to determine if the new basic feasible solution is optimal. If it is, the algorithm terminates, otherwise, the algorithm continues.

Example 2.9 Consider the standard form-version of the Electricity-Production Problem, which is introduced in Section 2.1.1 in matrix form. This matrix form is given by (2.20)–(2.22). Taking $x_B = (x_3, x_4, x_5, x_6)$ and $x_N = (x_1, x_2)$, we use the Simplex method to solve this LPP.

Using this starting partition of the variables into basic and non-basic variables, we can define:

$$c_B = \begin{pmatrix} 0 \\ 0 \\ 0 \\ 0 \end{pmatrix},$$

$$c_N = \begin{pmatrix} -1 \\ -1 \end{pmatrix},$$

$$B = \begin{bmatrix} 1 & 0 & 0 & 0 \\ 0 & -1 & 0 & 0 \\ 0 & 0 & 1 & 0 \\ 0 & 0 & 0 & 1 \end{bmatrix},$$

and:

$$N = \begin{bmatrix} 2/3 & 1 \\ 2 & 1 \\ 1 & 0 \\ 0 & 1 \end{bmatrix}.$$

Using Equations (2.24) and (2.25) we have:

$$\tilde{b} = B^{-1}b = \begin{bmatrix} 1 & 0 & 0 & 0 \\ 0 & -1 & 0 & 0 \\ 0 & 0 & 1 & 0 \\ 0 & 0 & 0 & 1 \end{bmatrix}^{-1} \begin{pmatrix} 18 \\ 8 \\ 12 \\ 16 \end{pmatrix} = \begin{pmatrix} 18 \\ -8 \\ 12 \\ 16 \end{pmatrix},$$

$$\tilde{N} = -B^{-1}N = -\begin{bmatrix} 1 & 0 & 0 & 0 \\ 0 & -1 & 0 & 0 \\ 0 & 0 & 1 & 0 \\ 0 & 0 & 0 & 1 \end{bmatrix}^{-1} \begin{bmatrix} 2/3 & 1 \\ 2 & 1 \\ 1 & 0 \\ 0 & 1 \end{bmatrix} = \begin{bmatrix} -2/3 & -1 \\ 2 & 1 \\ -1 & 0 \\ 0 & -1 \end{bmatrix},$$

$$\tilde{c}_0 = c_B^\top \tilde{b} = \begin{pmatrix} 0 & 0 & 0 & 0 \end{pmatrix} \begin{pmatrix} 18 \\ -8 \\ 12 \\ 16 \end{pmatrix} = 0,$$

and:

$$\tilde{c}^\top = c_B^\top \tilde{N} + c_N^\top = \begin{pmatrix} 0 & 0 & 0 & 0 \end{pmatrix} \begin{bmatrix} 2/3 & 1 \\ 2 & 1 \\ 1 & 0 \\ 0 & 1 \end{bmatrix} + \begin{pmatrix} -1 & -1 \end{pmatrix} = \begin{pmatrix} -1 & -1 \end{pmatrix}.$$

The starting tableau corresponding to this basic solution is shown in Table 2.14, where the 'B' and 'N' subscripts on the basic and non-basic variables are omitted. The starting basic solution has x_1 and x_2 (the variables in the original formulation given in Section 2.1.1) equal to zero and an objective-function value of zero. This starting basic solution is infeasible, however, because $\tilde{b}_4 = -8$ is negative, meaning that $x_4 = -8$. Figure 2.10 shows the feasible region of the LPP and the starting solution, further illustrating that the starting basic solution is infeasible.

Table 2.14 Starting tableau for Example 2.9

	1	x_1	x_2
z	0	-1	-1
x_3	18	$-2/3$	-1
x_4	-8	2	1
x_5	12	-1	0
x_6	16	0	-1

Fig. 2.10 Feasible region of the Electricity-Production Problem in Example 2.9 and starting basic solution

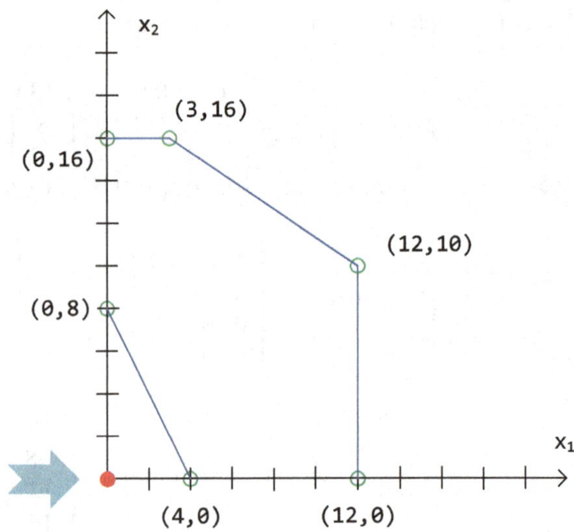

Because $\tilde{b}$ is not non-negative, a regularization step is needed at this point. To do this, we add a non-basic variable, which we denote x_7, and assign a value of $K = 2$, which is greater than all of the other values in the objective-function row. This selection of $K = 2$ is arbitrary. The regularization step can be conducted with any $K > -1$. At this point, the tableau is updated to that shown in Table 2.15. We next conduct a pivot operation, in which x_7 enters the basis. The basic variable that exits the basis is the one with index corresponding to:

$$\min \{\tilde{b}_3, \tilde{b}_4, \tilde{b}_5, \tilde{b}_6\} = \min \{18, -8, 12, 16\},$$

which is x_4.

Table 2.15 Tableau for Example 2.9 after the artificial variable, x_7, is added

	1	x_1	x_2	x_7
z	0	-1	-1	2
x_3	18	$-2/3$	-1	1
x_4	-8	2	1	1
x_5	12	-1	0	1
x_6	16	0	-1	1

Swapping x_7 and x_4 through a pivot operation gives the tableau shown in Table 2.16. Note that after conducting this regularization step, our new basic solution gives $(x_1, x_2) = (0, 0)$, which is infeasible in the original problem. This is consistent with our intuition in Section 2.5.1. Adding an artificial variable to an LPP is 'cheating' in the sense that we have added a new variable to find a starting basic solution in which all of the basic variables are non-negative. It is only after conducting Simplex iterations and (hopefully) driving the artificial variable, x_7, down to zero that we find a basic solution that is feasible in the original LPP. We will see this happen as we proceed with solving the problem.

Table 2.16 Tableau for Example 2.9 after regularization step is complete

	1	x_1	x_2	x_4
z	16	-5	-3	2
x_3	26	$-8/3$	-2	1
x_7	8	-2	-1	1
x_5	20	-3	-1	1
x_6	24	-2	-2	1

Now that we have a starting basic solution with non-negative values for the basic variables, we determine if the solution is optimal. This is done by examining the objective-function row of the tableau in Table 2.16. Seeing that both $\tilde{c}_1$ and $\tilde{c}_2$ are negative, we know that increasing either from zero will improve (decrease) the objective function. Because x_1 has a more negative objective-function coefficient, we chose x_1 to enter the basis. Note, however, that we could choose x_2 to enter the basis at the current iteration instead. The final solution that we find after finishing the Simplex method will be optimal regardless of the variable chosen to enter the basis. To determine the basic variable that exits the basis, we compute the ratios between $\tilde{b}$ and the column of negative $\tilde{N}$'s below x_1 in the tableau in Table 2.16. The basic variable to exit the basis is the one with index corresponding to:

$$\min_{\tilde{N}_{1,3}, \tilde{N}_{1,7}, \tilde{N}_{1,5}, \tilde{N}_{1,6} < 0} \left\{ -\frac{\tilde{b}_3}{\tilde{N}_{1,3}}, -\frac{\tilde{b}_7}{\tilde{N}_{1,7}}, -\frac{\tilde{b}_5}{\tilde{N}_{1,5}}, -\frac{\tilde{b}_6}{\tilde{N}_{1,6}} \right\} = \min \left\{ \frac{39}{4}, 4, \frac{20}{3}, 12 \right\},$$

which is x_7.

Thus, we conduct a pivot operation to swap x_1 and x_7, which gives the tableau shown in Table 2.17. This tableau gives the basic solution $(x_1, x_2) = (4, 0)$, which is feasible in the original problem. This can be verified by substituting these values of x_1 and x_2 into the original formulation given in Section 2.1.1. It can also be verified by observing that at this basic feasible solution we have the artificial variable x_7 equal to zero and $\tilde{b} > 0$. This means that the variable values satisfy the constraints of the original problem without needing the artificial variable any longer. Indeed, now that the artificial variable is equal to zero, we could drop it and its column from the tableau, as it will never again enter the basis. This is because we chose K in a way to ensure that the objective-function coefficient on the artificial variable never becomes negative again. The tableau also tells us that this basic feasible solution gives an objective function value of -4. Figure 2.11 shows the feasible region of the problem and our new basic solution after the first Simplex iteration, also illustrating that this solution is feasible.

Table 2.17 Tableau for Example 2.9 after completing one Simplex iteration

	1	x_7	x_2	x_4
z	-4	5/2	$-1/2$	$-1/2$
x_3	46/3	4/3	$-2/3$	$-1/3$
x_1	4	$-1/2$	$-1/2$	1/2
x_5	8	3/2	1/2	$-1/2$
x_6	16	1	-1	0

We now proceed by conducting another Simplex iteration using the tableau in Table 2.17. We first note that because the objective-function row of the tableau has negative values in it, the current basic feasible solution is not optimal. Increasing the values of either of x_2 or x_4 from zero improves the objective-function value. Moreover, both x_2 and x_4 have the same value in the objective-function row, thus we can arbitrarily choose either to enter the basis. We choose x_4 here. We next conduct the ratio test to determine which basic variable exits the basis. This will be the variable with index corresponding to:

$$\min_{\tilde{N}_{4,3},\tilde{N}_{4,1},\tilde{N}_{4,5},\tilde{N}_{4,6}<0}\left\{-\frac{\tilde{b}_3}{\tilde{N}_{4,3}},-\frac{\tilde{b}_1}{\tilde{N}_{4,1}},-\frac{\tilde{b}_5}{\tilde{N}_{4,5}},-\frac{\tilde{b}_6}{\tilde{N}_{4,6}}\right\}=\min\{46,/,18,/\},$$

where the slashes on the right-hand side of the equality indicate values of $\tilde{N}$ that are non-negative, and are, thus, excluded from consideration. Based on this test, x_5 is the variable to exit the basis. We then conduct a pivot operation to update the tableau to that shown in Table 2.18. Our new basic feasible solution has $(x_1, x_2) = (12, 0)$

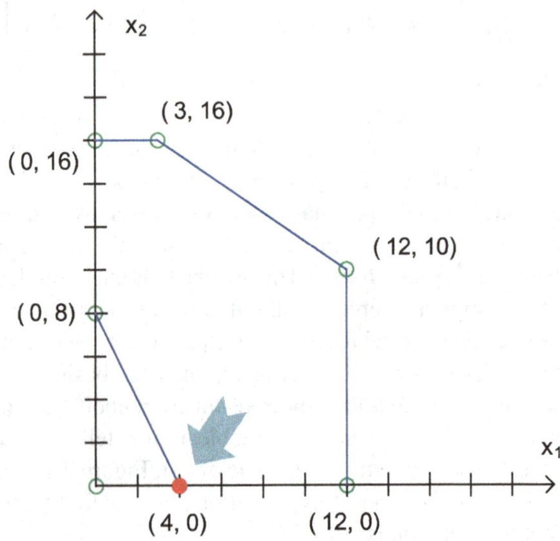

Fig. 2.11 Feasible region of the Electricity-Production Problem in Example 2.9 and basic feasible solution after completing one Simplex iteration

and gives an objective-function value of -12. Figure 2.12 shows the feasible region of the problem and the new basic feasible solution.

Table 2.18 Tableau for Example 2.9 after completing two Simplex iterations

	1	x_7	x_2	x_5
z	-12	1	-1	1
x_3	10	1/3	-1	2/3
x_1	12	1	0	-1
x_4	16	3	1	-2
x_6	16	1	-1	0

We again conduct another Simplex iteration using the tableau in Table 2.18. We note that the objective-function row is not non-negative and that increasing the value of x_2 from zero would improve the objective function. We next conduct the ratio test to determine which basic variable leaves the basis when x_2 enters it. The variable to exit the basis has index that corresponds to:

$$\min_{\tilde{N}_{2,3},\tilde{N}_{2,1},\tilde{N}_{2,4},\tilde{N}_{2,6}<0} \left\{ -\frac{\tilde{b}_3}{\tilde{N}_{2,3}}, -\frac{\tilde{b}_1}{\tilde{N}_{2,1}}, -\frac{\tilde{b}_4}{\tilde{N}_{2,3}}, -\frac{\tilde{b}_6}{\tilde{N}_{2,6}} \right\} = \min\{10, /, /, 16\},$$

which is x_3. We conduct a pivot operation to swap x_2 and x_3, which gives the updated tableau in Table 2.19. Our new basic feasible solution has $(x_1, x_2) = (12, 10)$ and an objective-function value of -22. Figure 2.13 shows the feasible region of the LPP and the new basic feasible solution found.

Fig. 2.12 Feasible region of the Electricity-Production Problem in Example 2.9 and basic feasible solution after completing two Simplex iterations

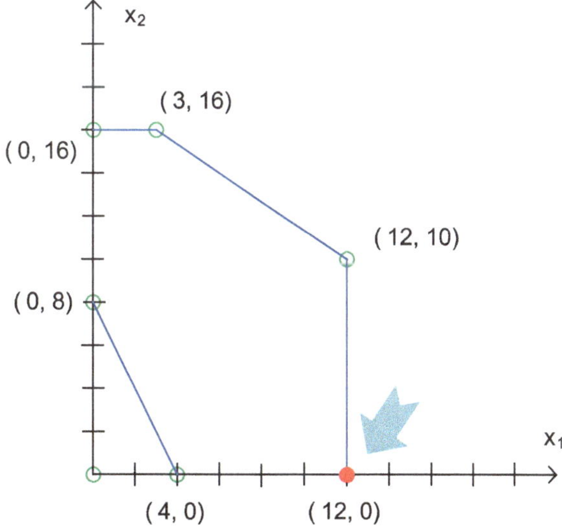

Table 2.19 Tableau for Example 2.9 after completing three Simplex iterations

	1	x_7	x_3	x_5
z	-22	2/3	1	1/3
x_2	10	1/3	-1	2/3
x_1	12	1	0	-1
x_4	26	10/3	-1	$-4/3$
x_6	6	2/3	1	$-2/3$

If we proceed to conduct an additional Simplex iteration using the tableau in Table 2.19, we find that the Simplex method terminates. This is because we now have $\tilde{c} \geq 0$ in the objective-function row, meaning that we cannot improve on the current solution. The point $(x_1, x_2) = (12, 10)$ is the same optimal solution to this problem found in Sections 2.1.1, 2.3.1, and 2.3.2. The tableau gives an optimal objective-function value of -22. However, recall that the problem was converted from a maximization to a minimization to put it into standard form. When the objective is converted back to a maximization, the objective-function value becomes 22, which is consistent with the discussion in Sections 2.1.1, 2.3.1, and 2.3.2.

Figure 2.14 shows the sequence of points that the Simplex method goes through to get from the starting basic solution, $(x_1, x_2) = (0, 0)$, to the final optimal basic feasible solution, $(x_1, x_2) = (12, 10)$.

$\square$

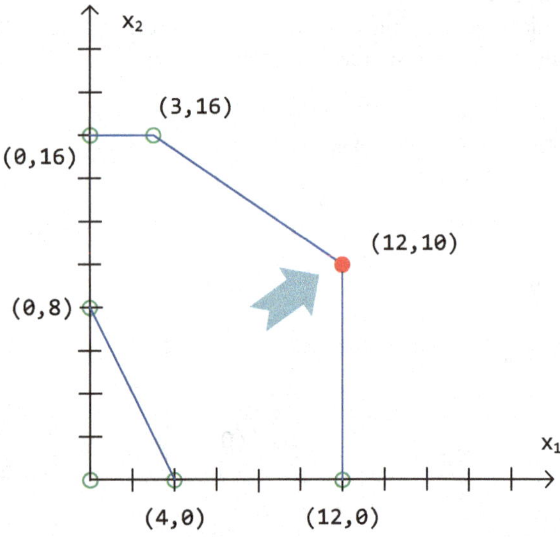

Fig. 2.13 Feasible region of the Electricity-Production Problem in Example 2.9 and basic feasible solution after completing three Simplex iterations

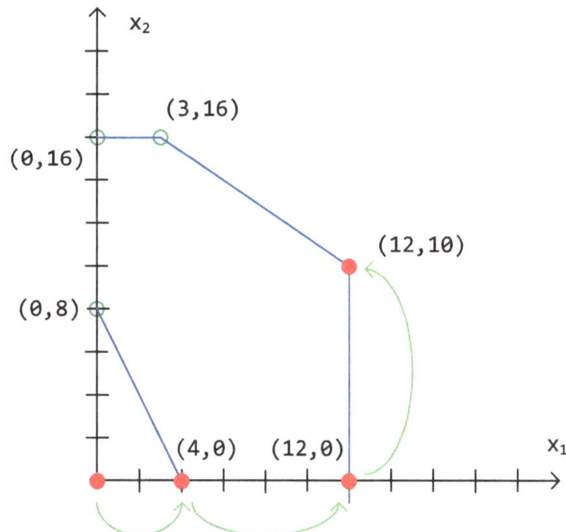

Fig. 2.14 Feasible region of the Electricity-Production Problem in Example 2.9 and the sequence of points that the Simplex method goes through

2.5.4 Convergence of the Simplex Method

The Simplex method is an iterative algorithm. As such, an important question is whether it is guaranteed to converge. That is to say, are we guaranteed to eventually find a basic feasible solution at which $\tilde{c} \geq 0$, which allows the Simplex method to terminate? If not, it is possible that we can get stuck in Steps 8 through 12 of the Simplex algorithm outlined in Section 2.5.3 without ever terminating.

To answer this question, we note that the Simplex method solves a linear optimization problem by going through extreme points of the feasible region. Because a linear optimization problem has a finite number of extreme points, this implies that the algorithm should eventually terminate. There is one added wrinkle to this, however. Certain problems can have multiple basic feasible solutions corresponding to a single extreme point. These occur because of what is called **degeneracy**. A degenerate basic solution is one in which one or more basic variables take on a value of zero when we solve for them using the structural equality constraints. Degenerate basic solutions normally arise because there are extra redundant constraints at an extreme point of the feasible region.

The difficulty that degeneracy raises is that the Simplex method may get 'stuck' at an extreme point by cycling through the same set of basic feasible solutions corresponding to that extreme point without ever moving. There is, however, a very easy way to ensure that the Simplex method does not get stuck at a degenerate extreme point. This is done by choosing the variable that enters the basis at each Simplex iterations based on their index number. That is to say, if both $\tilde{c}_i$ and $\tilde{c}_j$ are negative at a given Simplex iteration, then choose whichever has the smaller index (*i.e.*, the smaller of i or j) to be the variable entering the basis. One can show that using this rule to select the entering variable guarantees that the Simplex method leaves every degenerate extreme point after a finite number of Simplex iterations [9].

In practice, the Simplex method is not applied this way. Rather, the entering variable is chosen on the basis of which one has the most negative $\tilde{c}$. If the Simplex method spends multiple iterations at the same extreme point (suggesting a degeneracy problem), then the selection rule based on the index number is used instead until the Simplex method moves to a different extreme point.

2.5.5 Detecting Infeasible Linear Optimization Problems

The Simplex method detects that a linear optimization problem is infeasible based on the final value, after the Simplex algorithm terminates, of the artificial variable added in the regularization step. If the Simplex method terminates (*i.e.*, if $\tilde{c} \geq 0$) and the artificial variable has a non-zero value, this means that the starting linear optimization problem is infeasible.

To understand why, first note that if a linear optimization problem is infeasible, then the regularization step must be done at the beginning of the Simplex method. This is because for any starting partition of the variables into basic and non-basic variables, the resulting basic solution must be infeasible. Otherwise, if we can find a basic solution that is feasible then the linear optimization problem cannot be infeasible.

Next, recall that when the artificial variable is added in the regularization step, a comparatively high value of K is put in the objective-function row of the tableau. The purpose of K, as discussed in Section 2.5.1, is to make the objective function (which we seek to minimize) larger if the artificial variable takes on a positive value. Indeed, by making K larger than all of the other values in the objective-function row of the tableau, the cost on the artificial variable is higher than all of the other variables and the Simplex method seeks to make the artificial variable as small as possible. If the Simplex method terminates but the artificial variable is still positive, that means it is impossible to satisfy the constraints of the original problem without having the artificial variable allow for constraint violations. Thus, the original problem must be infeasible. Otherwise, if the Simplex method terminates and the artificial variable is equal to zero, the original problem is feasible.

2.5.6 Detecting Unbounded Linear Optimization Problems

The Simplex method detects that a linear optimization problem is unbounded through the ratio test conducted in Step 10 of the Simplex Method Algorithm, which is outlined in Section 2.5.3. Recall that the purpose of the ratio test is to determine how large the non-basic variable entering the basis can be made before causing one of the basic variables to become negative. If there is no such restriction, then the Simplex method would make the entering variable infinitely large because the negative $\tilde{c}$ value in the objective-function row of the tableau means that doing so would make the objective function go to $-\infty$.

We know that if at least one of the $\tilde{N}$ values in the column underneath the entering variable in the tableau is negative, then the ratio test gives a limit on how much the entering variable can increase. Otherwise, if all of the $\tilde{N}$ values are non-negative, then the ratio test allows the entering variable to become as large as possible. Thus, if at any point in the Simplex method there is an entering variable without any negative $\tilde{N}$ values in the tableau, the problem is known to be unbounded.

2.5.7 Detecting Multiple Optima

The Simplex method detects multiple optimal solutions based on the objective-function row of the final tableau. If all of the values in the objective-function row of the final tableau are strictly positive, this means that the optimal solution found is a unique optimal solution. Otherwise, if there are any zero values in the objective-function row of the final tableau, this means that there are multiple optimal solutions. The reason for this is that a zero in the objective-function row of the final tableau means that a non-basic variable can enter the basis without changing the objective-function value at all. Thus, there are additional optimal solutions in which non-basic variables with a zero in the objective-function row take on positive values (keeping in mind that the basic-variable values would have to be recomputed).

2.6 Sensitivity Analysis

The subject of **sensitivity analysis** answers the question of what effect changing a linear optimization problem has on the resulting optimal solution. Of course, one way to answer this question is to change a given problem, solve the new problem, and examine any resulting changes in the solution. This can be quite cumbersome and time-consuming, however. A large-scale linear optimization problem with millions of variables could take several hours to solve. Having to re-solve multiple versions of the same basic problem with different data can be impractical. Sensitivity analysis answers this question by using information from the optimal solution and the final tableau after applying the Simplex method.

Throughout this discussion we assume that we have a linear optimization problem that is already in standard form, which can be generically written as:

$$\min_{x} c^{\top} x \tag{2.46}$$

$$\text{s.t. } Ax = b \tag{2.47}$$

$$x \geq 0. \tag{2.48}$$

We also assume that we have solved the problem using the Simplex method and have an optimal set of decision-variable values, x^*. More specifically, we assume that we have an optimal basic feasible solution, meaning that x^* is partitioned into basic variables, x_B^*, and non-basic variables, $x_N^* = 0$. We, thus, also have the objective-function-coefficient vector partitioned into c_B and c_N and the A matrix partitioned into submatrices, B and N.

We use sensitivity analysis to examine the effect of three different types of changes: (i) changing the constants on the right-hand sides of the constraints (*i.e.*, the b vector in the structural equality constraints), (ii) changing the objective-function coefficients (*i.e.*, the c vector), and (iii) changing the coefficients on the left-hand sides of the constraints (*i.e.*, the A matrix).

2.6.1 Changing the Right-Hand Sides of the Constraints

We begin by considering the case in which the b vector is changed. To do this, recall that once we solve the original LPP, given by (2.46)–(2.48), the basic variable values are given by:

$$x_B^* = \tilde{b} = B^{-1}b.$$

Let us next examine the effect of changing the structural equality constraints from (2.47) to:

$$Ax = b + \Delta b,$$

on the basic feasible solution, x^*, that is optimal in the original problem. From (2.24) and (2.25) we have:

$$\hat{b} = B^{-1}(b + \Delta b),$$

$$\hat{N} = -B^{-1}N,$$

$$\hat{c}_0 = c_B^\top \hat{b},$$

and:

$$\hat{c}^\top = c_B^\top \hat{N} + c_N^\top,$$

where $\hat{b}$, $\hat{N}$, $\hat{c}_0$, and $\hat{c}$ denote the new values of the terms in the final tableau after the right-hand sides of the structural equality constraints are changed. Note that changing the right-hand sides of the structural equality constraints only changes the values of $\hat{b}$ and $\hat{c}_0$ in the final tableau. We have that $\hat{N} = \tilde{N}$ and $\hat{c} = \tilde{c}$.

From this observation we can draw the following important insight. If $\hat{b} \geq 0$, then the partition of x^* into basic and non-basic variables is still feasible when the equality constraints are changed. Moreover, the values in the objective-function row of the final tableau, $\hat{c} = \tilde{c}$, are not affected by the change in the right-hand side of the equality constraints. This means that if $\hat{b} \geq 0$ the partition of x^* into basic and non-basic variables is not only feasible but also optimal after the equality constraints are changed.

We can use this insight to first determine how much the right-hand side of the equality constraints can be changed before the partition of x^* into basic and non-basic variables becomes infeasible. We determine this bound from the requirement that $\hat{b} \geq 0$ as follows:

$$\hat{b} \geq 0$$
$$B^{-1} \cdot (b + \Delta b) \geq 0$$
$$B^{-1} b \geq -B^{-1} \Delta b$$
$$\tilde{b} \geq -B^{-1} \Delta b. \tag{2.49}$$

If Δb satisfies (2.49), then the optimal basis remains unchanged by the changes in the right-hand side of the equality constraints. Although the basis remains the same if (2.49) is satisfied, the values of the basic variables change. Specifically, we can compute the new values of the basic variables, $\hat{x}_B$, as:

$$\hat{x}_B = \hat{b} = B^{-1} \cdot (b + \Delta b) = B^{-1} b + B^{-1} \Delta b = \tilde{b} + B^{-1} \Delta b. \tag{2.50}$$

Equation (2.50) gives us an exact expression for how much the values of the basic variables change as a result of changing the right-hand sides of the equalities. Specifically, this change is $B^{-1} \Delta b$.

The next question is how much of an effect these changes in the values of the basic variables have on the objective-function value. From (2.25) we can write the objective-function value as:

$$z = c_B^\top \hat{x}_B + c_N^\top \hat{x}_N.$$

where $\hat{x}_N$ are the new non-basic variable values. We know, however, that the non-basic variables will still equal zero after the right-hand sides of the equality constraints are changed. Thus, using (2.50) we can write the objective-function value as:

$$z = c_B^\top \cdot (\tilde{b} + B^{-1} \Delta b)$$
$$= c_B^\top \tilde{b} + c_B^\top B^{-1} \Delta b$$
$$= z^* + \lambda^\top \Delta b, \tag{2.51}$$

where we define $\lambda^\top = c_B^\top B^{-1}$. The vector λ, which is called the **sensitivity vector**, gives the change in the optimal objective-function value of an LPP resulting from a sufficiently small change in the right-hand side of a structural equality constraint.

If Δb does not satisfy (2.49), then the basis that is optimal for the original LPP is no longer feasible after the constraints are changed. In such a case, one cannot directly compute the effect of changing the constraints on the solution. Rather, additional Simplex iterations must be conducted to find a new basis that is feasible and optimal. Note that when conducting the additional Simplex iterations, one can start with the final basis from solving the original LPP and conduct a regularization step to begin the additional Simplex iterations. Doing so can often decrease the number of Simplex iterations that must be conducted.

Example 2.10 Consider the standard form-version of the Electricity-Production Problem, which is introduced in Section 2.1.1. In Example 2.9 we find that the optimal solution has basis $x_B = (x_2, x_1, x_4, x_6)$ and $x_N = (x_3, x_5)$. We are excluding x_7 from the vector of non-basic variables, because this is an artificial variable added in the regularization step. However, one can list x_7 as a non-basic variable without affecting any of the results in this example. The optimal solution also has:

$$B = \begin{bmatrix} 1 & 2/3 & 0 & 0 \\ 1 & 2 & -1 & 0 \\ 0 & 1 & 0 & 0 \\ 1 & 0 & 0 & 1 \end{bmatrix},$$

$$\tilde{b} = \begin{pmatrix} 10 \\ 12 \\ 26 \\ 6 \end{pmatrix},$$

and:

$$c_B = \begin{pmatrix} -1 & -1 & 0 & 0 \end{pmatrix}.$$

We can compute the sensitivity vector as:

$$\lambda^\top = c_B^\top B^{-1} = \begin{pmatrix} -1 & 0 & -1/3 & 0 \end{pmatrix}.$$

Suppose that the structural equality constraints of the problem are changed from:

$$\begin{bmatrix} 2/3 & 1 & 1 & 0 & 0 & 0 \\ 2 & 1 & 0 & -1 & 0 & 0 \\ 1 & 0 & 0 & 0 & 1 & 0 \\ 0 & 1 & 0 & 0 & 0 & 1 \end{bmatrix} \begin{pmatrix} x_1 \\ x_2 \\ x_3 \\ x_4 \\ x_5 \\ x_6 \end{pmatrix} = \begin{pmatrix} 18 \\ 8 \\ 12 \\ 16 \end{pmatrix},$$

to:

$$\begin{bmatrix} 2/3 & 1 & 1 & 0 & 0 & 0 \\ 2 & 1 & 0 & -1 & 0 & 0 \\ 1 & 0 & 0 & 0 & 1 & 0 \\ 0 & 1 & 0 & 0 & 0 & 1 \end{bmatrix} \begin{pmatrix} x_1 \\ x_2 \\ x_3 \\ x_4 \\ x_5 \\ x_6 \end{pmatrix} = \begin{pmatrix} 18 \\ 8 \\ 12 \\ 16 \end{pmatrix} + \begin{pmatrix} 2 \\ -1 \\ 0 \\ 0 \end{pmatrix}.$$

We first determine whether the basis that is optimal in the original problem is still feasible after the constraints are changed. From (2.49) we know that the basis is still feasible if:

$$\tilde{b} \geq -B^{-1}\Delta b,$$

or if:

$$\begin{pmatrix} 10 \\ 12 \\ 26 \\ 6 \end{pmatrix} \geq - \begin{bmatrix} 1 & 2/3 & 0 & 0 \\ 1 & 2 & -1 & 0 \\ 0 & 1 & 0 & 0 \\ 1 & 0 & 0 & 1 \end{bmatrix}^{-1} \begin{pmatrix} 2 \\ -1 \\ 0 \\ 0 \end{pmatrix} = \begin{pmatrix} -2 \\ 0 \\ -3 \\ 2 \end{pmatrix}.$$

Seeing that it is, we can next determine the effect of the change in the constraints on the optimal objective-function value using the sensitivity vector as:

$$\lambda^\top \Delta b = -2.$$

$\square$

2.6.2 *Changing the Objective-Function Coefficients*

We next consider the case in which the c vector is changed. More specifically, let us suppose that the objective function changes from (2.46) to:

$$(c + \Delta c)^\top x.$$

We analyze the effect of this change following the same line of reasoning used to analyze changes in the b vector in Section 2.6.1. That is to say, we examine what effect it has on x^*, the optimal basic feasible solution of the original problem. From (2.24) and (2.25) we have that:

$$\hat{b} = B^{-1}b,$$

$$\hat{N} = -B^{-1}N,$$

$$\hat{c}_0 = (c_B + \Delta c_B)^\top \hat{b},$$

and:

$$\hat{c}^\top = (c_B + \Delta c_B)^\top \hat{N} + (c_N + \Delta c_N)^\top,$$

where $\hat{b}$, $\hat{N}$, $\hat{c}_0$, and $\hat{c}$ denote the new values of the terms in the final tableau after the c vector is changed and Δc_B and Δc_N partition the Δc vector according to the final partition of basic and non-basic variables. Note that changing the c vector only changes the values of $\hat{c}_0$ and $\hat{c}$ and that we have $\hat{b} = \tilde{b}$ and $\hat{N} = \tilde{N}$.

From this observation we can conclude that when the c vector is changed, the optimal basic feasible solution of the original problem, x^*, is still feasible in the new problem. The only question is whether this basic feasible solution remains optimal after the objective function is changed. We know that x^* remains optimal if $\hat{c} \geq 0$. From this, we can derive the following bound:

$$\hat{c} \geq 0$$
$$(c_B + \Delta c_B)^\top \hat{N} + (c_N + \Delta c_N)^\top \geq 0$$
$$(c_B + \Delta c_B)^\top \tilde{N} + (c_N + \Delta c_N)^\top \geq 0$$
$$c_B^\top \tilde{N} + c_N^\top + \Delta c_B^\top \tilde{N} + \Delta c_N^\top \geq 0$$
$$\tilde{c}^\top + \Delta c_B^\top \tilde{N} + \Delta c_N^\top \geq 0$$
$$\Delta c_B^\top \tilde{N} + \Delta c_N^\top \geq -\tilde{c}^\top, \tag{2.52}$$

on how much c can change and have x^* remain an optimal basic feasible solution. If (2.52) is satisfied, then x^* remains an optimal basic feasible solution. Of course, the objective-function value changes as a result of the c vector changing. However, it is straightforward to compute the new objective-function value as:

$$z = (c_B + \Delta c_B)^\top x_B^* + (c_N + \Delta c_N)^\top x_N^*$$
$$= (c_B + \Delta c_B)^\top x_B^*,$$

because $x_N^* = 0$. This then simplifies to:

$$z = z^* + \Delta c_B^\top x_B^*. \tag{2.53}$$

Thus, if Δc satisfies (2.52), the impact on the optimal objective-function value of changing the objective-function coefficient of a basic variable is equal to the basic variable's optimal value itself. Changing the objective-function value of a non-basic variable has no impact on the optimal objective-function value.

On the other hand, if (2.52) is not satisfied, then x^* is a basic feasible solution of the changed LPP but is not optimal in the new problem. As such, additional Simplex iterations would have to be conducted to find a new optimal basic feasible solution.

Example 2.11 Consider the standard form-version of the Electricity-Production Problem, which is introduced in Section 2.1.1. In Example 2.9 we find that:

$$\tilde{c} = \begin{pmatrix} 1 \\ 5/3 \end{pmatrix},$$

and:

$$\tilde{N} = \begin{bmatrix} -1 & -2/3 \\ 0 & -1 \\ -1 & -8/3 \\ 1 & 2/3 \end{bmatrix}.$$

Suppose we wish to change c_1 from -1 to -0.9. From (2.52) we know that x^* will remain an optimal solution so long as:

$$\Delta c_B^\top \tilde{N} + \Delta c_N^\top \geq -\tilde{c}^\top,$$

or so long as:

$$(0\ 0.1\ 0\ 0)\begin{bmatrix} -1 & -2/3 \\ 0 & -1 \\ -1 & -8/3 \\ 1 & 2/3 \end{bmatrix} \geq -(1\ 5/3),$$

or so long as:

$$(0\ -0.1) \geq (-1\ -5/3),$$

which holds true. Thus, based on (2.53) we know that the new objective-function value after this change is given by:

$$\Delta c_B^\top x_B^* = (0\ 0.1\ 0\ 0)\begin{pmatrix} 10 \\ 12 \\ 26 \\ 6 \end{pmatrix} = 1.2.$$

□

2.6.3 Changing the Left-Hand-Side Coefficients of the Constraints

We finally examine the effect of changing the coefficients multiplying the variables on the left-hand side of the structural equality constraints. More specifically, we examine the effect of changing (2.47) to:

$$(A + \Delta A)x = b.$$

We begin by partitioning ΔA into:

$$\Delta A = \begin{bmatrix} \Delta B & \Delta N \end{bmatrix}.$$

Using this and (2.24) and (2.25) we have that:

$$\hat{b} = (B + \Delta B)^{-1}b,$$

$$\hat{N} = -(B + \Delta B)^{-1}(N + \Delta N),$$

$$\hat{c}_0 = c_B^\top \hat{b},$$

and:

$$\hat{c}^\top = c_B^\top \hat{N} + c_N^\top,$$

where $\hat{b}$, $\hat{N}$, $\hat{c}_0$, and $\hat{c}$ denote the new values of the terms in the final tableau after the A matrix is changed.

We can draw two conclusions from these expressions. The first is that if ΔA changes values in the B matrix, then all of $\hat{b}$, $\hat{N}$, $\hat{c}_0$, and $\hat{c}$ are changed in the final tableau. Moreover, it is not straightforward to derive a bound on how large or small ΔB can be before x^* is no longer a feasible or optimal basic solution. This is because ΔB appears in matrix inversions in the expressions giving $\hat{b}$ and $\hat{c}$. Thus, we do not provide any such bounds on ΔB (because they are difficult to derive and work with).

We can, however, derive such bounds in the case in which ΔA changes values in the N matrix *only*. To do so, we note that if $\Delta B = 0$, then from (2.24) and (2.25) we have:

$$\hat{b} = B^{-1}b,$$

$$\hat{N} = -B^{-1} \cdot (N + \Delta N),$$

$$\hat{c}_0 = c_B^\top \hat{b},$$

and:

$$\hat{c}^\top = c_B^\top \hat{N} + c_N^\top.$$

Thus, we see that if the N matrix is changed, this only changes the values of $\hat{N}$ and $\hat{c}$ and that we have $\hat{b} = \tilde{b}$ and $\hat{c}_0 = \tilde{c}_0$. Thus, we know that x^* remains a basic feasible solution after the N matrix is changed. The only question is whether x^* remains an optimal basic feasible solution. We can derive the bound:

$$\hat{c} \geq 0$$
$$-c_B^\top B^{-1} \cdot (N + \Delta N) + c_N^\top \geq 0$$
$$-c_B^\top B^{-1} N - c_B^\top B^{-1} \Delta N + c_N^\top \geq 0$$
$$c_B^\top \tilde{N} + c_N^\top - c_B^\top B^{-1} \Delta N \geq 0$$
$$\tilde{c}^\top - c_B^\top B^{-1} \Delta N \geq 0$$
$$\tilde{c}^\top \geq c_B^\top B^{-1} \Delta N, \tag{2.54}$$

on ΔN, which ensures that x^* remains optimal. If (2.54) is satisfied, then x^* remains an optimal basic feasible solution. If so, we can determine the new objective-function value from (2.25) as:

$$z = \hat{c}_0 + \hat{c}^\top x_N^*$$
$$= \tilde{c}_0, \tag{2.55}$$

because we have that $\hat{c}_0 = \tilde{c}_0$ and $x_N^* = 0$. Thus, if (2.54) is satisfied, there is no change in the optimal objective-function value. Otherwise, if ΔN does not satisfy (2.54), then x^* is no longer an optimal basic feasible solution and additional Simplex iterations must be conducted to find a new optimal basic solution.

We finally conclude this discussion by noting that although we cannot derive a bound on the allowable change in the B matrix, we can approximate changes in the optimal objective-function value from changing the B matrix. We do this by supposing that one element of the A matrix, $A_{i,j}$, is changed to $A_{i,j} + \Delta A_{i,j}$. When this change is made, the ith structural equality constraint changes from:

$$A_{i,1}x_1^* + \cdots + A_{i,j-1}x_{j-1}^* + A_{i,j}x_j^* + A_{i,j+1}x_{j+1}^* + \cdots + A_{i,n}x_n^* = b_i,$$

to:

$$A_{i,1}x_1^* + \cdots + A_{i,j-1}x_{j-1}^* + (A_{i,j} + \Delta A_{i,j})x_j^* + A_{i,j+1}x_{j+1}^* + \cdots + A_{i,n}x_n^* = b_i.$$

The changed ith constraint can be rewritten as:

$$A_{i,1}x_1^* + \cdots + A_{i,j-1}x_{j-1}^* + A_{i,j}x_j^* + A_{i,j+1}x_{j+1}^* + \cdots + A_{i,n}x_n^* = b_i - \Delta A_{i,j}x_j^*.$$

Thus, one can view changing the A matrix as changing the right-hand side of the ith constraint by $-\Delta A_{i,j}x_j^*$. One can then use the results of Section 2.6.1 to estimate the change in the optimal objective-function value as:

$$\Delta z^* \approx -\lambda_i \Delta A_{i,j} x_j^*, \tag{2.56}$$

where λ_i is the ith component of the sensitivity vector derived in (2.51). Note that if x_j is a non-basic variable, then we have that the change in the optimal objective-function value is:

$$-\lambda_i \Delta A_{i,j} x_j^* = 0,$$

which is what is found in analyzing the case of changes in the N matrix in (2.55). However, Equation (2.56) can also be applied to approximate the change in the optimal objective-function value when changes are made to the B matrix. It is important to note, however, that (2.56) only approximates such changes (whereas all of the other sensitivity expressions are exact).

Example 2.12 Consider the standard form-version of the Electricity-Production Problem, which is introduced in Section 2.1.1, and suppose that $A_{1,2}$ changes from 1 to 1.01. Because this is making a change to a coefficient in the B matrix, we cannot exactly estimate the effect of this change on the objective function. We can, however, approximate this using (2.56) as:

$$-\lambda_1 \Delta A_{1,2} x_2^* = -1 \cdot (-0.01) \cdot 10 = 0.1.$$

To see that this is only an approximation, we can solve the problem with $A_{1,2} = 1.01$, which gives an optimal objective-function value of -21.900990. The exact change in the objective-function value is thus:

$$-21.900990 - (-22) = 0.099010 \approx 0.1.$$

$\square$

2.7 Duality Theory

Every linear optimization problem has an associated optimization problem called its **dual problem**. We show in this section that the starting linear optimization problem, which in the context of duality theory is called the **primal problem**, and its associated dual problem have some very important relationships. These relationships include the underlying structure of the primal and dual problems (*i.e.*, the objective function, constraints, and variables of the two problems) and properties of solutions of the two problems.

2.7.1 *Symmetric Duals*

We begin by looking at the simple case of a primal problem that is in canonical form. Recall from Section 2.2.2.2 that a generic primal problem in canonical form can be written compactly as:

$$\min_{x} z_P = c^\top x \tag{2.57}$$

$$\text{s.t.}\ Ax \geq b \tag{2.58}$$

$$x \geq 0. \tag{2.59}$$

The dual problem associated with this primal problem is:

$$\max_{y} z_D = b^\top y \tag{2.60}$$

$$\text{s.t.}\ A^\top y \leq c \tag{2.61}$$

$$y \geq 0. \tag{2.62}$$

As indicated above, we can note some relationships in the structures of the primal and dual problems. The vector of objective-function coefficients in the primal problem, c, is the vector of the right-hand side constants in the structural constraints in the dual problem. The vector of the right-hand side constants in the structural constraints in the primal problem, b, is the vector of objective-function coefficients in the dual problem. We also see that the coefficients multiplying the variables in the left-hand sides of the constraints (which are in the A matrix) are the same in the dual problem as in the primal, except that the coefficient matrix is transposed in the dual. We finally note that whereas the primal problem is a minimization, the dual problem is a maximization.

We can also observe a one-to-one relationship between the variables and constraints in the primal and dual problems. Recall that for a generic primal problem in canonical form, A is an $m \times n$ matrix. This means that the primal problem has m structural constraints and n variables (*i.e.*, $x \in \mathbb{R}^n$). Because $A^\top$ is an $n \times m$ matrix, this means that the dual problem has n structural constraints and m variables (*i.e.*, $y \in \mathbb{R}^m$). Thus, we can say that each primal constraint has an associated dual variable and each primal variable has an associated dual constraint.

We can show a further symmetry between the primal and dual problems. By this we mean that each dual constraint has an associated primal variable and each dual variable has an associated primal constraint. This is because if we start with a primal problem in canonical form and find the dual of its dual problem, we get back the original primal problem. To see this, we note from the discussion above that if we have the primal problem:

$$\min_{x} c^\top x$$
$$\text{s.t. } Ax \geq b$$
$$x \geq 0,$$

its dual problem is:

$$\max_{y} b^\top y$$
$$\text{s.t. } A^\top y \leq c$$
$$y \geq 0.$$

If we want to find the dual of the dual problem, we must first convert it to canonical form, which is:

$$\min_{y} -b^\top y$$
$$\text{s.t. } -A^\top y \geq -c$$
$$y \geq 0.$$

The dual of this problem is:

$$\max_{w} -c^\top w$$
$$\text{s.t. } (-A^\top)^\top w \leq -b$$
$$w \geq 0,$$

where we let w be a vector of variables. This problem can be rewritten as:

$$\min_{w} c^{\mathsf{T}} w$$
$$\text{s.t. } Aw \geq b$$
$$w \geq 0,$$

which is identical to the starting problem, except that the variables are now called w instead of x.

Example 2.13 Consider the Electricity-Production Problem, which is introduced in Section 2.1.1. In Example 2.4 we show the canonical form of this problem to be:

$$\min_{x} z_P = 5x_{1,1} + 4x_{1,2} + 3x_{2,1} + 6x_{2,2}$$
$$\text{s.t. } -x_{1,1} - x_{1,2} \geq -7$$
$$-x_{2,1} - x_{2,2} \geq -12$$
$$-x_{1,1} - x_{2,1} \geq -10$$
$$x_{1,1} + x_{2,1} \geq 10$$
$$-x_{1,2} - x_{2,2} \geq -8$$
$$x_{1,2} + x_{2,2} \geq 8$$
$$x_{i,j} \geq 0, \forall\, i = 1, 2; \; j = 1, 2.$$

This can be written as:

$$\min_{x} z_P = c^{\mathsf{T}} x$$
$$\text{s.t. } Ax \geq b$$
$$x \geq 0,$$

where:

$$x = \begin{pmatrix} x_{1,1} \\ x_{1,2} \\ x_{2,1} \\ x_{2,2} \end{pmatrix},$$

$$c = \begin{pmatrix} 5 \\ 4 \\ 3 \\ 6 \end{pmatrix},$$

$$A = \begin{bmatrix} -1 & -1 & 0 & 0 \\ 0 & 0 & -1 & -1 \\ -1 & 0 & -1 & 0 \\ 1 & 0 & 1 & 0 \\ 0 & -1 & 0 & -1 \\ 0 & 1 & 0 & 1 \end{bmatrix},$$

and:

$$b = \begin{pmatrix} -7 \\ -12 \\ -10 \\ 10 \\ -8 \\ 8 \end{pmatrix}.$$

The dual problem is:

$$\max_{y} z_D = b^\top y$$

$$\text{s.t. } A^\top y \le c$$

$$y \ge 0,$$

which can be expanded out to:

$$\max_{y} z_D = \begin{pmatrix} -7 & -12 & -10 & 10 & -8 & 8 \end{pmatrix} \begin{pmatrix} y_1 \\ y_2 \\ y_3 \\ y_4 \\ y_5 \\ y_6 \end{pmatrix}$$

$$\text{s.t. } \begin{bmatrix} -1 & 0 & -1 & 1 & 0 & 0 \\ -1 & 0 & 0 & 0 & -1 & 1 \\ 0 & -1 & -1 & 1 & 0 & 0 \\ 0 & -1 & 0 & 0 & -1 & 0 \end{bmatrix} \begin{pmatrix} y_1 \\ y_2 \\ y_3 \\ y_4 \\ y_5 \\ y_6 \end{pmatrix} \le \begin{pmatrix} 5 \\ 4 \\ 3 \\ 6 \end{pmatrix}$$

$$\begin{pmatrix} y_1 \\ y_2 \\ y_3 \\ y_4 \\ y_5 \\ y_6 \end{pmatrix} \ge \begin{pmatrix} 0 \\ 0 \\ 0 \\ 0 \\ 0 \\ 0 \end{pmatrix},$$

or, by simplifying, to:

$$\max_{y} z_D = -7y_1 - 12y_2 - 10y_3 + 10y_4 - 8y_5 + 8y_6$$

$$\text{s.t.} \quad -y_1 - y_3 + y_4 \leq 5$$
$$-y_1 - y_5 + y_6 \leq 4$$
$$-y_2 - y_3 + y_4 \leq 3$$
$$-y_2 - y_5 \leq 6$$
$$y_1, y_2, y_3, y_4, y_5, y_6 \geq 0.$$

$\square$

2.7.2 Asymmetrical Duals

Now consider a generic LPP in standard form:

$$\min_{x} z_P = c^\top x$$
$$\text{s.t.} \quad Ax = b$$
$$x \geq 0.$$

To find the dual of this problem, we can convert it to canonical form, by replacing the structural equality constraints with the pair of inequalities:

$$\min_{x} z_P = c^\top x$$
$$\text{s.t.} \quad Ax \geq b$$
$$Ax \leq b$$
$$x \geq 0,$$

and then converting the less-than-or-equal-to constraint into a greater-than-or-equal-to constraint:

$$\min_{x} z_P = c^\top x$$
$$\text{s.t.} \quad Ax \geq b$$
$$-Ax \geq -b$$
$$x \geq 0.$$

If we let u and v denote the dual variables associated with the two structural inequality constraints, the dual of this problem is then:

$$\max_{u,v} z_D = b^\top u - b^\top v$$

$$\text{s.t. } A^\top u - A^\top v \le c$$

$$u, v \ge 0.$$

If we define y as the difference between the dual-variable vectors:

$$y = u - v,$$

then the dual can be written as:

$$\max_y z_D = b^\top y$$

$$\text{s.t. } A^\top y \le c,$$

where there is no sign restriction on y because it is the difference of two non-negative vectors. This analysis of a primal problem with equality constraints gives an important conclusion, which is that the dual variable associated with an equality constraint has no sign restriction.

2.7.3 Duality Conversion Rules

Example 2.13 demonstrates a straightforward, but often tedious way of finding the dual of any linear optimization problem. This method is to first convert the problem into canonical form, using the rules outlined in Section 2.2.2.2. Then, one can use the symmetric dual described in Section 2.7.1 to find the dual of the starting primal problem. While straightforward, this is cumbersome, as it requires the added step of first converting the primal problem to canonical form. Here we outline a standard set of rules that allow us to directly find a dual problem without first converting the primal problem to canonical form. We demonstrate these rules using the problem introduced in Example 2.1, which is:

$$\max_x \ 3x_1 + 5x_2 - 3x_3 + 1.3x_4 - x_5$$

$$\text{s.t. } x_1 + x_2 - 4x_4 \le 10$$

$$x_2 - 0.5x_3 + x_5 = -1$$

$$x_3 \ge 5$$

$$x_1 \ge 0$$

$$x_2 \ge 0$$

$$x_4 \le 0.$$

We then demonstrate that the dual we find using the rules outlined is equivalent to the dual we would obtain if we first convert the primal problem to canonical form and use the symmetric dual introduced in Section 2.7.1.

2.7.3.1   Step 1: Determine Number of Dual Variables

The first step is to determine the number of dual variables, which is based on the number of structural constraints in the primal problem. Our example problem has three structural constraints, which are:

$$x_1 + x_2 - 4x_4 \leq 10$$

$$x_2 - 0.5x_3 + x_5 = -1$$

$$x_3 \geq 5.$$

The three other constraints:

$$x_1 \geq 0$$

$$x_2 \geq 0$$

$$x_4 \leq 0,$$

are not structural constraints but are rather non-negativity and non-positivity restrictions. These are handled differently than structural constraints when writing the dual problem. Because there are three structural constraints, there will be three dual variables, which we refer to as y_1, y_2, and y_3. To help in the subsequent steps of writing the dual problem, we write each dual variable next to its associated primal constraint as follows:

$$
\begin{aligned}
\max_{x} \quad & 3x_1 + 5x_2 - 3x_3 + 1.3x_4 - x_5 \\
\text{s.t.} \quad & x_1 + x_2 - 4x_4 \leq 10 && (y_1) \\
& x_2 - 0.5x_3 + x_5 = -1 && (y_2) \\
& x_3 \geq 5 && (y_3) \\
& x_1 \geq 0 \\
& x_2 \geq 0 \\
& x_4 \leq 0.
\end{aligned}
$$

2.7.3.2 Step 2: Determine Objective of the Dual Problem

The next step is to determine the objective function of the dual problem. First, the direction of the optimization of the dual problem is always opposite to that of the primal problem. Because the primal problem in our example is a maximization, the dual will be a minimization. Next, to determine the objective function itself, we multiply each dual variable with the constant on the right-hand side of its associated primal constraint and sum the products. Thus, in our example we multiply y_1 by 10, y_2 by -1 and y_3 by 5 giving:

$$\min_{y}\ 10y_1 - y_2 + 5y_3.$$

2.7.3.3 Step 3: Determine Number of Structural Constraints in the Dual Problem

There is one structural constraint in the dual problem for each primal variable. Moreover, there is a one-to-one correspondence between dual constraints and primal variables. Thus, just as we associate dual variables with primal constraints in Step 1, we associate primal variables with dual constraints here. Our example problem has five primal variables. Thus, the dual problem will have five structural constraints, and we can associate the constraints and variables as follows:

$$\min_{y}\ 10y_1 - y_2 + 5y_3$$

$$\text{s.t.} \qquad\qquad\qquad\qquad (x_1)$$
$$(x_2)$$
$$(x_3)$$
$$(x_4)$$
$$(x_5)$$

2.7.3.4 Step 4: Determine Right-Hand Sides of Structural Constraints in the Dual Problem

We next determine the right-hand side of each structural constraint in the dual problem. These are given by the coefficient multiplying the associated primal variable in the objective function of the primal problem. In our example, the five structural constraints in the dual problem correspond to x_1, x_2, x_3, x_4, and x_5, respectively, which have objective-function coefficients in the primal problem of 3, 5, -3, 1.3, and -1, respectively. Thus, these will be the right-hand sides of the associated structural constraints in the dual:

$$\min_{y} \ 10y_1 - y_2 + 5y_3$$

$$
\begin{array}{llr}
\text{s.t.} & 3 & (x_1) \\
 & \quad\ 5 & (x_2) \\
 & \qquad -3 & (x_3) \\
 & \quad 1.3 & (x_4) \\
 & \ -1. & (x_5)
\end{array}
$$

2.7.3.5   Step 5: Determine Left-Hand Sides of Structural Constraints in the Dual Problem

The left-hand sides of the structural constraints in the dual problem have the dual variables multiplied by the transpose of the coefficient matrix defining the structural constraints in the primal problem. The result of this is that the coefficient multiplying the primal variable associated with each structural dual constraint is multiplied by the dual variable associated with the structural primal constraint.

To illustrate this rule, take the first structural dual constraint, which is associated with x_1. If we examine the structural constraints in the primal problem, we see that x_1 has coefficients of 1, 0, and 0 in each of the structural constraints in the primal problem. Moreover, these three primal constraints are associated with dual variables y_1, y_2, and y_3. Multiplying these coefficients with the associated dual variables and summing the products gives:

$$1y_1 + 0y_2 + 0y_3,$$

as the left-hand side of the first structural constraint in the dual problem. Next, take the second structural dual constraint, which is associated with x_2. The primal variable x_2 has coefficients 1, 1, and 0 in the three structural constraints of the primal, which are associated with dual variables y_1, y_2, and y_3. Thus, the left-hand side of the second structural constraint of the dual problem is:

$$1y_1 + 1y_2 + 0y_3.$$

Repeating this process three more times gives the following left-hand sides of the structural constraints in the dual problem:

$$\min_{y} \ 10y_1 - y_2 + 5y_3$$

$$
\begin{array}{llr}
\text{s.t.} & y_1 \quad 3 & (x_1) \\
 & y_1 + y_2 \quad 5 & (x_2) \\
 & -0.5y_2 + y_3 \quad -3 & (x_3) \\
 & \ -4y_1 \quad 1.3 & (x_4) \\
 & y_2 \quad -1. & (x_5)
\end{array}
$$

2.7.3.6 Step 6: Determine the Types of Structural Constraints in the Dual Problem

The types of structural constraints in the dual problem (*i.e.*, greater-than-or-equal-to, less-than-or-equal-to, or equality constraints) is determined by the sign restrictions on the associated primal variables. To determine the type of each constraint, we refer to the symmetric duals introduced in Section 2.7.1.

Recall that the canonical primal:

$$\min_{x} c^{\top} x$$

$$\text{s.t. } Ax \geq b$$

$$x \geq 0,$$

has:

$$\max_{y} b^{\top} y$$

$$\text{s.t. } A^{\top} y \leq c$$

$$y \geq 0,$$

as its dual. Recall, also, that there is a further symmetry between these two problems, in that the dual of the dual problem is the original primal problem. Because of this, we begin by first determining which of these two canonical problems matches the type of problem that our primal is. In our example, the primal is a maximization problem. Thus, we examine the problem:

$$\max_{y} b^{\top} y$$

$$\text{s.t. } A^{\top} y \leq c$$

$$y \geq 0.$$

Specifically, we notice that each variable in this problem is non-negative and that the associated structural constraints in the problem:

$$\min_{x} c^{\top} x$$

$$\text{s.t. } Ax \geq b$$

$$x \geq 0,$$

are greater-than-or-equal-to inequalities. Thus, in our example, each structural constraint in the dual that is associated with a primal variable that is non-negative in the primal problem will be a greater-than-or-equal-to constraint. Specifically, in our example, x_1 and x_2 are non-negative, and so their associated dual constraints will be greater-than-or-equal-to inequalities, as follows:

$$\min_{y} \ 10y_1 - y_2 + 5y_3$$

$$
\begin{aligned}
\text{s.t.} \quad & y_1 \geq 3 && (x_1) \\
& y_1 + y_2 \geq 5 && (x_2) \\
& -0.5y_2 + y_3 \quad -3 && (x_3) \\
& -4y_1 \quad 1.3 && (x_4) \\
& y_2 \quad -1. && (x_5)
\end{aligned}
$$

Next, we examine the variable x_4 in our primal problem. Notice that the sign restriction on this variable is opposite sign restriction (2.62) in the canonical maximization problem. Thus, its associated dual structural constraint will have the opposite direction of the canonical minimization problem (*i.e.*, opposite greater-than-or-equal-to constraint (2.58) in the canonical minimization problem). This means that the dual constraint associated with x_4 will be a less-than-or-equal-to constraint, as follows:

$$\min_{y} \ 10y_1 - y_2 + 5y_3$$

$$
\begin{aligned}
\text{s.t.} \quad & y_1 \geq 3 && (x_1) \\
& y_1 + y_2 \geq 5 && (x_2) \\
& -0.5y_2 + y_3 \quad -3 && (x_3) \\
& -4y_1 \leq 1.3 && (x_4) \\
& y_2 \quad -1. && (x_5)
\end{aligned}
$$

Finally, for the variables x_3 and x_5, which are unrestricted in sign in the primal problem, their associated structural constraints in the dual problem are equalities, as follows:

$$\min_{y} \ 10y_1 - y_2 + 5y_3$$

$$
\begin{aligned}
\text{s.t.} \quad & y_1 \geq 3 && (x_1) \\
& y_1 + y_2 \geq 5 && (x_2) \\
& -0.5y_2 + y_3 = -3 && (x_3) \\
& -4y_1 \leq 1.3 && (x_4) \\
& y_2 = -1. && (x_5)
\end{aligned}
$$

When finding the dual of a minimization problem, one would reverse the roles of the two canonical problems (*i.e.*, the roles of problems (2.60)–(2.62) and (2.57)–(2.59)). Specifically, one would have less-than-or-equal-to structural inequality constraints in the dual for primal variables that are non-negative, greater-than-or-equal-to inequalities for primal variables that are non-positive, and equality constraints for primal variables that are unrestricted in sign.

2.7.3.7   Step 7: Determine Sign Restrictions on the Dual Variables

The final step is to determine sign restrictions on the dual variables. The sign restrictions depend on the type of structural constraints in the primal problem. The method of determining the sign restrictions is analogous to determining the types of structural constraints in the dual problem in Step 6. We again rely on the two symmetric dual problems introduced in Section 2.7.1.

Because the primal problem in our example is a maximization, we focus on the canonical problem:

$$\max_{y} b^{\top} y$$

$$\text{s.t. } A^{\top} y \leq c$$

$$y \geq 0,$$

and its dual:

$$\min_{x} c^{\top} x$$

$$\text{s.t. } Ax \geq b$$

$$x \geq 0.$$

Specifically, we begin by noting that the first primal constraint:

$$x_1 + x_2 - 4x_4 \leq 10,$$

is consistent with constraint (2.61) in the canonical maximization problem. Thus, the dual variable associated with this constraint, which is y_1, is non-negative, which is consistent with sign restriction (2.59) in the canonical minimization problem. We next examine the third primal structural constraint:

$$x_3 \geq 5.$$

This is the opposite type of constraint compared to constraint (2.61) in the canonical maximization problem. Thus, its associated dual variable, which is y_3, will have a sign restriction that is opposite to non-negativity constraint (2.59) in the canonical minimization problem. Finally, we note that the second structural constraint in the primal problem is an equality. As such, its associated dual variable, which is y_2, is unrestricted in sign. Taking these sign restrictions together, the dual problem is:

$$\min_{y} \ 10y_1 - y_2 + 5y_3$$

$$
\begin{aligned}
\text{s.t.} \quad & y_1 \geq 3 && (x_1) \\
& y_1 + y_2 \geq 5 && (x_2) \\
& -0.5y_2 + y_3 = -3 && (x_3) \\
& -4y_1 \leq 1.3 && (x_4) \\
& y_2 = -1 && (x_5) \\
& y_1 \geq 0 \\
& y_3 \leq 0.
\end{aligned}
$$

One can verify these rules by taking different starting primal problems, converting them to canonical form, and examining the resulting dual. We now show, in the following example, that the dual problem found by applying these rules is identical to what would be obtained if the primal problem is first converted to canonical form.

Example 2.14 Consider the linear optimization problem:

$$\max_{x} \ 3x_1 + 5x_2 - 3x_3 + 1.3x_4 - x_5$$

$$
\begin{aligned}
\text{s.t.} \quad & x_1 + x_2 - 4x_4 \leq 10 \\
& x_2 - 0.5x_3 + x_5 = -1 \\
& x_3 \geq 5 \\
& x_1, x_2 \geq 0 \\
& x_4 \leq 0,
\end{aligned}
$$

which is introduced in Example 2.1. In Example 2.3 we find that the canonical form of this problem is:

$$\min_{x} \ -3x_1 - 5x_2 + 3x_3^+ - 3x_3^- + 1.3\tilde{x}_4 + x_5^+ - x_5^-$$

$$
\begin{aligned}
\text{s.t.} \quad & -x_1 - x_2 - 4\tilde{x}_4 \geq -10 \\
& -x_2 + 0.5x_3^+ - 0.5x_3^- - x_5^+ + x_5^- \geq 1 \\
& x_2 - 0.5x_3^+ + 0.5x_3^- + x_5^+ - x_5^- \geq -1 \\
& x_3^+ - x_3^- \geq 5 \\
& x_1, x_2, x_3^-, x_3^+, \tilde{x}_4, x_5^-, x_5^+ \geq 0,
\end{aligned}
$$

which can be written as:

$$\min_{x} \ c^{\mathsf{T}} x$$

$$
\begin{aligned}
\text{s.t.} \quad & Ax \geq b \\
& x \geq 0,
\end{aligned}
$$

where we have:

$$
x = \begin{pmatrix} x_1 \\ x_2 \\ x_3^+ \\ x_3^- \\ \tilde{x}_4 \\ x_5^+ \\ x_5^- \end{pmatrix},
$$

$$
c = \begin{pmatrix} -3 \\ -5 \\ 3 \\ -3 \\ 1.3 \\ 1 \\ -1 \end{pmatrix},
$$

$$
A = \begin{bmatrix} -1 & -1 & 0 & 0 & -4 & 0 & 0 \\ 0 & -1 & 0.5 & -0.5 & 0 & -1 & 1 \\ 0 & 1 & -0.5 & 0.5 & 0 & 1 & -1 \\ 0 & 0 & 1 & -1 & 0 & 0 & 0 \end{bmatrix},
$$

and:

$$
b = \begin{pmatrix} -10 \\ 1 \\ -1 \\ 5 \end{pmatrix}.
$$

Thus, the dual problem is:

$$
\max_{w} \; \begin{pmatrix} -10 & 1 & -1 & 5 \end{pmatrix} \begin{pmatrix} w_1 \\ w_2 \\ w_3 \\ w_4 \end{pmatrix}
$$

$$
\text{s.t.} \quad \begin{bmatrix} -1 & 0 & 0 & 0 \\ -1 & -1 & 1 & 0 \\ 0 & 0.5 & 0.5 & 1 \\ 0 & -0.5 & 0.5 & -1 \\ -4 & 0 & 0 & 0 \\ 0 & -1 & 1 & 0 \\ 0 & 1 & -1 & 0 \end{bmatrix} \begin{pmatrix} w_1 \\ w_2 \\ w_3 \\ w_4 \end{pmatrix} \leq \begin{pmatrix} -3 \\ -5 \\ 3 \\ -3 \\ 1.3 \\ 1 \\ -1 \end{pmatrix}
$$

$$\begin{pmatrix} w_1 \\ w_2 \\ w_3 \\ w_4 \end{pmatrix} \geq \begin{pmatrix} 0 \\ 0 \\ 0 \\ 0 \end{pmatrix},$$

which simplifies to:

$$\max_{w} \; -10w_1 + w_2 - w_3 + 5w_4$$

$$\text{s.t.} \quad -w_1 \leq -3$$
$$-w_1 - w_2 + w_3 \leq -5$$
$$0.5w_2 - 0.5w_3 + w_4 \leq 3$$
$$-0.5w_2 + 0.5w_3 - w_4 \leq -3$$
$$-4w_1 \leq 1.3$$
$$-w_2 + w_3 \leq 1$$
$$w_2 - w_3 \leq -1$$
$$w_1, w_2, w_3, w_4 \geq 0.$$

Note that if we change the direction of optimization, this problem becomes:

$$\min_{w} \; 10w_1 - w_2 + w_3 - 5w_4$$

$$\text{s.t.} \quad -w_1 \leq -3$$
$$-w_1 - w_2 + w_3 \leq -5$$
$$0.5w_2 - 0.5w_3 + w_4 \leq 3$$
$$-0.5w_2 + 0.5w_3 - w_4 \leq -3$$
$$-4w_1 \leq 1.3$$
$$-w_2 + w_3 \leq 1$$
$$w_2 - w_3 \leq -1$$
$$w_1, w_2, w_3, w_4 \geq 0.$$

Next, note that if we define the variables $y_1 = w_1$, $y_2 = w_2 - w_3$, and $y_3 = -w_4$, this problem can be written as:

$$\min_{y} \; 10y_1 - y_2 + 5y_3$$

$$\text{s.t.} \quad -y_1 \leq -3$$
$$-y_1 - y_2 \leq -5$$
$$0.5y_2 - y_3 \leq 3 \tag{2.63}$$
$$-0.5y_2 + y_3 \leq -3 \tag{2.64}$$
$$-4y_1 \leq 1.3$$

$$-y_2 \leq 1 \qquad (2.65)$$
$$y_2 \leq -1 \qquad (2.66)$$
$$y_1 \geq 0$$
$$y_3 \leq 0.$$

Note that because $y_1 = w_1$ and $w_1 \geq 0$ we have $y_1 \geq 0$. Similarly, $y_3 = -w_4$ and $w_4 \geq 0$, implying that $y_3 \leq 0$. Because y_2 is defined as the difference between two non-negative variables, it can be either positive or negative in sign.

We next note that (2.63) and (2.64) can be rewritten as:

$$-0.5w_2 + 0.5w_3 + w_4 \geq -3,$$

and:

$$-0.5w_2 + 0.5w_3 + w_4 \leq -3,$$

which together imply:

$$-0.5w_2 + 0.5w_3 + w_4 = -3.$$

Similarly, (2.65) and (2.66) imply:

$$y_2 = -1.$$

Making these substitutions and multiplying some of the other structural inequalities through by -1 gives the following dual problem:

$$\min_{y} 10y_1 - y_2 + 5y_3$$
$$\text{s.t.} \ \ y_1 \geq 3$$
$$y_1 + y_2 \geq 5$$
$$-0.5y_2 + y_3 = -3$$
$$-4y_1 \leq 1.3$$
$$y_2 = -1$$
$$y_1 \geq 0$$
$$y_3 \leq 0,$$

which is identical to the dual problem found directly by applying the conversion rules.

$\square$

Example 2.15 Consider the Gasoline-Mixture Problem, which is introduced in Section 2.1.3. The primal problem is formulated as:

$$\min_{x} 200x_1 + 220x_2$$

$$\text{s.t. } 0.7x_1 + 0.6x_2 \geq 0.65$$

$$0.9x_1 + 0.8x_2 \leq 0.85$$

$$x_1 + x_2 = 1$$

$$x_1, x_2 \geq 0.$$

To find the dual of this problem directly, we first observe that this problem has three structural constraints, thus the dual will have three variables, which we call y_1, y_2, and y_3. We associate these dual variables with the primal constraints as follows:

$$\min_{x} 200x_1 + 220x_2$$

$$\text{s.t. } 0.7x_1 + 0.6x_2 \geq 0.65 \qquad (y_1)$$

$$0.9x_1 + 0.8x_2 \leq 0.85 \qquad (y_2)$$

$$x_1 + x_2 = 1 \qquad (y_3)$$

$$x_1, x_2 \geq 0.$$

We next determine the objective function of the dual problem, which is:

$$\max_{y} \ 0.65y_1 + 0.85y_2 + y_3,$$

based on the direction of optimization of the primal problem and the right-hand side constants of the structural constraints. We also know that because the primal problem has two variables the dual problem has two structural constraints, each of which we associate with a primal variable as follows:

$$\max_{y} 0.65y_1 + 0.85y_2 + y_3$$

$$\text{s.t.} \qquad (x_1)$$

$$(x_2)$$

We next determine the right-hand sides of the structural constraints in the dual problem based on the objective-function coefficients in the primal problem, which gives:

$$\max_{y} 0.65y_1 + 0.85y_2 + y_3$$

$$\text{s.t.} \qquad 200 \qquad (x_1)$$

$$220. \qquad (x_2)$$

Next we determine the left-hand sides of the structural constraints in the dual problem using the coefficients on the left-hand sides of the structural constraints of the primal problem, giving:

$$\max_{y} 0.65y_1 + 0.85y_2 + y_3$$

$$\text{s.t. } 0.7y_1 + 0.9y_2 + y_3 \quad 200 \qquad (x_1)$$
$$0.6y_1 + 0.8y_2 + y_3 \quad 220. \qquad (x_2)$$

Next, to determine the types of structural constraints in the dual problem, we note that the primal problem is a minimization problem and each of x_1 and x_2 are non-negative. Based on the canonical minimization problem given by (2.57)–(2.59) and its dual, we observe that a minimization problem with non-negative constraints has a dual problem with less-than-or-equal-to structural constraints. Thus, both of the structural constraints will be less-than-or-equal-to constraints, which gives:

$$\max_{y} 0.65y_1 + 0.85y_2 + y_3$$

$$\text{s.t. } 0.7y_1 + 0.9y_2 + y_3 \le 200 \qquad (x_1)$$
$$0.6y_1 + 0.8y_2 + y_3 \le 220. \qquad (x_2)$$

Finally, we must determine the sign restrictions on the dual variables. Again, we examine the canonical minimization problem given by (2.57)–(2.59) and its dual. Note that the first primal constraint in the Gasoline-Mixture Problem is a greater-than-or-equal-to constraint, which is consistent with (2.58). Thus, the associated dual variable, y_1, is non-negative. Because the second primal constraint is inconsistent with (2.58), the associated variable, y_2, is non-positive. The third primal constraint is an equality, meaning that the associated variable, y_3, is unrestricted in sign. Thus, the dual problem is:

$$\max_{y} 0.65y_1 + 0.85y_2 + y_3$$

$$\text{s.t. } 0.7y_1 + 0.9y_2 + y_3 \le 200 \qquad (x_1)$$
$$0.6y_1 + 0.8y_2 + y_3 \le 220 \qquad (x_2)$$
$$y_1 \ge 0$$
$$y_2 \le 0.$$

$\square$

2.7.3.8 Summary of Duality Conversion Rules

From the discussion in this section, we can summarize the following duality conversion rules. We use the terms 'primal' and 'dual' in these rules, however the two words can be interchanged. This is because there is a symmetry between primal and dual problems (*i.e.*, the primal problem is the dual of the dual).

For a primal problem that is a minimization we have that:

- the dual problem is a maximization;
- dual variables associated with equality constraints in the primal problem are unrestricted in sign;
- dual variables associated with greater-than-or-equal-to constraints in the primal problem are non-negative;
- dual variables associated with less-than-or-equal-to constraints in the primal problem are non-positive;
- dual constraints associated with non-negative primal variables are less-than-or-equal-to inequalities;
- dual constraints associated with non-positive primal variables are greater-than-or-equal-to inequalities; and
- dual constraints associated with primal variables that are unrestricted in sign are equalities.

For a dual problem that is a maximization we have that:

- the primal problem is a minimization;
- primal variables associated with equality constraints in the dual problem are unrestricted in sign;
- primal variables associated with less-than-or-equal-to constraints in the dual problem are non-negative;
- primal variables associated with greater-than-or-equal-to constraints in the dual problem are non-positive;
- primal constraints associated with non-negative dual variables are greater-than-or-equal-to inequalities;
- primal constraints associated with non-positive dual variables are less-than-or-equal-to inequalities; and
- primal constraints associated with dual variables that are unrestricted in sign are equalities.

2.7.4 Weak- and Strong-Duality Properties

We saw in Section 2.7.1 that any linear optimization problem and its dual have structural relationships. We now show some further relationships between a primal problem and its dual, focusing specifically on the objective-function values of points that are feasible in a primal problem and its dual. We show these relationships for the special case of the symmetric duals introduced in Section 2.7.1. However, these results extend to any primal problem (even if it is not in canonical form) [9]. This is because any primal problem can be converted to canonical form and then its symmetric dual can be found.

Weak-Duality Property: Consider the primal problem:

$$\min_{x} c^{\top} x$$

$$\text{s.t. } Ax \geq b$$

$$x \geq 0,$$

and its dual:

$$\max_{y} b^{\top} y$$

$$\text{s.t. } A^{\top} y \leq c$$

$$y \geq 0.$$

If x is feasible in the primal problem and y is feasible in the dual problem, then we have that:

$$c^{\top} x \geq b^{\top} y.$$

To show this we first note that if x is feasible in the primal problem then we must have:

$$b \leq Ax. \tag{2.67}$$

Next, we know that if y is feasible in the dual problem, we must have $y \geq 0$. Thus, multiplying both sides of (2.67) by y gives:

$$y^{\top} b \leq y^{\top} Ax. \tag{2.68}$$

We also know that if y is feasible in the dual problem, then:

$$A^{\top} y \leq c,$$

which can be rewritten as:

$$y^{\top} A \leq c^{\top},$$

by transposing the expressions on both sides of the inequality. Combining this inequality with (2.68) gives:

$$y^{\top} b \leq y^{\top} Ax \leq c^{\top} x,$$

or simply:

$$y^\top b \le c^\top x,$$

which is the weak-duality inequality.

The weak-duality property says that the objective-function value of any point that is feasible in the dual problem provides a bound on the objective-function value of any point that is feasible in the primal problem and *vice versa*. One can take this weak-duality relationship further. That is because any point that is feasible in the dual problem provides a bound on the objective-function of the *optimal* objective-function value of the primal problem. For a given pair of feasible solutions to the primal and dual problems, the difference:

$$c^\top x - y^\top b,$$

is called the **duality gap**. One way that the weak-duality property can be used is to determine how close to optimal a feasible solution to the primal problem is. One can estimate this by finding a solution that is feasible in the dual problem and computing the duality gap.

As noted above, the weak-duality property applies to any primal problem and its dual (not only to a primal problem in canonical form). However, the direction of the inequality between the objective functions of the two problems may change, depending on the directions of the inequalities, signs of the variables, and the direction of optimization in the primal and dual problems. Thus, for the purposes of applying the weak-duality property, it is often beneficial to convert the primal problem into canonical form.

Primal and dual problems have a further related property, known as the strong-duality equality.

Strong-Duality Property: Consider the primal problem:

$$\min_{x} c^\top x$$
$$\text{s.t. } Ax \ge b$$
$$x \ge 0,$$

and its dual:

$$\max_{y} b^\top y$$
$$\text{s.t. } A^\top y \le c$$
$$y \ge 0.$$

> If x^* is optimal in the primal problem and y^* is optimal in the dual problem, then we have that:
>
> $$c^\top x^* = b^\top y^*.$$

The strong-duality equality says that the objective-function value of the dual problem evaluated at an optimal solution is equal to the optimal primal objective-function value. Luenberger and Ye [9] provide a formal proof of this so-called strong-duality theorem. In some sense, the primal problem pushes its objective-function value down (because it is a minimization problem) while the dual problem pushes its objective-function value up (as it is a maximization problem). The two problems 'meet' at an optimal solution where they have the same objective-function value. Figure 2.15 illustrates this concept by showing the duality gap for a given pair of solutions that are feasible in the primal and dual problems (in the left-hand side of the figure) and the gap being reduced to zero as the two problems are solved to optimality (in the right-hand side of the figure).

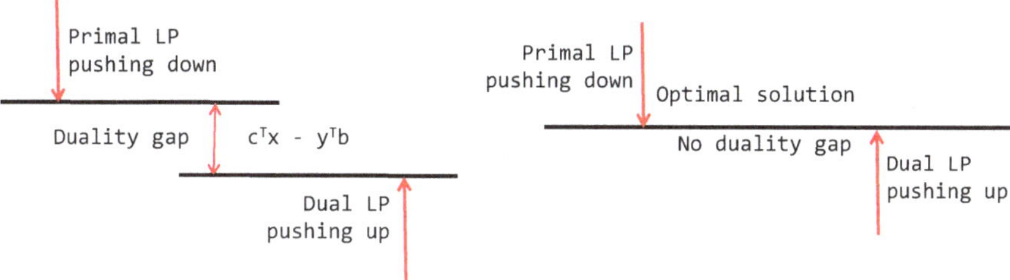

Fig. 2.15 Illustration of the duality gap being reduced as the primal and dual problems are solved to optimality

The strong-duality equality also gives us another way to solve a linear optimization problem. As an illustrative example, consider a primal problem in canonical form:

$$\min_{x} c^\top x$$
$$\text{s.t. } Ax \geq b$$
$$x \geq 0,$$

and its dual:

$$\max_{y} b^\top y$$
$$\text{s.t. } A^\top y \leq c$$
$$y \geq 0.$$

Instead of solving the primal problem directly, one can solve the following system of equalities and inequalities:

$$Ax \geq b$$

$$A^\top y \leq c$$

$$c^\top x = b^\top y$$

$$x \geq 0$$

$$y \geq 0,$$

for x and y. This system of equalities and inequalities consist of the constraints of the primal problem (*i.e.*, $Ax \geq b$ and $x \geq 0$), the constraints of the dual problem (*i.e.*, $A^\top y \leq c$ and $y \geq 0$), and the strong-duality equality (*i.e.*, $c^\top x = b^\top y$). If we find a pair of vectors, x^* and y^*, that satisfy all of these conditions, then from the strong-duality property we know that x^* is optimal in the primal problem and y^* is optimal is the dual problem.

These strong-duality properties (including the alternate method of solving a linear optimization problem) apply to any primal problem and its dual (regardless of whether the primal problem is in canonical form). This is because the strong-duality property is an equality, meaning that the result is the same regardless of the form of the primal problem. This is demonstrated in the following example.

Example 2.16 Consider the Electricity-Production Problem, which is introduced in Section 2.1.1. This problem is formulated as:

$$\max_{x} x_1 + x_2 \tag{2.69}$$

$$\text{s.t. } \frac{2}{3}x_1 + x_2 \leq 18 \tag{2.70}$$

$$2x_1 + x_2 \geq 8 \tag{2.71}$$

$$x_1 \leq 12 \tag{2.72}$$

$$x_2 \leq 16 \tag{2.73}$$

$$x_1, x_2 \geq 0. \tag{2.74}$$

Taking this is as the primal problem, its dual is:

$$\min_{y} 18y_1 + 8y_2 + 12y_3 + 16y_4 \tag{2.75}$$

$$\text{s.t. } \frac{2}{3}y_1 + 2y_2 + y_3 \geq 1 \tag{2.76}$$

$$y_1 + y_2 + y_4 \geq 1 \tag{2.77}$$

$$y_1, y_3, y_4 \geq 0 \tag{2.78}$$

$$y_2 \leq 0. \tag{2.79}$$

We can also find the dual of this problem by first converting it to canonical form, which is:

$$\min_{x} \ -x_1 - x_2 \tag{2.80}$$

$$\text{s.t.} \ -\frac{2}{3}x_1 - x_2 \geq -18 \tag{2.81}$$

$$2x_1 + x_2 \geq 8 \tag{2.82}$$

$$-x_1 \geq -12 \tag{2.83}$$

$$-x_2 \geq -16 \tag{2.84}$$

$$x_1, x_2 \geq 0, \tag{2.85}$$

and has the dual problem:

$$\max_{y} \ -18y_1 + 8y_2 - 12y_3 - 16y_4 \tag{2.86}$$

$$\text{s.t.} \ -\frac{2}{3}y_1 + 2y_2 - y_3 \leq -1 \tag{2.87}$$

$$-y_1 + y_2 - y_4 \leq -1 \tag{2.88}$$

$$y_1, y_2, y_3, y_4 \geq 0. \tag{2.89}$$

Consider the solution $(x_1, x_2) = (4, 0)$, which is feasible but not optimal in the canonical-form primal problem, which is given by (2.80)–(2.85). Substituting this solution into objective function (2.80) gives a value of -4. Next, consider the solution $(y_1, y_2, y_3, y_4) = (1.5, 0, 0, 0)$, which is feasible but not optimal in the symmetric dual problem, which is given by (2.86)–(2.89). Substituting this solution into objective function (2.86) gives a value of -27. We see that the dual objective-function value is less than the primal value, verifying that the weak-duality inequality is satisfied.

These solutions are also feasible in the original primal problem, given by (2.69)–(2.74), and its dual, given by (2.75)–(2.79). However, substituting these solutions gives a primal objective-function value of 4 and a dual value of 27. We see that in this case the weak-duality inequality is reversed, because the original primal problem is not in canonical form.

Next, consider the primal solution $(x_1^*, x_2^*) = (12, 10)$ and the dual solution $(y_1^*, y_2^*, y_3^*, y_4^*) = (1, 0, 1/3, 0)$. It is straightforward to verify that these solutions are feasible in both forms of the primal and dual problems. Moreover, note that if we substitute x^* into primal objective function (2.69) we obtain a value of 22. Similarly, substituting y^* into (2.75) yields a dual objective-function value of 22. Thus, by the strong-duality property, we know that x^* and y^* are optimal in their respective problems.

We also obtain the same results from examining the objective functions of the canonical-form primal and its dual, except that the signs are reversed. More specifically, substituting x^* into (2.80) yields a value of -22 and substituting y^* into (2.86) also gives -22.

□

2.7.5 *Duality and Sensitivity*

In this section we show yet another important relationship between a primal problem and its dual, specifically focusing on the relationship between dual variables and the sensitivity vector of the primal problem. We assume that we are given a linear optimization problem in standard form:

$$\min_{x} c^\top x$$

$$\text{s.t. } Ax = b$$

$$x \geq 0.$$

Moreover, we assume that we have an optimal basic feasible solution x^*, which is partitioned into basic and nonbasic variables, $x^* = (x_B^*, x_N^*)$. Thus, we also have the objective-function and constraint coefficients partitioned as $c = (c_B, c_N)$ and $A = [B, N]$. Because the nonbasic variables are equal to zero, the objective function can be written as:

$$c^\top x^* = c_B^\top x_B^*.$$

We also know that the basic-variable vector can be written as:

$$x_B^* = B^{-1}b,$$

thus we can write the optimal objective-function value as:

$$c^\top x^* = c_B^\top x_B^* = c_B^\top B^{-1}b. \tag{2.90}$$

At the same time, the strong-duality equality gives us:

$$c^\top x^* = b^\top y^*, \tag{2.91}$$

where y^* is a dual-optimal solution. Combining (2.90) and (2.91) gives:

$$y^{*\top} b = c_B^\top B^{-1}b,$$

or:

$$y^{*\top} = c_B^\top B^{-1},$$

which provides a convenient way of computing dual-variable values from the final tableau after solving the primal problem using the Simplex method. We next recall sensitivity result (2.51):

$$c_B^\top B^{-1}\Delta b = \lambda^\top \Delta b,$$

which implies:

$$y^{*\top} = c_B^\top B^{-1} = \lambda^\top,$$

or that the sensitivity vector is equal to the dual variables.

2.7.6 Complementary Slackness

In this section we show a final relationship between a primal problem and its dual. We do not prove this relationship formally, but rather rely on the interpretation of dual variables as providing sensitivity information derived in Section 2.7.5. We, again, consider the case of a primal problem in canonical form:

$$\min_x c^\top x$$
$$\text{s.t. } Ax \geq b$$
$$x \geq 0,$$

and its symmetric dual:

$$\max_y b^\top y$$
$$\text{s.t. } A^\top y \leq c$$
$$y \geq 0.$$

The discussion in Section 2.7.5 leads to the conclusion that the dual variables, y, provide sensitivity information for the structural constraints in the primal problem. Moreover, we know from Section 2.7.1 that the primal problem is the dual of the dual. Thus, we can also conclude that the primal variables, x, provide sensitivity information for the structural constraints in the dual problem.

Before proceeding, we first define what it means for an inequality constraint in a linear optimization to be binding as opposed to non-binding. For this discussion, let us consider the jth structural constraint in the primal problem, which can be written as:

$$A_{j,.}x \geq b_j,$$

where $A_{j,.}$ is the jth row of the A matrix and b_j the jth element of the b vector. Note, however, that these definitions can be applied to any inequality constraint (of any direction) in any problem. This constraint is said to be **non-binding** at the point $\hat{x}$ if:

$$A_{j,.}\hat{x} > b_j.$$

That is to say, the inequality constraint is non-binding if there is a difference or slack between the left- and right-hand sides of the constraint when we substitute the values of $\hat{x}$ into the constraint. Conversely, we say that this constraint is **binding** at the point $\hat{x}$ if:

$$A_{j,\cdot}\hat{x} = b_j.$$

The constraint is binding if there is no difference or slack between its two sides at $\hat{x}$.

Now that we have these two definitions, let us consider the effect of changing the right-hand side of the jth structural constraint in the primal problem, which is:

$$A_{j,\cdot}x \geq b_j,$$

to:

$$A_{j,\cdot}x \geq b_j + \Delta b_j,$$

where Δb_j is sufficiently close to zero to satisfy condition (2.49). We know from Sections 2.6 and 2.7.5 that the change in the primal objective-function value can be computed as:

$$\Delta b_j y_j^*,$$

where y_j^* is the optimal dual-variable value associated with the jth primal structural constraint.

Let us next intuitively determine what the effect of changing this constraint is. First, consider the case in which the original constraint:

$$A_{j,\cdot}x \geq b_j,$$

is not binding. In this case, we can conclude that changing the constraint to:

$$A_{j,\cdot}x \geq b_j + \Delta b_j,$$

will have no effect on the optimal solution, because the constraint is already slack at the point x^*. If changing the constraint causes the feasible region to decrease in size, x^* will still be feasible so long as $|\Delta b_j|$ is not too large. Similarly, if changing the constraint causes the feasible region to increase in size, x^* will still remain optimal because loosening the jth constraint should not result in the optimal solution changing. Based on this intuition, we conclude that if the jth constraint is non-binding then $\Delta b_j y_j^* = 0$ (because the optimal solution does not change) and, thus, $y_j^* = 0$.

We can also draw the converse conclusion for the case of a binding constraint. If the jth constraint is binding, then changing the constraint to:

$$A_{j,\cdot}x \geq b_j + \Delta b_j,$$

may have an effect on the optimal solution. This is because if changing the constraint causes the feasible region to decrease in size, x^* may change causing the primal objective-function value to increase (get worse). If changing the constraint causes the feasible region to increase in size, x^* may change causing the primal objective-function value to get better. Thus, we conclude that if the jth constraint is binding then $\Delta b_j y_j^*$ can be non-zero, meaning that y_j^* may be non-zero.

These conclusions give what is known as the **complementary-slackness** condition between the primal constraints and their associated dual variables. The Primal Complementary-Slackness Property is written explicitly in the following.

Primal Complementary-Slackness Property: Consider the primal problem:

$$\min_{x} c^\top x$$

$$\text{s.t. } Ax \geq b$$

$$x \geq 0,$$

and its dual:

$$\max_{y} b^\top y$$

$$\text{s.t. } A^\top y \leq c$$

$$y \geq 0.$$

If x^* is optimal in the primal problem and y^* is optimal in the dual problem, then for each primal constraint, $j = 1, \ldots, m$, we have that either:

1. $A_{j,.}x^* = b_j$,
2. $y_j^* = 0$, or
3. both.

The third case in the Primal Complementary-Slackness Property implies some level of redundancy in the constraints. This is because in Case 3 there is at least one inequality constraint that is binding, but which has a sensitivity value of zero.

This complementary-slackness property can also be written more compactly as:

$$(A_{j,.}x^* - b_j)y_j^* = 0, \forall j = 1, \ldots, m.$$

This is because for the product, $(A_{j,.}x^* - b_j)y_j^*$, to equal zero for a given j we must either have $(A_{j,.}x^* - b_j) = 0$, which is the first complementary slackness condition, or $y_j^* = 0$, which is the second. Indeed, instead of writing:

$$(A_{j,.}x^* - b_j)y_j^* = 0,$$

for each j, one can also write the complementary slackness condition even more compactly as:

$$(Ax^* - b)^\top y = 0,$$

which enforces complementary slackness between all of the primal constraints and dual variables in a single equation.

We can carry out a similar analysis of changing the right-hand sides of the dual constraints, which gives rise to the following Dual Complementary-Slackness Property.

Dual Complementary-Slackness Property: Consider the primal problem:

$$\min_{x} c^\top x$$

$$\text{s.t.}\ \ Ax \geq b$$

$$x \geq 0,$$

and its dual:

$$\max_{y} b^\top y$$

$$\text{s.t.}\ \ A^\top y \leq c$$

$$y \geq 0.$$

If x^* is optimal in the primal problem and y^* is optimal in the dual problem, then for each dual constraint, $i = 1, \ldots, n$, we have that either:

1. $A_{.,i}^\top y^* = c_i$,
2. $x_i^* = 0$, or
3. both;

where $A_{.,i}$ is the ith column of the A matrix.

As with the Primal Complementary-Slackness Property, the third case in the Dual Complementary-Slackness Property also implies some level of redundancy in the constraints. This is because in Case 3 there is at least one inequality constraint that is binding, but which has a sensitivity value of zero.

As with the Primal Complementary-Slackness Property, the Dual Complementary-Slackness Property can be written more compactly as:

$$(A_{.,i}^\top y^* - c_i)x_i^* = 0,\ \forall i = 1, \ldots, n;$$

or as:

$$(A^\top y^* - c)^\top x^* = 0.$$

The Primal and Dual Complementary-Slackness Properties can be extended to structural constraints that are different from the canonical form. The same result is obtained, if there is an inequality in one of the primal or dual problem either it must be binding or the associated variable in the other problem must equal zero. One can also write complementary slackness involving equality constraints, however these conditions do not provide any useful information. This is because an equality constraint must always be binding. Thus, the variable associated with an equality constraint can take on any value (*i.e.*, it is never restricted to equal zero).

One of the benefits of the complementary-condition properties is that it provides another means to recover an optimal solution to either of the primal or dual problem from an optimal solution to the other. This is demonstrated in the following example.

Example 2.17 Consider the Electricity-Production Problem, which is introduced in Section 2.1.1. This problem is formulated as:

$$\max_{x} x_1 + x_2$$

$$\begin{aligned}
\text{s.t.} \quad & \frac{2}{3}x_1 + x_2 \leq 18 && (y_1) \\
& 2x_1 + x_2 \geq 8 && (y_2) \\
& x_1 \leq 12 && (y_3) \\
& x_2 \leq 16 && (y_4) \\
& x_1, x_2 \geq 0,
\end{aligned}$$

and its dual is:

$$\min_{y} 18y_1 + 8y_2 + 12y_3 + 16y_4$$

$$\begin{aligned}
\text{s.t.} \quad & \frac{2}{3}y_1 + 2y_2 + y_3 \geq 1 && (x_1) \\
& y_1 + y_2 + y_4 \geq 1 && (x_2) \\
& y_1, y_3, y_4 \geq 0 \\
& y_2 \leq 0,
\end{aligned}$$

where the variable associations are denoted in the parentheses. We know that $(x_1^*, x_2^*) = (12, 10)$ is optimal in the primal problem.

Substituting these values into the primal constraints, we see that the second and fourth one are non-binding. Thus, the Primal Complementary-Slackness Property tells us that $y_2^* = 0$ and $y_4^* = 0$. Furthermore, because both x_1^* and x_2^* are non-zero, we know that their associated dual constraints must be binding (*i.e.*, we can write them as equalities). Doing so and substituting in the values found for $y_2^* = 0$ and $y_4^* = 0$ gives:

$$\frac{2}{3}y_1 + y_3 = 1$$

$$y_1 = 1,$$

which have: $y_1^* = 1$ and $y_3^* = 1/3$ as solutions. This dual solution, $(y_1^*, y_2^*, y_3^*, y_4^*) = (1, 0, 1/3, 0)$ coincides with the dual-optimal solution found in Example 2.16.

$\square$

2.8 Final Remarks

This chapter introduces the Simplex method, which was initially proposed by Dantzig [5], as an algorithm to solve linear optimization problems. The Simplex method works by navigating around the boundary of the polytope that describes the feasible region of the problem, jumping from one extreme point (or basic feasible solution) to another until reaching an optimal corner. An obvious question that this raises is whether it would be beneficial to navigate through the interior of the polytope instead of around its exterior. For extremely large problems, this may be the case. Interested readers are referred to more advanced textbooks, which introduce such **interior-point** algorithms [11]. Additional information on LPPs, their formulation, properties, and solutions algorithms, can be found in a number of other advanced textbooks [1, 2, 9, 12]. Modeling issues are treated extensively by Castillo *et al.* [4].

2.9 GAMS Codes

This final section provides GAMS [3] codes for the main problems considered in this chapter. GAMS can use a variety of different software packages, among them CPLEX [8] and GUROBI [7], to actually solve an LPP.

2.9.1 Electricity-Production Problem

The Electricity-Production Problem, which is introduced in Section 2.1.1, has the following GAMS formulation:

```
1   variable z;
2   positive variables x1, x2;
3   equations of, eq1, eq2, eq3, eq4;
4   of .. z =e= x1+x2;
5   eq1 .. (2/3)*x1+x2 =l= 18;
6   eq2 .. 2*x1+x2 =g= 8;
7   eq3 .. x1 =l= 12;
8   eq4 .. x2 =l= 16;
9   model ep /all/;
10  solve ep using lp maximizing z;
```

Lines 1 and 2 are variable declarations, Line 3 gives names to the equations (*i.e.*, the objective function, equalities, and inequalities) of the model, and Lines 5–8 define these equations (*i.e.*, the objective function and constraints). The double-dot separates the name of an equation from its definition. "=e=" indicates an equality, "=l=" a less-than-or-equal-to inequality, and "=g=" a greater-than-or-equal-to inequality. Line 9 gives a name to the model and indicates that all equations should be considered. Finally, Line 10 directs GAMS to solve the problem using an LP solver while minimizing z. All lines end with a semicolon.

The GAMS output that provides information about the optimal solution is:

		LOWER	LEVEL	UPPER	MARGINAL
----	VAR z	-INF	22.000	+INF	.
----	VAR x1	.	12.000	+INF	.
----	VAR x2	.	10.000	+INF	.

2.9.2 Natural Gas-Transportation Problem

The Natural Gas-Transportation Problem, which is introduced in Section 2.1.2, has the following GAMS formulation:

```
variable z;
positive variables x11, x12, x21, x22;
equations of, s1, s2, d1, d2;
of .. z =e= 5*x11+4*x12+3*x21+6*x22;
s1 .. x11+x12 =l= 7;
s2 .. x21+x22 =l= 12;
d1 .. x11+x21 =e= 10;
d2 .. x12+x22 =e= 8;
model ng /all/;
solve ng using lp minimizing z;
```

Lines 1 and 2 declare variables, Line 3 gives names to the model equations, Line 4 defines the objective function, Lines 5–8 specify the constraints, Line 9 defines the model, and Line 10 directs GAMS to solve it.

The GAMS output that provides information about the optimal solution is:

		LOWER	LEVEL	UPPER	MARGINAL
----	VAR z	-INF	64.000	+INF	.
----	VAR x11	.	.	+INF	4.000
----	VAR x12	.	7.000	+INF	.
----	VAR x21	.	10.000	+INF	.
----	VAR x22	.	1.000	+INF	.

2.9.3 Gasoline-Mixture Problem

The Gasoline-Mixture Problem, which is introduced in Section 2.1.3, has the following GAMS formulation:

```
1   variable z;
2   positive variables x1, x2;
3   equations of, eq1, eq2, eq3;
4   of .. z =e= 200*x1+220*x2;
5   eq1 .. 0.7*x1+0.6*x2 =g= 0.65;
6   eq2 .. 0.9*x1+0.8*x2 =l= 0.85;
7   eq3 .. x1+x2=e=1;
8   model mg /all/;
9   solve mg using lp minimizing z;
```

Lines 1 and 2 declare variables, Line 3 gives names to the model equations, Line 4 defines the objective function, Lines 5–7 specify the constraints, Line 8 defines the model, and Line 9 directs GAMS to solve it.

The GAMS output that provides information about the optimal solution is:

```
1                   LOWER      LEVEL       UPPER      MARGINAL

3   ---- VAR z       -INF     210.000      +INF          .
4   ---- VAR x1        .         0.500      +INF          .
5   ---- VAR x2        .         0.500      +INF          .
```

2.9.4 Electricity-Dispatch Problem

The Electricity-Dispatch Problem, which is introduced in Section 2.1.4, has the following GAMS formulation:

```
1    variable z, theta1, theta2;
2    positive variables x1, x2;
3    equations of, ba1, ba2, ba3, bo1, bo2;
4    of .. z =e= x1+2*x2;
5    ba1 .. x1 =e= (theta1-theta2)+(theta1-0);
6    ba2 .. x2 =e= (theta2-theta1)+(theta2-0);
7    ba3 .. 10 =e= (theta1-0) +(theta2-0);
8    bo1 .. x1 =l= 6;
9    bo2 .. x2 =l= 8;
10   model ed /all/;
11   solve ed using lp minimizing z;
```

Lines 1 and 2 declare variables, Line 3 gives names to the equations of the model, Line 4 defines the objective function, Lines 5–9 specify the constraints, Line 10 defines the model, and Line 11 directs GAMS to solve it.

The GAMS output that provides information about the optimal solution is:

		LOWER	LEVEL	UPPER	MARGINAL
----	VAR z	-INF	14.000	+INF	.
----	VAR theta1	-INF	5.333	+INF	.
----	VAR theta2	-INF	4.667	+INF	.
----	VAR x1	.	6.000	+INF	.
----	VAR x2	.	4.000	+INF	.

2.10 Exercises

2.1 Jose builds electrical cable using two types of metallic alloys. Alloy 1 is 55% aluminum and 45% copper, while alloy 2 is 75% aluminum and 25% copper. Market prices for alloys 1 and 2 are \$5 and \$4 per ton, respectively. Formulate a linear optimization problem to determine the cost-minimizing quantities of the two alloys that Jose should use to produce 1 ton of cable that is at least 30% copper.

2.2 Transform the linear optimization problem:

$$\max_{x_1, x_2} z = x_1 + 2x_2$$
$$\text{s.t. } 2x_1 + x_2 \leq 12$$
$$x_1 - x_2 \geq 2$$
$$x_1, x_2 \geq 0,$$

to standard and canonical forms.

2.3 Consider the tableau shown in Table 2.20. Conduct a regularization step and pivot operation to make the $\tilde{b}$ vector non-negative.

Table 2.20 Tableau for Exercise 2.3

	1	x_1	x_2
z	0	-1	-1
x_3	12	-2	-1
x_4	-3	1	0

2.4 Consider the tableau shown in Table 2.21. Conduct Simplex iterations to solve the associated linear optimization problem. What are the optimal solution found and sensitivity vector?

Table 2.21 Tableau for Exercise 2.4

	1	x_5	x_2	x_4
z	-3	21	-1	$-1/2$
x_3	6	3	-1	-2
x_1	3	-1	0	1

2.5 Find the dual of the primal problem:

$$\min_{x_1, x_2} z_P = 3x_1 + 4x_2$$

$$\text{s.t. } 2x_1 + 3x_2 \geq 4$$
$$3x_1 + 4x_2 \leq 10$$
$$x_1 + x_2 = 5$$
$$x_1 \geq 0$$
$$x_2 \leq 0.$$

2.6 The optimal solution to the primal problem in Exercise 2.5 is $(x_1^*, x_2^*) = (14/3, 8/3)$. The optimal solution to the dual problem has 1 as the value of the dual variable associated with the first primal constraint and -1 as the value of the dual variable associated with the second. Using this information, answer the following three questions.

1. What is the change in the primal objective-function value if the right-hand side of the first primal constraint changes to 12.1?
2. What is the change in the primal objective-function value if the right-hand side of the second primal constraint changes to 1.9?
3. What is the change in the primal objective-function value if the right-hand side of the first primal constraint changes to 12.1 and the right-hand side of the second primal constraint changes to 1.9?

2.7 Write a GAMS code for the model formulated in Exercise 2.1.

References

1. Bazaraa MS, Jarvis JJ, Sherali HD (2009) Linear programming and network flows, 4th edn. Wiley, New York
2. Bertsimas D, Tsitsiklis JN (1997) Introduction to linear optimization. Athena scientific series in optimization and neural computation, vol 6, third printing edn. Athena Scientific, Belmont
3. Brook A, Kendrick D, Meeraus, A (1988) GAMS – a user's guide. ACM, New York. www.gams.com
4. Castillo E, Conejo AJ, Pedregal P, García R, Alguacil N (2002) Building and solving mathematical programming models in engineering and science. Pure and applied mathematics series. Wiley, New York
5. Dantzig GB (1963) Linear programming and extensions. Princeton University Press, Princeton

6. Eaton JW, Bateman D, Hauberg S, Wehbring R (2016) GNU octave version 4.2.0 manual: a high-level interactive language for numerical computations. www.gnu.org/software/octave
7. Gurobi Optimization, Inc. (2010) Gurobi Optimizer Reference Manual, Version 3.0. Houston, Texas. www.gurobi.com
8. IBM ILOG CPLEX Optimization Studio (2016) CPLEX User's Manual, Version 12 Release 7. IBM Corp. www.cplex.com
9. Luenberger DG, Ye Y (2008) Linear and nonlinear programming. International series in operations research & management science, vol 116, 3rd edn. Springer, New York
10. MATLAB (2016) The MathWorks, Inc., Natick, Massachusetts. www.mathworks.com/products/matlab
11. Vanderbei RJ (2001) Linear programming: foundations and extensions. Springer, New York
12. Wright SJ (1987) Primal-dual interior-point methods. Society for industrial and applied mathematics, Philadelphia, Pennsylvania

Chapter 3
Mixed-Integer Linear Optimization

In this chapter, we study mixed-integer linear optimization problems. Also known as mixed-integer linear programming problems (MILPPs), these are problems with an objective function and constraints that are all linear in the decision variables. What sets MILPPs apart from linear programming problems (LPPs) is that at least some of the variables in MILPPs are constrained to take on integer values. LPPs, conversely, have no such constraints and all of the variables can take on any continuous value.

This chapter begins by providing a number of illustrative examples to show the practical significance of MILPPs. Then, a general formulation of the MILPP is provided. Next, we demonstrate the use of a special type of integer variable known as a binary variable. We show that binary variables can be used to model a number of types of nonlinearities and discontinuities while maintaining a linear model structure. This use of binary variables is, in some sense, the true power of mixed-integer linear optimization models. Two solution techniques for MILPPs are then introduced. We first discuss the use of a branch-and-bound method to solve general MILPPs. We next introduce a cutting-plane algorithm for MILPPs in which *all* of the variables are constrained to take on integer values. Problems with this structure are commonly referred to as pure-integer linear optimization problems or pure-integer linear programming problems (PILPPs). This chapter closes with some final remarks, GAMS codes for the illustrative examples, and a number of end-of-chapter exercises.

3.1 Motivating Examples

This introductory section provides and explains a number of illustrative examples to show the practical value of MILPPs. These examples pertain to the energy sector and cover most of the standard 'classes' of MILPPs.

© Springer International Publishing AG 2017

R. Sioshansi and A.J. Conejo, *Optimization in Engineering*,

Springer Optimization and Its Applications 120, DOI 10.1007/978-3-319-56769-3_3

3.1.1 *Photovoltaic Panel-Repair Problem*

A spacecraft has a series of photovoltaic (PV) solar panels installed that must be repaired. There are two types of repair units, type A and B, that can be used to repair the PV panels. One type-A unit has a mass of 17 kg and occupies 32 m^3 of space while one type-B unit has a mass of 32 kg and occupies 15 m^3. The shuttle that is available to transport repair units to the spacecraft can only accommodate up to 136 kg and up to 120 m^3. How many units of each type (A and B) should be transported to maximize the total number of repair units sent to the spacecraft?

There are two decision variables in this problem. We let x_1 denote the number of type-A units transported to the spacecraft and x_2 the number of type-B units transported.

The objective of this problem is to maximize the total number of repair units transported:

$$\max_{x_1,x_2} \; x_1 + x_2.$$

There are four sets of constraints in this problem. First, we must impose the mass constraint, which is:

$$17x_1 + 32x_2 \leq 136.$$

Second, we have the volume constraint:

$$32x_1 + 15x_2 \leq 120.$$

Third, we must constrain each of x_1 and x_2 to be non-negative, because transporting a negative number of repair units is physically meaningless:

$$x_1, x_2 \geq 0.$$

Finally, we impose a constraint that each of the variables, x_1 and x_2, must take on integer values. The most compact way to express this is:

$$x_1, x_2 \in \mathbb{Z},$$

as $\mathbb{Z}$ is the standard notation for the set of integers.

When we impose this fourth constraint, the two variables x_1 and x_2 become **integer variables**. This should be contrasted with all of the variables used in formulating LPPs in Chapter 2. We have no restriction in any of the models formulated in Chapter 2 that the variables take on integer values. Indeed, the Simplex method, which is used to solve LPPs, has no general guarantee that any of the optimal decision variable values obtained have integer values. It is straightforward to verify that one can formulate a multitude of LPPs that do not yield integer-valued decision variables.

Taking all of these elements together, the problem can be formulated as:

$$\max_{x_1,x_2} z = x_1 + x_2 \tag{3.1}$$

$$\text{s.t. } 17x_1 + 32x_2 \le 136 \tag{3.2}$$

$$32x_1 + 15x_2 \le 120 \tag{3.3}$$

$$x_1, x_2 \ge 0 \tag{3.4}$$

$$x_1, x_2 \in \mathbb{Z}. \tag{3.5}$$

Because problem (3.1)–(3.5) includes integer variables, we refer to it as a **mixed-integer linear optimization problem**. Indeed, because all of the variables in this particular problem are restricted to take on integer values, we can refer to it more specifically as a **pure-integer linear optimization problem**. The distinction between a mixed- and pure-integer optimization problem is that the former can include a mix of variables that are and are not restricted to take on integer values. The latter *only* includes variables restricted to take on integer values.

Before proceeding, we finally note that we can express this optimization problem in the even more compact form:

$$\max_{x_1,x_2} z = x_1 + x_2$$

$$\text{s.t. } 17x_1 + 32x_2 \le 136$$

$$32x_1 + 15x_2 \le 120$$

$$x_1, x_2 \in \mathbb{Z}^+,$$

where $\mathbb{Z}^+$ is the standard notation for the set of non-negative integers.

The feasible region of problem (3.1)–(3.5) is shown in Figure 3.1. All of the feasible (integer) solutions are indicated in this figure by small circles and we see

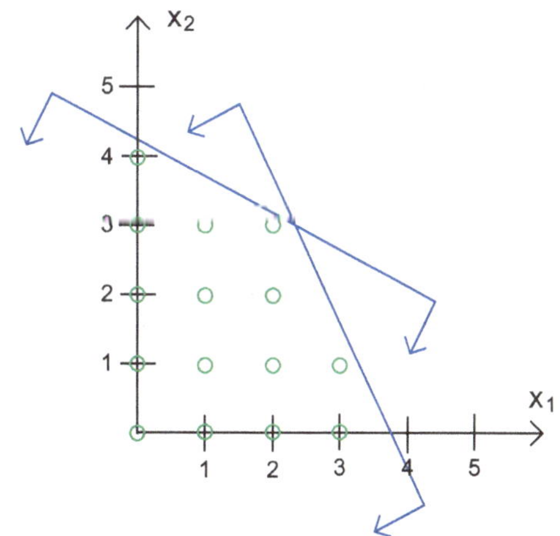

Fig. 3.1 Geometrical representation of the feasible region of the Photovoltaic Panel-Repair Problem

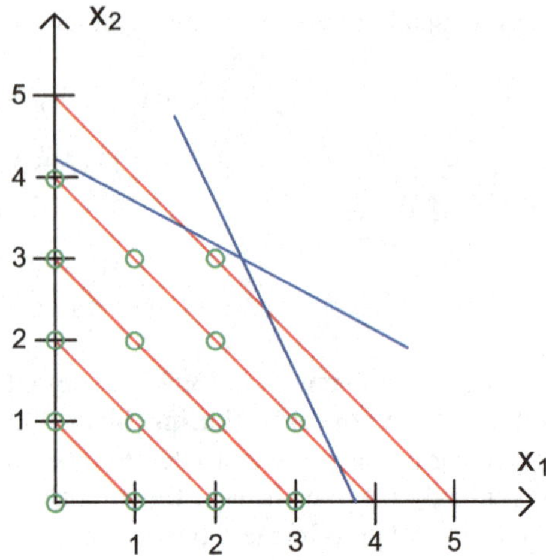

Fig. 3.2 Geometrical representation of the feasible region of the Photovoltaic Panel-Repair Problem with objective-function contour plot overlaid

that there are a total of 15 feasible points: (x_1, x_2) can feasibly equal any of $(0, 0)$, $(0, 1)$, $(0, 2)$, $(0, 3)$, $(0, 4)$, $(1, 0)$, $(1, 1)$, $(1, 2)$, $(1, 3)$, $(2, 0)$, $(2, 1)$, $(2, 2)$, $(2, 3)$, $(3, 0)$, or $(3, 1)$.

Figure 3.2 overlays the contour plot of the objective function on the feasible region. Visual inspection of this figure shows that of the 15 feasible solutions, $(x_1, x_2) = (2, 3)$ is optimal because it allows the most (five) repair units to be sent to the spacecraft.

We finally note that the Photovoltaic Panel-Repair Problem is a simplified instance of what is known as a knapsack problem.

3.1.2 Natural Gas-Storage Problem

A natural gas supplier needs to build gas-storage facilities with which to supply customers in two cities. The company has three possible sites where storage facilities can be built. Table 3.1 indicates how much revenue is earned (in \$ billion) by the company per GJ of natural gas supplied from each of the three prospective facilities to each of the two cities.

Table 3.2 summarizes the building cost (in \$ billion) of each of the three prospective gas-storage facilities. It also lists the capacity of each of the facilities, in GJ, if that facility is built. Note that each of the costs is incurred if the corresponding gas-storage facility is built, irrespective of how much gas is ultimately stored there.

Table 3.1 Revenue earned from each potential gas-storage facility [$ billion/GJ] in the Natural Gas-Storage Problem

	City 1	City 2
Storage Facility 1	1	6
Storage Facility 2	2	5
Storage Facility 3	3	4

Table 3.2 Building cost and capacity of gas-storage facilities in the Natural Gas-Storage Problem

Storage Facility	Cost [$ billion]	Capacity [GJ]
1	8	7
2	9	8
3	7	9

Finally, Table 3.3 summarizes the demand for natural gas (in GJ) in each city, which must be exactly met. The company would like to determine which (if any) of the three gas-storage facilities to build and how much gas to supply from each facility built to each city to maximize its profit.

Table 3.3 Demand for natural gas in each city in the Natural Gas-Storage Problem

City	Demand [GJ]
1	10
2	6

To formulate this problem, we introduce two sets of variables. The first, which we denote x_1, x_2, and x_3, represent the company's decision of whether to build each of the three storage facilities. We model these using a specific type of integer variable, called a **binary variable**. As the name suggests, a binary variable is restricted to take on two values, which are almost always 0 and 1. Binary variables are almost always used to model logical conditions or decisions. In this example, we define each of the three x variables to represent the logical decision of whether to build each of the three storage facilities. Thus, we define each of these variables as:

$$x_1 = \begin{cases} 1, & \text{if gas-storage facility 1 is built,} \\ 0, & \text{if gas-storage facility 1 is not built,} \end{cases}$$

$$x_2 = \begin{cases} 1, & \text{if gas-storage facility 2 is built,} \\ 0, & \text{if gas-storage facility 2 is not built,} \end{cases}$$

and:

$$x_3 = \begin{cases} 1, & \text{if gas-storage facility 3 is built,} \\ 0, & \text{if gas-storage facility 3 is not built.} \end{cases}$$

We next define an additional set of six variables, which we denote $y_{i,j}$ where $i = 1, 2, 3$ and $j = 1, 2$. The variable $y_{i,j}$ is defined as the amount of natural gas (in GJ) supplied by storage facility i to city j.

The company's objective is to maximize its profit, which is defined as revenue less cost. Using the data reported in Table 3.1, the company's revenue is defined as:

$$1y_{1,1} + 2y_{2,1} + 3y_{3,1} + 6y_{1,2} + 5y_{2,2} + 4y_{3,2}.$$

The company incurs costs for building gas-storage facilities, which are summarized in Table 3.2. Indeed, we can compute the company's cost by multiplying the cost data reported in Table 3.2 by each of the corresponding x variables. This gives:

$$8x_1 + 9x_2 + 7x_3,$$

as the company's cost. To understand this expression, note that if gas-storage facility i is not built, then $x_i = 0$. In that case, the product of x_i and the corresponding cost will be zero, meaning that the company incurs no cost for building storage facility i. If, on the other hand, facility i is built, then $x_i = 1$ and the product of x_i and the corresponding cost gives the correct cost. Taking the revenue and cost expressions together, the company's objective function is:

$$\max_{x,y} \; y_{1,1} + 2y_{2,1} + 3y_{3,1} + 6y_{1,2} + 5y_{2,2} + 4y_{3,2} - (8x_1 + 9x_2 + 7x_3).$$

This problem has four types of constraints. The first imposes the requirement that the demand in each city is met:

$$y_{1,1} + y_{2,1} + y_{3,1} = 10,$$

and:

$$y_{1,2} + y_{2,2} + y_{3,2} = 6.$$

We next need constraints that ensure that none of the three gas-storage facilities operate above maximum capacities. One may be tempted to write these constraints as:

$$y_{1,1} + y_{1,2} \le 7,$$

$$y_{2,1} + y_{2,2} \le 8,$$

and:

$$y_{3,1} + y_{3,2} \le 9.$$

However, it is actually preferable to write these constraints as:

$$y_{1,1} + y_{1,2} \le 7x_1,$$

$$y_{2,1} + y_{2,2} \leq 8x_2,$$

and:

$$y_{3,1} + y_{3,2} \leq 9x_3.$$

The reason for this is that the second set of constraints, with the x's on the right-hand sides, impose an additional restriction that a gas-storage facility cannot operate (meaning that its maximum capacity is zero) if the facility is not built. To understand this, let us examine the first of the three constraints:

$$y_{1,1} + y_{1,2} \leq 7x_1.$$

If gas-storage facility 1 is built, then $x_1 = 1$ and this constraint simplifies to:

$$y_{1,1} + y_{1,2} \leq 7,$$

which is what we want (the gas-storage facility can hold at most 7 GJ of gas). Otherwise, if the facility 1 is not built, then $x_1 = 0$ and the constraint simplifies to:

$$y_{1,1} + y_{1,2} \leq 0.$$

This is, again, what we want in this case because if facility 1 is *not* built, then it can hold 0 GJ of gas. These three constraints are an example of **logical constraints**. This is because they encode a logical condition. In this case, a gas-storage facility only has capacity available if it is built. Otherwise, it has zero capacity available. We next require the y variables to be non-negative:

$$y_{1,1}, y_{1,2}, y_{2,1}, y_{2,2}, y_{3,1}, y_{3,2} \geq 0.$$

We finally require the x variables to be binary:

$$x_1, x_2, x_3 \in \{0, 1\}.$$

Taking all of these elements together, the MILPP for this problem is:

$$\max_{x,y} z = y_{1,1} + 2y_{2,1} + 3y_{3,1} + 6y_{1,2} + 5y_{2,2} + 4y_{3,2} - 8x_1 - 9x_2 - 7x_3$$

$$\text{s.t. } y_{1,1} + y_{2,1} + y_{3,1} = 10$$

$$y_{1,2} + y_{2,2} + y_{3,2} = 6$$

$$y_{1,1} + y_{1,2} \leq 7x_1 \tag{3.6}$$

$$y_{2,1} + y_{2,2} \leq 8x_2 \tag{3.7}$$

$$y_{3,1} + y_{3,2} \leq 9x_3 \tag{3.8}$$

$$y_{i,j} \geq 0, \forall\, i = 1, 2, 3;\, j = 1, 2$$

$$x_i \in \{0, 1\}, \forall\, i = 1, 2, 3.$$

The Natural Gas-Storage Problem is a simplified version of a facility-location problem.

3.1.3 Electricity-Scheduling Problem

An electricity producer needs to supply 50 kW of power to a remote village over the next hour. To do so, it has three generating units available. The producer must decide whether to switch each of the three generating units on or not.

If it does switch a generating unit on, the producer must pay a fixed cost to operate the unit, which is independent of how much energy the unit produces. In addition to this fixed cost, there is a variable cost that the producer must pay for each kWh produced by each unit.

If a unit is switched on, it must produce any amount of energy between a minimum and maximum production level. Otherwise, its production level must equal zero. Table 3.4 summarizes the operating costs and production limits of the electricity-production units. The electricity producer would like to schedule the three production units to minimize the cost of exactly serving the 50 kW demand of the village.

Table 3.4 Operating costs and production limits of electricity-production units in the Electricity-Scheduling Problem

Production Unit	Cost		Production Limits [kW]	
	Variable [$/kWh]	Fixed [$/hour]	Minimum	Maximum
1	2	40	5	20
2	5	50	6	40
3	1	35	4	35

To formulate this problem, we introduce two sets of variables. We first have variables p_1, p_2 and p_3, where we define p_i as the kW produced by unit i. We next define three binary variables, x_1, x_2, and x_3, where we define x_i as:

$$x_i = \begin{cases} 1, & \text{if unit } i \text{ is switched on,} \\ 0, & \text{if unit } i \text{ is not switched on.} \end{cases}$$

The producer's objective function, which is to minimize cost, can be written as:

$$\min_{p,x} 2p_1 + 5p_2 + 1p_3 + 40x_1 + 50x_2 + 35x_3.$$

The problem requires three types of constraints. We first need a constraint that ensures that the 50 kW of demand is exactly met:

$$p_1 + p_2 + p_3 = 50.$$

We next must impose the production limits of the units, which can be written as:

$$5x_1 \leq p_1 \leq 20x_1,$$

$$6x_2 \leq p_2 \leq 40x_2,$$

and:

$$4x_3 \leq p_3 \leq 35x_3.$$

Note that these three production-limit constraints have the logical constraint involving the decision to switch each unit on or not embedded within them. To see this, let us look more closely at the first of these three constraints:

$$5x_1 \leq p_1 \leq 20x_1.$$

If unit 1 is switched on, then $x_1 = 1$ and this constraint simplifies to:

$$5 \leq p_1 \leq 20.$$

This is the constraint that we want in this case, because if unit 1 is switched on, then it must produce between its minimum (5 kW) and maximum (20 kW) production levels. Otherwise, if unit 1 is not switched on then $x_1 = 0$ and the production-limit constraint becomes:

$$0 \leq p_1 \leq 0,$$

or simply:

$$p_1 = 0.$$

This is, again, correct because if unit 1 is not switched on, its production must be equal to zero. We finally have a constraint that the x variables be binary:

$$x_1, x_2, x_3 \in \{0, 1\}.$$

Thus, the MILPP for this problem is:

$$\min_{p,x} z = 2p_1 + 5p_2 + p_3 + 40x_1 + 50x_2 + 35x_3$$

$$\text{s.t. } p_1 + p_2 + p_3 = 50$$

$$5x_1 \leq p_1 \leq 20x_1 \tag{3.9}$$

$$6x_2 \leq p_2 \leq 40x_2 \tag{3.10}$$

$$4x_3 \leq p_3 \leq 35x_3 \tag{3.11}$$

$$x_i \in \{0, 1\}, \forall\, i = 1, 2, 3.$$

Although it is physically meaningless for any of the p_1, p_2, or p_3 to take on a negative values, we do not need to include explicit non-negativity constraints for these variables. This is because the left-hand sides of double-sided inequalities (3.9)–(3.11) ensure that each of the p variables take on non-negative values. We could include non-negativity constraints for the p variables, however, they would be redundant in this problem.

We finally note that this example is a simplified instance of a unit-scheduling problem.

3.1.4 Oil-Transmission Problem

An oil company owns two oil wells that are connected via three pipelines to one another and to a refinery, as shown in Figure 3.3. The refinery has a fixed demand for 30 t of oil. The company is considering expanding the capacity of the two pipelines that directly connect the wells with the refinery, as depicted in the figure.

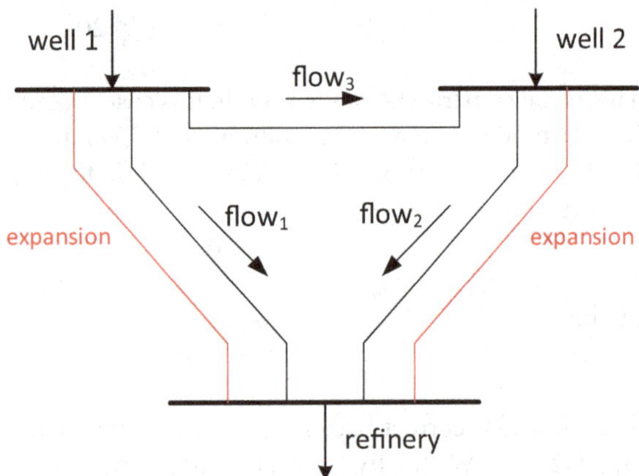

Fig. 3.3 Network in the Oil-Transmission Problem

The company earns a profit of $2000 per ton of oil sold from well 1 to the refinery and a profit of $3000/t for oil sold from well 2 to the refinery.

The pipeline that directly connects well 1 with the refinery can carry 12 t of oil. If it is expanded, which costs $50000, then this capacity increases by 11 t to 23 t total. Similarly, the pipeline directly connecting well 2 with the refinery can carry 11 t of oil. This capacity increases by 12 t if the pipeline is expanded, which costs $55000 to do. The pipeline directly connecting wells 1 and 2 can carry at most 10 t and this pipeline *cannot* be expanded.

The company would like to determine which (if any) of the pipelines to expand and how to ship oil from its wells to the refinery to maximize its profit.

To formulate this problem, we introduce three sets of variables. First, we let p_1 and p_2 denote the amount of oil extracted from each of wells 1 and 2, respectively. We next define f_1, f_2, and f_3 as the flows along the three pipelines, as shown in Figure 3.3. We use the convention that if f_i is positive, that means there is a net flow in the direction of the arrow depicted in the figure. We finally define two binary variables, x_1 and x_2, representing the pipeline-expansion decisions. More specifically, we define x_i as:

$$x_i = \begin{cases} 1, & \text{if pipeline } i \text{ is expanded,} \\ 0, & \text{if pipeline } i \text{ is not expanded.} \end{cases}$$

The company's objective function, which maximizes profit, is:

$$\max_{p,f,x} \ 2000p_1 + 3000p_2 - 50000x_1 - 55000x_2.$$

This problem has five types of constraints. First, we must ensure that we deliver exactly 30 t of oil to the refinery:

$$p_1 + p_2 = 30.$$

This constraint ensures that a total of exactly 30 t of oil is extracted from the two wells. Next, we must ensure that the oil extracted at the two wells is not left at the two wells but is instead delivered to the refinery through the pipeline network. We write these constraints as:

$$p_1 = f_1 + f_3,$$

and:

$$p_2 = f_2 - f_3.$$

These two constraints require that the total amount of oil extracted at each well (which is on the left-hand sides of the constraints) exactly equals the amount of oil that leaves each well through the pipelines (which is on the right-hand sides of the constraints). We next impose the flow constraints on the three pipelines:

$$-12 - 11x_1 \le f_1 \le 12 + 11x_1,$$

$$-11 - 12x_2 \le f_2 \le 11 + 12x_2,$$

and:

$$-10 \le f_3 \le 10.$$

We next require that the amount produced by each well be non-negative:

$$p_1, p_2 \ge 0,$$

and that the x variables be binary:

$$x_1, x_2 \in \{0, 1\}.$$

Putting these elements together, this problem is formulated as:

$$\max_{p,f,x} z = 2000p_1 + 3000p_2 - 50000x_1 - 55000x_2$$

$$\text{s.t. } p_1 + p_2 = 30$$
$$p_1 = f_1 + f_3$$
$$p_2 = f_2 - f_3$$
$$-12 - 11x_1 \leq f_1 \leq 12 + 11x_1$$
$$-11 - 12x_2 \leq f_2 \leq 11 + 12x_2$$
$$-10 \leq f_3 \leq 10$$
$$p_i \geq 0, \forall i = 1, 2$$
$$x_i \in \{0, 1\}, \forall i = 1, 2.$$

This problem is a simple version of a transmission-expansion problem.

3.1.5 Charging-Station Problem

To serve electric vehicle (EV) owners in four neighborhoods, a city needs to identify
which (if any) of three potential EV charging stations to build. The city's goal is
to minimize the total cost of building the stations, while properly serving the EV-
charging needs of the four neighborhoods.

Table 3.5 summarizes which of the four neighborhoods can be served by each of
the three potential EV-charging-station locations. An entry of 1 in the table means
that the neighborhood can be served by a station at the location while an entry of 0
means that it cannot be. Table 3.6 summarizes the cost incurred for building each of
the three potential EV charging stations.

Table 3.5 Neighborhoods that can use each potential EV-charging-station location in the Charging-Station Problem

	Location 1	Location 2	Location 3
Neighborhood 1	1	0	1
Neighborhood 2	0	1	0
Neighborhood 3	1	1	0
Neighborhood 4	0	0	1

Table 3.6 Cost of building EV-charging stations in the Charging-Station Problem

	Cost [$ million]
Location 1	10
Location 2	12
Location 3	13

To formulate this problem, we define three binary variables, x_1, x_2 and x_3, where we define x_i as:

$$x_i = \begin{cases} 1, & \text{if EV-charging station } i \text{ is built,} \\ 0, & \text{if EV-charging station } i \text{ is not built.} \end{cases}$$

The city's objective function is to minimize cost, which is given by:

$$\min_x \ 10x_1 + 12x_2 + 13x_3,$$

where the objective is measured in millions of dollars.

There are two types of constraints in this problem. First, we must ensure that the EV charging stations built are capable of serving EV owners in each of the four neighborhoods. These constraints take the form:

$$1x_1 + 0x_2 + 1x_3 \geq 1,$$

$$0x_1 + 1x_2 + 0x_3 \geq 1,$$

$$1x_1 + 1x_2 + 0x_3 \geq 1,$$

and:

$$0x_1 + 0x_2 + 1x_3 \geq 1.$$

To understand the logic behind each of these constraints, let us examine the first one:

$$1x_1 + 0x_2 + 1x_3 \geq 1,$$

more closely. We know, from Table 3.5 that EV owners in neighborhood 1 can only be served by stations 1 or 3. Thus, at least one of those two stations must be built. The constraint imposes this requirement by multiplying each of x_1 and x_3 by 1. The sum:

$$1x_1 + 0x_2 + 1x_3 = x_1 + x_3,$$

measures how many of the stations that can serve EV owners in neighborhood 1 are built. The constraint requires that at least one such station be built. An analysis of the other three constraints have the same interpretation. We must also impose a constraint that the x variables be binary:

$$x_1, x_2, x_3 \in \{0, 1\}.$$

Taking these together, the problem is formulated as:

$$\min_{x} z = 10x_1 + 12x_2 + 13x_3$$

$$\text{s.t. } x_1 + x_3 \geq 1$$

$$x_2 \geq 1$$

$$x_1 + x_2 \geq 1$$

$$x_3 \geq 1$$

$$x_i \in \{0, 1\}, \forall\, i = 1, 2, 3.$$

The Charging-Station Problem is an example of an area-covering or set-covering problem.

3.1.6 Wind Farm-Maintenance Problem

A wind-generation company must perform annual maintenance on its three wind farms. The company has three maintenance teams to carry out this work. Each maintenance team must be assigned to exactly one wind farm and each wind farm must have exactly one maintenance team assigned to it. Table 3.7 lists the costs of assigning the three maintenance teams to each wind farm. The company would like to determine the assignments to minimize its total maintenance costs.

Table 3.7 Cost of assigning maintenance crews to wind farms in the Wind Farm-Maintenance Problem

	Wind Farm 1	Wind Farm 2	Wind Farm 3
Maintenance Team 1	10	12	14
Maintenance Team 2	9	8	15
Maintenance Team 3	10	5	15

To formulate this problem, we define one set of variables, which we denote $x_{i,j}$ with $i = 1, 2, 3$ and $j = 1, 2, 3$, where we define $x_{i,j}$ as:

$$x_{i,j} = \begin{cases} 1, & \text{if maintenance crew } i \text{ is assigned to wind farm } j, \\ 0, & \text{otherwise.} \end{cases}$$

The company's objective function is to minimize cost, which is given by:

$$\min_{x} 10x_{1,1} + 12x_{1,2} + 14x_{1,3} + 9x_{2,1} + 8x_{2,2} + 15x_{2,3} + 10x_{3,1} + 5x_{3,2} + 15x_{3,3}.$$

This problem has three sets of constraints. First, we must ensure that each maintenance crew is assigned to exactly one wind farm, giving the following constraints:

$$x_{1,1} + x_{1,2} + x_{1,3} = 1,$$

$$x_{2,1} + x_{2,2} + x_{2,3} = 1,$$

and:

$$x_{3,1} + x_{3,2} + x_{3,3} = 1.$$

We must next ensure that each wind farm has exactly one maintenance crew assigned to it:

$$x_{1,1} + x_{2,1} + x_{3,1} = 1,$$

$$x_{1,2} + x_{2,2} + x_{3,2} = 1,$$

and:

$$x_{1,3} + x_{2,3} + x_{3,3} = 1.$$

We must finally ensure that all of the variables are binary:

$$x_{i,j} \in \{0, 1\}, \forall\, i = 1, 2, 3;\, j = 1, 2, 3.$$

Taking these elements together, the company's problem is:

$$
\begin{aligned}
\max_{x} z = {}& 10x_{1,1} + 12x_{1,2} + 14x_{1,3} + 9x_{2,1} + 8x_{2,2} + 15x_{2,3} \\
& + 10x_{3,1} + 5x_{3,2} + 15x_{3,3} & (3.12) \\
\text{s.t. } & x_{1,1} + x_{1,2} + x_{1,3} = 1 & (3.13) \\
& x_{2,1} + x_{2,2} + x_{2,3} = 1 & (3.14) \\
& x_{3,1} + x_{3,2} + x_{3,3} = 1 & (3.15) \\
& x_{1,1} + x_{2,1} + x_{3,1} = 1 & (3.16) \\
& x_{1,2} + x_{2,2} + x_{3,2} = 1 & (3.17) \\
& x_{1,3} + x_{2,3} + x_{3,3} = 1 & (3.18) \\
& x_{i,j} \in \{0, 1\}, \forall\, i = 1, 2, 3;\, j = 1, 2, 3. & (3.19)
\end{aligned}
$$

This problem is an example of an assignment problem.

3.2 Types of Mixed-Integer Linear Optimization Problems

Mixed-integer linear optimization problems can take a number of different more specific forms. We have discussed some examples of these, such as pure-integer linear optimization problems. We describe some of these forms in more detail here.

3.2.1 General Mixed-Integer Linear Optimization Problems

A general mixed-integer linear optimization problem or MILPP has the form:

$$
\min_{x_1,\ldots,x_n} \quad c_0 + \sum_{i=1}^{n} c_i x_i
$$

$$
\text{s.t.} \quad \sum_{i=1}^{n} A_{j,i}^e x_i = b_j^e, \qquad \forall \, j = 1, \ldots, m_e
$$

$$
\sum_{i=1}^{n} A_{j,i}^g x_i \geq b_j^g, \qquad \forall \, j = 1, \ldots, m_g
$$

$$
\sum_{i=1}^{n} A_{j,i}^l x_i \leq b_j^l, \qquad \forall \, j = 1, \ldots, m_l
$$

$$
x_i \in \mathbb{Z}, \qquad \text{for some } i = 1, \ldots, n
$$

$$
x_i \in \mathbb{R}, \qquad \text{for the remaining } i = 1, \ldots, n,
$$

where m_e, m_g, and m_l are the numbers of equal-to, greater-than-or-equal-to, and less-than-or-equal to constraints. Thus, $m = m_e + m_g + m_l$ is the total number of constraints. The coefficients, $A_{j,i}^e, \forall i = 1, \ldots, n, j = 1, \ldots, m_e$, $A_{j,i}^g, \forall i = 1, \ldots, n, j = 1, \ldots, m_g$, and $A_{j,i}^l, \forall i = 1, \ldots, n, j = 1, \ldots, m_l$, the terms on the right-hand sides of the constraints, $b_j^e, \forall j = 1, \ldots, m_e, b_j^g, \forall j = 1, \ldots, m_g$, and $b_j^l, \forall j = 1, \ldots, m_l$, and the coefficients, $c_0, \ldots, c_n$, in the objective function are all constants.

Some subset of the variables are restricted to take on integer values:

$$
x_i \in \mathbb{Z}, \text{ for some } i = 1, \ldots, n,
$$

where:

$$
\mathbb{Z} = \{\ldots, -2, -1, 0, 1, 2, \ldots\},
$$

is standard notation for the set of integers. These are called the integer variables. The remaining variables have no such restriction and can take on any non-integer values that satisfy the remaining constraints.

Of the examples given in Section 3.1, the Natural Gas-Storage, Electricity-Scheduling and Oil-Transmission Problems, which are introduced in Sections 3.1.2, 3.1.3 and 3.1.4, respectively, are general MILPPs.

3.2.2 Pure-Integer Linear Optimization Problems

A pure-integer linear optimization problem or PILPP is a special case of a general MILPP in which *all* of the variables are restricted to take on integer values. A PILPP has the generic form:

$$\min_{x_1,\ldots,x_n} c_0 + \sum_{i=1}^{n} c_i x_i$$

$$\text{s.t.} \sum_{i=1}^{n} A_{j,i}^e x_i = b_j^e, \qquad \forall\, j = 1,\ldots, m_e$$

$$\sum_{i=1}^{n} A_{j,i}^g x_i \geq b_j^g, \qquad \forall\, j = 1,\ldots, m_g$$

$$\sum_{i=1}^{n} A_{j,i}^l x_i \leq b_j^l, \qquad \forall\, j = 1,\ldots, m_l$$

$$x_i \in \mathbb{Z}, \qquad \forall\, i = 1,\ldots, n,$$

where m_e, m_g, m_l, $A_{j,i}^e$, $A_{j,i}^g$, $A_{j,i}^l$, b_j^e, b_j^g, b_j^l, $c_0, \ldots, c_n$, and $\mathbb{Z}$ have the same interpretations as in the generic MILPP, which is given in Section 3.2.1. Thus, the only distinction between the formulation of a general MILPP and a PILPP is that *all* of the variables in a PILPP are restricted to take on integer values. Conversely, a general MILPP can include a combination of variables with and without such a restriction.

Among the examples introduced in Section 3.1, the Photovoltaic Panel-Repair Problem, which is introduced in Section 3.1.1, is a PILPP.

3.2.3 Mixed-Binary Linear Optimization Problems

A **mixed-binary linear optimization problem** is a special case of a general MILPP in which the variables that are restricted to take on integer values are actually further restricted to take on binary values. With rare exceptions, these binary variables are restricted to take on the values of 0 and 1 and are often used to model logical decisions or constraints.

A mixed-binary linear optimization problem has the generic form:

$$\min_{x_1,\ldots,x_n} \; c_0 + \sum_{i=1}^{n} c_i x_i$$

$$\text{s.t.} \quad \sum_{i=1}^{n} A_{j,i}^{e} x_i = b_j^{e}, \qquad \forall \, j = 1, \ldots, m_e$$

$$\sum_{i=1}^{n} A_{j,i}^{g} x_i \geq b_j^{g}, \qquad \forall \, j = 1, \ldots, m_g$$

$$\sum_{i=1}^{n} A_{j,i}^{l} x_i \leq b_j^{l}, \qquad \forall \, j = 1, \ldots, m_l$$

$$x_i \in \{0, 1\}, \qquad \text{for some } i = 1, \ldots, n$$

$$x_i \in \mathbb{R}, \qquad \text{for the remaining } i = 1, \ldots, n,$$

where $m_e, m_g, m_l, A_{j,i}^{e}, A_{j,i}^{g}, A_{j,i}^{l}, b_j^{e}, b_j^{g}, b_j^{l}$, and $c_0, \ldots, c_n$ have the same interpretations as in the generic MILPP, which is given in Section 3.2.1.

Of the examples given in Section 3.1, the Natural Gas-Storage, Electricity-Scheduling, and Oil-Transmission Problems, which are introduced in Sections 3.1.2, 3.1.3 and 3.1.4, respectively, are mixed-binary linear optimization problems.

3.2.4 Pure-Binary Linear Optimization Problems

We finally have the case of a **pure-binary linear optimization problem**, which is a special case of a mixed-binary linear optimization problem in which all of the variables are restricted to being binary variables. Such a problem has the form:

$$\min_{x_1,\ldots,x_n} \; c_0 + \sum_{i=1}^{n} c_i x_i$$

$$\text{s.t.} \quad \sum_{i=1}^{n} A_{j,i}^{e} x_i = b_j^{e}, \qquad \forall \, j = 1, \ldots, m_e$$

$$\sum_{i=1}^{n} A_{j,i}^{g} x_i \geq b_j^{g}, \qquad \forall \, j = 1, \ldots, m_g$$

$$\sum_{i=1}^{n} A_{j,i}^{l} x_i \leq b_j^{l}, \qquad \forall \, j = 1, \ldots, m_l$$

$$x_i \in \{0, 1\}, \qquad \forall \, i = 1, \ldots, n,$$

where $m_e, m_g, m_l, A_{j,i}^{e}, A_{j,i}^{g}, A_{j,i}^{l}, b_j^{e}, b_j^{g}, b_j^{l}, c_0, \ldots, c_n$, and $\mathbb{Z}$ have the same interpretations as in the generic MILPP, which is given in Section 3.2.1.

Of the examples introduced in Section 3.1, the Charging-Station and Wind Farm-Maintenance Problems, which are discussed in Sections 3.1.5 and 3.1.6, respectively, are examples pure-binary linear optimization problems.

3.3 Linearizing Nonlinearities Using Binary Variables

This section introduces one of the most powerful uses of integer optimization techniques. This is the use of integer, and in particular binary, variables to linearize complex nonlinearities and discontinuities and to formulate logical constraints in optimization problems. This is particularly useful, because it is considerably easier to solve optimization problems in which the objective function and constraints are linear in the decision variables.

Some of these linearizations are employed in formulating the examples that are discussed in Section 3.1.

3.3.1 Variable Discontinuity

On occasion a variable may have a discontinuity in the sense that we would like it to be within one of two intervals. To more concretely explain this, suppose that we have a variable, x, and we would like x to either be between l_1 and u_1 or between l_2 and u_2. We could write this restriction on x as the following logical condition:

$$l_1 \leq x \leq u_1 \text{ or } l_2 \leq x \leq u_2.$$

This is not a valid linear constraint, however, because constraints in optimization problems cannot have logical statements (*i.e.*, the 'or') in them.

We can model this type of a restriction as a valid linear constraint by introducing a new binary variable, which we denote y. The restriction on x is then written as:

$$l_1 y + l_2 \cdot (1 - y) \leq x \leq u_1 y + u_2 \cdot (1 - y). \tag{3.20}$$

Note that if $y = 0$, then (3.20) simplifies to:

$$l_2 \leq x \leq u_2,$$

and that if $y = 1$, then it becomes:

$$l_1 \leq x \leq u_1.$$

A common situation in which this type of variable discontinuity arises is when we are modeling the production of a facility that must be switched on or off. If it is switched off, its production level must equal 0. Otherwise, if it is switched on, its production level must be between some lower and upper limits, which we denote $x^{\min}$ and $x^{\max}$. This type of a production discontinuity is illustrated in Figure 3.4. We can model this type of a restriction using (3.20) by letting $l_2 = u_2 = 0$, $l_1 = x^{\min}$, and $u_1 = x^{\max}$. In this case, Constraint (3.20) becomes:

$$x^{\min} y \le x \le x^{\max} y. \tag{3.21}$$

Fig. 3.4 Production discontinuity

This technique is used in both the Natural Gas-Storage and Electricity-Scheduling Problems, which are introduced in Sections 3.1.2 and 3.1.3, respectively. More specifically, Constraints (3.6)–(3.8) in the Natural Gas-Storage Problem impose the capacities of the storage facilities and only allow each facility to store gas if it is built. Constraints (3.9)–(3.11) in the Electricity-Scheduling Problem impose the minimum and maximum production levels of the units. These constraints also restrict each production unit to produce power only if it is switched on.

It is important to stress that these types of variable discontinuities can be applied to other settings besides the modeling of production processes and facilities.

3.3.2 Fixed Activity Cost

Some systems have a cost or other term in the objective function that is incurred when an activity occurs. As an example of this, consider the modeling of a production facility that must be switched on or off, as discussed in Section 3.3.1. Suppose that if the production facility is switched on, there is a fixed cost, c_0, that is incurred that does not depend on how many units the facility produces. In addition to this fixed cost, the facility incurs a cost of m per unit produced. Thus, the total cost of operating the facility can be written as:

$$\text{cost} = \begin{cases} c_0 + mx, & \text{if facility is switched on,} \\ 0, & \text{otherwise,} \end{cases} \tag{3.22}$$

where x represents the facility's production level. Figure 3.5 illustrates such a cost function where we further impose minimum and maximum production levels on the facility, in line with the example that is given in Section 3.3.1.

Fig. 3.5 Fixed activity cost

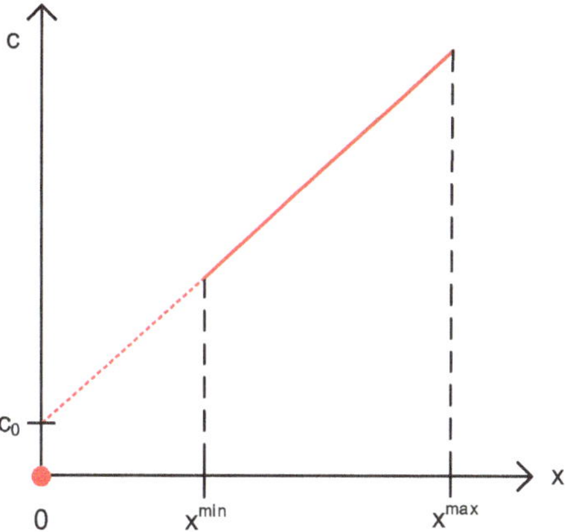

Expression (3.22) is not a valid term to include in the objective function of an optimization problem, because it includes logical 'if' statements. To model this type of cost function, we introduce a binary variable, y, which is defined as:

$$y = \begin{cases} 1, & \text{if facility is switched on,} \\ 0, & \text{otherwise.} \end{cases}$$

The facility's cost is then modeled as:

$$\text{cost} = c_0 y + mx.$$

In nearly all cases, we include a discontinuity constraint similar to (3.21), which 'links' the decision to switch the production facility on or off (*i.e.*, the y variable) to the production decision (*i.e.*, the x variable). This is because if a constraint similar to (3.21) is not included, the model will allow the facility to produce units while being switched off. If there is no explicit upper bound on how many units the facility can produce, one can set x^{max} to an arbitrarily high number in (3.21). In practice, however, a real-world production facility almost invariably has an upper limit on its output.

Both the Natural Gas-Storage and Electricity-Scheduling Problems, which are introduced in Sections 3.1.2 and 3.1.3, respectively, employ this technique to model a fixed activity cost. In the Natural Gas-Storage Problem, there are binary variables representing the decision to build a storage facility. There is a fixed cost associated with building each facility that does not depend on how much gas is ultimately stored in it. Similarly, there are binary variables in the Electricity-Scheduling Problem representing the decision to switch each of the three production units on or not. Switching the production units on imposes a fixed cost that is independent of how much power is produced.

3.3.3   Non-convex Piecewise-Linear Cost

Another type of cost structure that can arise are so-called piecewise-linear costs. Figure 3.6 illustrates an example piecewise-linear cost function. The figure shows that the first b_1 units produced cost \$$m_1$ per unit to produce. After the first b_1 units, the next $(b_2 - b_1)$ units cost \$$m_2$ per unit. Finally, any remaining units after the first b_2 units cost \$$m_3$ per unit.

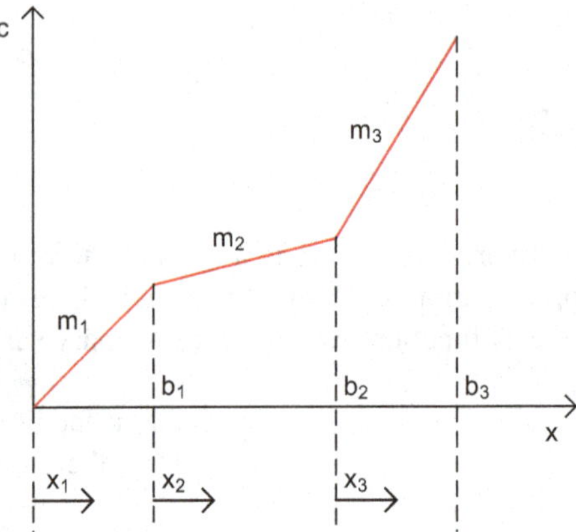

Fig. 3.6 Non-convex piecewise-linear cost

The idea of a piecewise-linear function is that the function is linear between two breakpoints. For instance, the costs shown in Figure 3.6 are linear between 0 and b_1, between b_1 and b_2, and between b_2 and b_3. However, the entire function is not linear because of the 'kinks' at the breakpoints. One can write the cost function shown in Figure 3.6 as:

$$\text{cost} = \begin{cases} m_1 x, & \text{if } 0 \le x \le b_1, \\ m_1 b_1 + m_2 \cdot (x - b_1), & \text{if } b_1 < x \le b_2, \\ m_1 b_1 + m_2 b_2 + m_3 \cdot (x - b_2), & \text{otherwise,} \end{cases} \tag{3.23}$$

where we let x denote the production level of the facility. This is not a valid term to include in the objective function of an optimization problem, however, because it employs logical 'if' statements.

We can linearize (3.23) by introducing three new continuous variables, x_1, x_2, and x_3. As shown in Figure 3.6, x_1 is defined as the number of units produced at a cost of m_1. The variables, x_2 and x_3, are similarly defined as the number of units produced at costs of m_2 and m_3, respectively. We next define two binary variables, y_1 and y_2, as:

$$y_i = \begin{cases} 1, & \text{if the maximum units that can be produced at a cost of } m_i \text{ are produced,} \\ 0, & \text{otherwise.} \end{cases}$$

We then model the production cost as:

$$\text{cost} = m_1 x_1 + m_2 x_2 + m_3 x_3,$$

and include the constraints:

$$x = x_1 + x_2 + x_3 \tag{3.24}$$

$$b_1 y_1 \leq x_1 \leq b_1 \tag{3.25}$$

$$(b_2 - b_1) y_2 \leq x_2 \leq (b_2 - b_1) y_1 \tag{3.26}$$

$$0 \leq x_3 \leq (b_3 - b_2) y_2 \tag{3.27}$$

$$y_1 \leq y_2$$

$$y_1, y_2 \in \{0, 1\},$$

in the optimization model.

To understand the logic behind this set of constraints, first note that because of the constraint $y_1 \leq y_2$ we only have three possible cases for the values of the y variables. In the first, in which $y_1 = y_2 = 0$, (3.24)–(3.27) reduce to:

$$x = x_1 + x_2 + x_3$$

$$0 \leq x_1 \leq b_1$$

$$0 \leq x_2 \leq 0$$

$$0 \leq x_3 \leq 0,$$

meaning that we have $x_2 = x_3 = 0$ and $x = x_1$ can take on any value between 0 and b_1.

Next, if $y_1 = 1$ and $y_2 = 0$, then (3.24)–(3.27) become:

$$x = x_1 + x_2 + x_3$$

$$b_1 \leq x_1 \leq b_1$$

$$0 \leq x_2 \leq b_2 - b_1$$

$$0 \leq x_3 \leq 0,$$

meaning that we have $x_1 = b_1$, $x_3 = 0$, and x_2 can take on any value between 0 and $(b_2 - b_1)$.

Finally, in the case in which $y_1 = y_2 = 1$, Constraints (3.24)–(3.27) become:

$$x = x_1 + x_2 + x_3$$

$$b_1 \leq x_1 \leq b_1$$

$$b_2 - b_1 \leq x_2 \leq b_2 - b_1$$

$$0 \leq x_3 \leq b_3 - b_2,$$

meaning that we have $x_1 = b_1$, $x_2 = b_2$, and that x_3 can take on any value between 0 and $(b_3 - b_2)$.

Note that we can further generalize this technique to model a piecewise-linear function with any arbitrary number of pieces (unlike the three pieces assumed in Figure 3.6). To see this, suppose that x measures the total production level being modeled and that per-unit production costs $m_1, m_2, \ldots, m_N$ apply to different production levels with breakpoints at $b_1, b_2, \ldots, b_N$. We would model the production cost by introducing N variables, denoted $x_1, x_2, \ldots, x_N$, and $(N - 1)$ binary variables, denoted $y_1, y_2, \ldots, y_{N-1}$. We would model the production cost as:

$$\text{cost} = \sum_{i=1}^{N} m_i x_i,$$

and add the constraints:

$$x = \sum_{i=1}^{N} x_i$$

$$b_1 y_1 \leq x_1 \leq b_1$$

$$(b_2 - b_1) y_2 \leq x_2 \leq (b_2 - b_1) y_1$$

$$(b_3 - b_2) y_3 \leq x_3 \leq (b_3 - b_2) y_2$$

$$\vdots$$

$$(b_{N-1} - b_{N-1}) y_{N-1} \leq x_{N-1} \leq (b_{N-1} - b_{N-1}) y_{N-2}$$

$$0 \leq x_N \leq (b_N - b_{N-1}) y_{N-1}$$

$$y_1 \leq y_2 \leq y_3 \leq \cdots \leq y_{N-1}$$

$$y_i \in \{0, 1\}, \forall i = 1, \ldots, N_1,$$

to the model.

We finally note that this use of binary variables is generally needed if one is modeling a piecewise-linear function that is *not* convex (*cf.* Appendix B for further discussion of convex functions). The piecewise-linear function depicted in Figure 3.6 is non-convex because the per-unit costs, m_1, m_2, and m_3, are not non-decreasing (*i.e.*, we do not have $m_1 \leq m_2 \leq m_3$). Otherwise, if the per-unit costs are non-decreasing, the piecewise-linear cost function is convex. In that case, if the convex piecewise-linear cost function is being minimized, binary variables are not needed.

To see this, return to the general case in which x measures the total production level being modeled and that per-unit production costs $m_1, m_2, \ldots, m_N$ apply to different production levels with breakpoints at $b_1, b_2, \ldots, b_N$. Moreover, assume that we have $m_1 \leq m_2 \leq \cdots \leq m_N$. We would model the production cost by introducing N variables, denoted $x_1, x_2, \ldots, x_N$. The production cost is then computed as:

$$\text{cost} = \sum_{i=1}^{N} m_i x_i,$$

and we would add the constraints:

$$x = \sum_{i=1}^{N} x_i$$

$$0 \leq x_1 \leq b_1$$

$$0 \leq x_2 \leq b_2 - b_1$$

$$\vdots$$

$$0 \leq x_N \leq b_N - b_{N-1},$$

to the model.

We do not need to use binary variables in this case because it is optimal to fully exhaust production in level i before using any production in level $(i + 1)$. The reason for this is that production in level $(i + 1)$ (and all subsequent levels) is more costly than production in level i. The example that is given in Figure 3.6, with the non-convex cost, does not exhibit this property. Without the binary variables included, the model would choose to use production level 2 (at a per-unit cost of m_2) before production level 2 (with a higher per-unit cost of m_1) is used.

3.3.4 Alternative Constraints

In some situations, we may be interested in enforcing either one constraint:

$$\sum_{i=1}^{n} a_{1,i} x_i \leq b_1, \tag{3.28}$$

or an alternative one:

$$\sum_{i=1}^{n} a_{2,i} x_i \leq b_2, \tag{3.29}$$

but not both. As discussed in Section 3.3.1, the constraint:

$$\sum_{i=1}^{n} a_{1,i} x_i \leq b_1 \text{ or } \sum_{i=1}^{n} a_{2,i} x_i \leq b_2, \tag{3.30}$$

is not valid in an optimization problem, because it includes a logical 'or' statement. We can, however, linearize the 'or' statement through the use of a binary variable.

To do this, we define the binary variable, y, as:

$$y = \begin{cases} 1, & \text{if the constraint } \sum_{i=1}^{n} a_{1,i} x_i \leq b_1 \text{ is enforced,} \\ 0, & \text{otherwise.} \end{cases}$$

We then replace (3.30) with:

$$\sum_{i=1}^{n} a_{1,i} x_i \leq b_1 + M_1 \cdot (1 - y) \tag{3.31}$$

$$\sum_{i=1}^{n} a_{2,i} x_i \leq b_2 + M_2 y, \tag{3.32}$$

where M_1 and M_2 are sufficiently large constants.

To see how this formulation works, let us suppose that we have $y = 1$. If so, then (3.31) becomes:

$$\sum_{i=1}^{n} a_{1,i} x_i \leq b_1,$$

meaning that Constraint (3.28) is being enforced on the optimization problem. On the other hand, Constraint (3.32) becomes:

$$\sum_{i=1}^{n} a_{2,i} x_i \leq b_2 + M_2.$$

Note that if M_2 is sufficiently large, then any values for x will satisfy this constraint, meaning that Constraint (3.29) is not being imposed on the problem. In the case in which $y = 0$, Constraint (3.31) becomes:

$$\sum_{i=1}^{n} a_{1,i} x_i \leq b_1 + M_1,$$

meaning that Constraint (3.28) is not being imposed, so long as M_1 is sufficiently large. On the other hand, (3.32) becomes:

$$\sum_{i=1}^{n} a_{2,i} x_i \leq b_2,$$

meaning that Constraint (3.29) is being enforced.

3.3.5 *Product of Two Variables*

A very powerful use of binary variables is to linearize the product of two variables in an optimization problem. We discuss here how to linearize the product of variables in three different cases. We first discuss linearizing the product of a real and binary variable and then the product of two binary variables, both of which can be linearized exactly. We then discuss the use of a technique, known as binary expansion, to linearly approximate the product of two real variables.

3.3.5.1   Product of a Real and Binary Variable

Here we demonstrate how to linearize the product of a real and binary variable. Suppose that $x \in \mathbb{R}$ is a real variable while $y \in \{0, 1\}$ is a binary variable. If we define $p = xy$ as the product of these two variables, then we know that p can take on only the following two values:

$$p = \begin{cases} 0, & \text{if } y = 0, \\ x, & \text{otherwise.} \end{cases} \tag{3.33}$$

To come up with a linear expression for p, we must further assume that x is bounded:

$$-l < x < u.$$

If there are no explicit bounds on x, one can impose bounds on x in an optimization problem by making l and u sufficiently small and large, respectively. We must also add a new real variable, which we denote as z. We can then define p through the following constraints:

$$p = x - z \tag{3.34}$$

$$-ly \leq p \leq uy \tag{3.35}$$

$$-l \cdot (1 - y) \leq z \leq u \cdot (1 - y). \tag{3.36}$$

To see how this linearization works, first consider the case in which $y = 0$. If so, Constraints (3.34)–(3.36) simplify to:

$$p = x - z$$

$$0 \leq p \leq 0$$

$$-l \leq z \leq u,$$

meaning that $p = 0$, which is consistent with (3.33), and that $z = x$. The variable z is essentially playing the role of a slack variable in this formulation. Otherwise, if $y = 1$, then (3.34)–(3.36) become:

$$p = x - z$$

$$-l \leq p \leq u$$

$$0 \leq z \leq 0,$$

in which case $z = 0$ and $p = x$, as required by (3.33).

3.3.5.2 Product of Two Binary Variables

We now consider the case in which two binary variables are being multiplied. Let us suppose that $x, y \in \{0, 1\}$ are two binary variables and define $p = xy$ as their product. We know that p can take on one of two values:

$$p = \begin{cases} 1, & \text{if } x = 1 \text{ and } y = 1, \\ 0, & \text{otherwise.} \end{cases}$$

We can express p linearly through the following set of constraints:

$$p \leq x \tag{3.37}$$

$$p \leq y \tag{3.38}$$

$$p \geq 0 \tag{3.39}$$

$$p \geq x + y - 1. \tag{3.40}$$

To see how this linearization works, first consider the case in which both $x = 1$ and $y = 1$. If so, then (3.37)–(3.40) become:

$$p \leq 1$$

$$p \leq 1$$

$$p \geq 0$$

$$p \geq 1,$$

which forces $p = 1$. Otherwise, if at least one of x or y is equal to zero, then (3.37)–(3.40) force $p = 0$.

3.3.5.3 Product of Two Real Variables

The product of two real variables cannot be linearized exactly. However, there is a technique, known as **binary expansion**, that can be used to linearly *approximate* the product of two real variables. To demonstrate the concept of binary expansion, let us define $x, y \in \mathbb{R}$ as two real variables, and $p = xy$ as their product.

To linearize the product, we must approximate one of the two variables as taking on one of a finite number of values. Let us suppose that we approximate y as taking on one of the N values $y_1, y_2, \ldots, y_N$, where $y_1, y_2, \ldots, y_N$ are fixed constants. To conduct the binary expansion, we introduce N binary variables, which we denote $z_1, z_2, \ldots, z_N$. We then approximate y using the following:

$$y \approx \sum_{i=1}^{N} y_i z_i$$

$$\sum_{i=1}^{N} z_i = 1.$$

Note that because exactly one of the z_i's must equal 1, y is approximated as taking on the corresponding value of y_i.

Using this approximation of y, we can then approximate the product of x and y by adding the following constraints to the optimization problem:

$$p = \sum_{i=1}^{N} y_i z_i x \tag{3.41}$$

$$\sum_{i=1}^{N} z_i = 1$$

$$z_i \in \{0, 1\}, \forall i = 1, \ldots, N.$$

Because the y_i's are constants, the right-hand side of (3.41) involves the product of a binary and real variable. This must then be linearized using the technique that is outlined in Section 3.3.5.1.

It is important to stress once again that binary expansion does not exactly represent the product of two real variables. Rather, it approximates the product. Nevertheless, it can be a very useful technique. How good of an approximation binary expansion provides depends on whether the 'true' value of y is close to one of the y_i's. If so, the approximation will be better. Because we do not typically know *a priori* what the 'true' value of y is, we deal with this issue by using a large number of y_i's. Doing so increases the size of the problem, however, because more binary variables (*i.e.*, z_i's) must be introduced into the model. Increasing the size of the problem invariably makes it more difficult to solve.

3.4 Relaxations

This section introduces a very important concept in solving mixed-integer linear optimization problems. This is the idea of a **relaxation**. A relaxation of an optimization problem is a problem in which one or more of the constraints are loosened, relaxed, or entirely removed. When the constraint is relaxed, the feasible region of the problem grows in size. As a result of this, there are some important properties linking a problem and a relaxation that are useful when we solve MILPPs.

All of this discussion assumes that we have a generic MILPP, which can be written as:

$$\min_{x_1,\ldots,x_n} c_0 + \sum_{i=1}^{n} c_i x_i \tag{3.42}$$

$$\text{s.t.} \ \sum_{i=1}^{n} A^e_{j,i} x_i = b^e_j, \qquad\qquad \forall \ j = 1, \ldots, m_e$$

$$\sum_{i=1}^{n} A^g_{j,i} x_i \geq b^g_j, \qquad\qquad \forall \ j = 1, \ldots, m_g$$

$$\sum_{i=1}^{n} A^l_{j,i} x_i \leq b^l_j, \qquad\qquad \forall \ j = 1, \ldots, m_l \tag{3.43}$$

$$x_i \in \mathbb{Z}, \qquad\qquad\qquad \text{for some } i = 1, \ldots, n$$

$$x_i \in \mathbb{R}, \qquad\qquad\qquad \text{for the remaining } i = 1, \ldots, n.$$

This generic form captures all of the special cases of MILPPs that are introduced in Section 3.2. Moreover, we know from the discussion in Section 2.2.2.1 that a MILPP that is a maximization can be converted to a minimization simply by multiplying the objective function by -1.

We now introduce what is known as the linear relaxation of this MILPP, which is:

$$\min_{x_1,\ldots,x_n} c_0 + \sum_{i=1}^{n} c_i x_i$$

$$\text{s.t.} \quad \sum_{i=1}^{n} A^e_{j,i} x_i = b^e_j, \qquad \forall\, j = 1,\ldots,m_e$$

$$\sum_{i=1}^{n} A^g_{j,i} x_i \geq b^g_j, \qquad \forall\, j = 1,\ldots,m_g$$

$$\sum_{i=1}^{n} A^l_{j,i} x_i \leq b^l_j, \qquad \forall\, j = 1,\ldots,m_l$$

$$x_i \in \mathbb{R}, \qquad \forall\, i = 1,\ldots,n.$$

The only difference between the original MILPP and its linear relaxation is that Constraint (3.43) is relaxed, because we allow *all* of the variables to be real-valued in the relaxed problem. This can be contrasted with the original MILPP, in which some of the variables are restricted to take on only integer values.

It is important to stress that there are many ways in which an optimization problem can be relaxed. When writing the linear relaxation, we remove the constraints that the variables be integer-valued. However, one could, for example, relax the MILPP by removing the equality and greater-than-or-equal-to constraints, which would give the following relaxed problem:

$$\min_{x_1,\ldots,x_n} c_0 + \sum_{i=1}^{n} c_i x_i$$

$$\text{s.t.} \quad \sum_{i=1}^{n} A^l_{j,i} x_i \leq b^l_j, \qquad \forall\, j = 1,\ldots,m_l$$

$$x_i \in \mathbb{Z}, \qquad \text{for some } i = 1,\ldots,n$$

$$x_i \in \mathbb{R}, \qquad \text{for the remaining } i = 1,\ldots,n.$$

Moreover, one can relax constraints in any type of optimization problem (including linear optimization problems).

We can now show three useful relationships between the original problem and its relaxation. Note that these relationships apply to any problem and a relaxation. However, to make the results more clear, we will focus on the generic MILPP and its linear relaxation.

Relaxation-Optimality Property: The optimal objective-function value of the linear relaxation is less than or equal to the optimal objective-function value of the original MILPP.

Suppose that x^* is an optimal solution to the original MILPP. By definition, x^* is feasible in the MILPP. x^* is also feasible in the linear relaxation. This is because the linear relaxation has the same equality and inequality constraints as the original MILPP, but does not have the integrality constraints. There may, however, be a solution, $\tilde{x}$, that is feasible in the linear relaxation that gives a smaller objective-function value than x^*. $\tilde{x}$ is not feasible in the original MILPP (if it is, then x^* would not be an optimal solution to the MILPP). Thus, the linear relaxation may have an optimal solution that is not feasible in the original MILPP and gives a smaller objective-function value than x^*.

Relaxation-Optimality Corollary: If the optimal solution of the linear relaxation satisfies the constraints of the original MILPP, then this solution is also optimal in the original MILPP.

This result comes immediately from the Relaxation-Optimality Property. Suppose that x^* is optimal in the linear relaxation and that it is feasible in the original MILPP. We know from the Relaxation-Optimality Property that the optimal objective-function value of the original MILPP can be no lower than the objective-function value given by x^*. Combining this observation with the fact that x^* is feasible in the original MILPP tells us that x^* is also optimal in the original MILPP.

Relaxation-Feasibility Property: If the linear relaxation of the original MILPP is infeasible, then the original MILPP is also infeasible.

We show this by contradiction. To do so, we suppose that the property is not true. This would mean that the linear relaxation is infeasible but the original MILPP is feasible. If so, then there is a solution, $\hat{x}$, that is feasible in the original MILPP. $\hat{x}$ must also be feasible in the linear relaxation, however. This is because the linear relaxation has the same equality and inequality constraints as the original MILPP but does not have the integrality constraints. Thus, it is impossible for the Relaxation-Feasibility Property to not hold.

We use these three properties in Section 3.5 to outline an effective algorithm for solving general MILPPs.

3.5 Solving Mixed-Integer Linear Optimization Problems Using Branch and Bound

We describe in this section the **Branch-and-Bound Algorithm**, which can be used to efficiently solve MILPPs. We begin by first providing a high-level motivation behind the algorithm. We next outline the steps of the algorithm and then illustrate its use with a simple example. We finally provide a more formal outline of the algorithm.

All of this discussion assumes that we have a generic MILPP that is in the form:

$$\min_{x_1,\ldots,x_n} \quad c_0 + \sum_{i=1}^{n} c_i x_i$$

$$\text{s.t.} \quad \sum_{i=1}^{n} A_{j,i}^e x_i = b_j^e, \qquad \forall\, j = 1, \ldots, m_e$$

$$\sum_{i=1}^{n} A_{j,i}^g x_i \geq b_j^g, \qquad \forall\, j = 1, \ldots, m_g$$

$$\sum_{i=1}^{n} A_{j,i}^l x_i \leq b_j^l, \qquad \forall\, j = 1, \ldots, m_l$$

$$x_i \in \mathbb{Z}, \qquad \text{for some } i = 1, \ldots, n$$

$$x_i \in \mathbb{R}, \qquad \text{for the remaining } i = 1, \ldots, n.$$

3.5.1 Motivation

At a high level, the Branch-and-Bound Algorithm works through the following four steps.

1. The algorithm starts by solving the linear relaxation of the original MILPP.
2. Based on the solution of the linear relaxation, the algorithm generates a sequence of additional optimization problems in which constraints are added to the linear relaxation.
3. As this sequence of additional optimization problems is solved, the algorithm establishes upper and lower bounds on the optimal objective-function value of the original MILPP. The upper bound progressively decreases while the lower bound increases.
4. *Ideally*, this process continues until the full sequence of additional optimization problems are solved, which gives an optimal solution. In practice, we may stop once we have found a solution that is feasible in the original MILPP and that has upper and lower bounds that are close enough to one another (indicating that the feasible solution is very close to optimal).

The lower bound on the optimal objective-function value is found by appealing to the Relaxation-Optimality Property, which is discussed in Section 3.4. We know from the Relaxation-Optimality Property that the solution to the linear relaxation provides a lower bound on the optimal objective-function value of the original MILPP.

We obtain upper bounds whenever solving one of the additional optimization problems gives a solution that is feasible in the MILPP. We can reason that such a solution gives an upper bound on the optimal objective-function value. This is because of the definition of an optimal solution as being a feasible solution that gives the best objective-function value. If we find a solution that is feasible, it may be optimal. However, it may not be, in which case the objective-function value of the feasible solution is greater than the optimal objective-function value.

3.5.2 *Outline of Branch-and-Bound Algorithm*

Building off of the high-level motivation given in Section 3.5.1, we now outline the major steps of the Branch-and-Bound Algorithm in further detail.

3.5.2.1 Step 0: Initialization

We begin the Branch-and-Bound Algorithm by initializing it. Throughout the algorithm, we let z^l denote the current lower bound and z^u the current upper bound on the optimal objective-function value. We also keep track of a set, which we denote as Ξ, of linear optimization problems that remain to be solved. As we solve the sequence of linear optimization problems in Ξ, we also keep track of the best solution (in the sense of giving the smallest objective-function value) that has been found so far that satisfies all of the integrality constraints of the original MILPP. We denote this by x^b.

We initialize the bounds by letting $z^l \leftarrow -\infty$ and $z^u \leftarrow +\infty$. The reasoning behind this is that we have, as of yet, not done any work to solve the original MILPP. Thus, all we know is that its optimal objective-function value is between $-\infty$ and

$+\infty$. We also set $\varXi \leftarrow \emptyset$, because we have not generated any linear optimization problems to solve yet. We do not initialize x^b because we have not yet found a solution that satisfies the integrality constraints of the original MILPP.

We next solve the linear relaxation of the MILPP. When we solve the linear relaxation, one of following four outcomes are possible.

1. The linear relaxation may be infeasible. If so, then based on the Relaxation-Feasibility Property, the original MILPP is infeasible as well. As such, we terminate the Branch-and-Bound Algorithm and report that the original MILPP is infeasible.
2. Solving the linear relaxation may give a solution that satisfies all of the integrality constraints of the original MILPP. If so, then based on the Relaxation-Optimality Corollary, the optimal solution of the linear relaxation is also optimal in the original MILPP. Thus, we terminate the Branch-and-Bound Algorithm and report the solution found as being optimal.
3. Solving the linear relaxation may give a solution that does not satisfy all of the integrality constraints of the original MILPP. If so, we know from the Relaxation-Optimality Property that the optimal objective-function value of the linear relaxation is a lower bound on the optimal objective-function value of the original MILPP. Thus, we update the current lower bound as $z^l \leftarrow z^0$, where z^0 denotes the optimal objective-function value of the linear relaxation. We then continue with Step 1 of the algorithm.
4. The linear relaxation may be unbounded. If so, we cannot necessarily conclude whether the original MILPP is unbounded or not. We can only conclude that the original MILPP is unbounded if we can find a solution that is feasible in the original MILPP and that makes the objective function go to $-\infty$. If so, we terminate the algorithm and report that the original MILPP is unbounded. Otherwise, we continue with Step 1 of the algorithm.

3.5.2.2 Step 1: Initial Branching

Let x^0 denote the optimal solution to the linear relaxation, found in Step 0. Pick one of the variables that is supposed to be integer-valued in the original MILPP but has a non-integer value in x^0. We hereafter call this chosen variable x_i.

We generate two new linear optimization problems in which we add a new constraint to the linear relaxation. The first one, which we denote LPP_1, consists of the linear relaxation with the added constraint:

$$x_i \leq \left\lfloor x_i^0 \right\rfloor.$$

The notation $\left\lfloor x_i^0 \right\rfloor$ denotes the floor of x_i^0, which means that we round x_i^0 down to the nearest integer less than x_i^0, which is the value for x_i in the optimal solution to the linear relaxation. The second linear optimization problem, which we denote LPP_2, consists of the linear relaxation with the added constraint:

$$x_i \geq \lceil x_i^0 \rceil.$$

The notation $\lceil x_i^0 \rceil$ denotes the ceiling of x_i^0, meaning that we round x_i^0 up to the nearest integer greater than x_i^0.

After generating these two new linear optimization problems, we update the set of linear optimization problems that remain to be solved as:

$$\Xi \leftarrow \Xi \cup \mathrm{LPP}_1 \cup \mathrm{LPP}_2.$$

These two new problems, LPP_1 and LPP_2, cover the entire feasible region of the original MILPP. The reason for this is that all we have done in generating LPP_1 and LPP_2 is created two new problems, with x_i being constrained to take on values less than $\lfloor x_i^0 \rfloor$ in one and values greater than $\lceil x_i^0 \rceil$ in the other. Thus, the only points that are no longer feasible in these two LPPs are those in which x_i takes on a value strictly between $\lfloor x_i^0 \rfloor$ and $\lceil x_i^0 \rceil$. However, such points are infeasible in the original MILPP, because there are no integer values between $\lfloor x_i^0 \rfloor$ and $\lceil x_i^0 \rceil$ and x_i is restricted to taking on integer values in the original MILPP.

3.5.2.3   Step 2: Solving

Select a linear optimization problem in Ξ, solve it, and remove it from Ξ. Let $\hat{x}$ denote the optimal solution to this problem and let $\hat{z}$ denote its optimal objective-function value.

3.5.2.4   Step 3: Bound Updating and Branching

When we solve the linear optimization problem from Ξ chosen in Step 2, one of the four following outcomes are possible. In this step, we can update the bounds and may need to add more linear optimization problems to Ξ, based on the different possible outcomes.

1. $\hat{x}$ may satisfy all of the integrality constraints of the original MILPP. This means that $\hat{x}$ is feasible in the original MILPP. We further know, from the discussion in Section 3.5.1, that $\hat{z}$ provides an upper bound on the optimal objective-function value of the original MILPP.

   If $\hat{z} < z^u$, then this means that $\hat{x}$ is the best solution that is feasible in the original MILPP that we have found thus far. In this case, we update the upper bound $z^u \leftarrow \hat{z}$ and the best feasible solution found thus far $x^b \leftarrow \hat{x}$.

   Otherwise, if $\hat{z} \geq z^u$, the upper bound cannot be updated and we proceed to Step 4.

2. The problem solved in Step 2 may be infeasible. In this case, the bounds cannot be updated and we proceed to Step 4.

3. $\hat{x}$ may not satisfy all of the integrality constraints of the original MILPP. If so, $\hat{z}$ may provide an updated lower bound on the optimal objective-function value of the original MILPP. We may also need to generate two new linear optimization problems to add to Ξ.

The lower bound on the optimal objective-function value of the original MILPP can be updated if $z^l < \hat{z} \leq z^u$ and $z^u < +\infty$. If both of these conditions are met, then we can update the lower bound as $z^l \leftarrow \hat{z}$. Otherwise, we cannot update the lower bound. We cannot update the bound in this case because the upper bound is still $+\infty$, which is not a valid bounding reference.

Moreover, if $\hat{z} < z^u$, then we must also generate two new linear optimization problems. This is done in a manner similar to the Initial Branching in Step 1. To do this, we select one of the variables that is supposed to be integer-valued in the original MILPP but has a non-integer value in $\hat{x}$. We hereafter call this chosen variable, x_i. We generate two new linear optimization problems, in which we add a new constraint to the linear optimization problem most recently solved in Step 2 (*i.e.*, the problem that has $\hat{x}$ as an optimal solution). The first problem has the constraint:

$$x_i \leq \lfloor \hat{x}_i \rfloor,$$

while the second has the constraint:

$$x_i \geq \lceil \hat{x}_i \rceil,$$

added. These two problems are added to Ξ. We then proceed to Step 4. Note that for the same reason as with the Initial Branching step, these two new problems that are added to Ξ cover the entire feasible region of the linear optimization problem most recently solved in Step 2.

If $\hat{z} \geq z^u$, then we do not need to add any problems to Ξ and we proceed to Step 4. The reason we do not add any problems to Ξ is that no better solution than the current best one (*i.e.*, x^b) can be found from the problems generated from the linear optimization problem most recently solved in Step 2.

4. The problem solved in Step 2 may be unbounded. If so, we cannot necessarily conclude that the original MILPP is unbounded. We can only conclude that the original MILPP is unbounded if we can find a solution that is feasible in the original MILPP and that makes the objective function go to $-\infty$. If so, we terminate the algorithm and report that the original MILPP is unbounded. Otherwise, we generate two new linear optimization problems, following the process outlined in Case 3 and then proceed to Step 4.

3.5.2.5 Step 4: Optimality Check

We finally, in this step, determine if the Branch-and-Bound Algorithm has more problems to solve or if the algorithm can terminate. If $\Xi \neq \emptyset$, that means that there are more linear optimization problems to be solved. In this case, we return to Step 2, select a new problem from Ξ to solve, and continue with the algorithm.

If $\Xi = \emptyset$ and a value has been assigned to x^b, this means that we have found a feasible solution in the original MILPP and that there are no remaining feasible solutions in the original MILPP that give a better objective-function value than x^b. In this case, we terminate the algorithm and report that x^b is an optimal solution to the original MILPP.

The final case is that $\Xi = \emptyset$ and no value has been assigned to x^b. This means that we are not able to find a feasible solution to the MILPP and that the MILPP is infeasible. Thus, we terminate the algorithm and report this.

We now demonstrate the use of the Branch-and-Bound Algorithm in the following example.

Example 3.1 Consider the Photovoltaic Panel-Repair Problem, which is introduced in Section 3.1.1. This problem can be formulated as:

$$\min_{x_1,x_2} z = -x_1 - x_2$$

$$\text{s.t. } 17x_1 + 32x_2 \leq 136$$

$$32x_1 + 15x_2 \leq 120$$

$$x_1, x_2 \geq 0$$

$$x_1, x_2 \in \mathbb{Z},$$

where the objective function has been changed to a minimization by multiplying through by -1.

To solve this problem using the Branch-and-Bound Algorithm, we first initialize:

$$z^l \leftarrow -\infty,$$

$$z^u \leftarrow +\infty,$$

and:

$$\Xi = \emptyset.$$

We next solve the linear relaxation of the original MILPP, which is:

$$\min_{x_1,x_2} z = -x_1 - x_2$$

$$\text{s.t. } 17x_1 + 32x_2 \leq 136$$

$$32x_1 + 15x_2 \leq 120$$

$$x_1, x_2 \geq 0.$$

The optimal solution to this problem is $x^0 = (2.341, 3.007)^\top$ with $z^0 = -5.347$. This solution to the linear relaxation falls under Case 3 of the initialization step of the algorithm (we have a bounded feasible solution to the linear relaxation that *does not* satisfy the integrality constraints of the original MILPP). Thus, we update the lower bound:

$$z^l \leftarrow z^0 = -5.347.$$

This initialization step and the relaxation of the original MILPP into its linear relaxation are illustrated in Figure 3.7, where we denote the linear relaxation as 'LPP$_0$'.

$$\text{MILPP} \longrightarrow \text{LPP}_0$$

Fig. 3.7 Initialization of the Branch-and-Bound Algorithm

We next go to the Initial Branching step. To do this, we pick one of the variables that has a non-integer value in x^0 but must be integer-valued in the original MILPP. In this case, we can choose either of x_1 or x_2 to branch on, and arbitrarily pick x_1 for this example. We form two new linear optimization problems from the linear relaxation. The first one will have the constraint:

$$x_1 \leq \lfloor x_1^0 \rfloor = 2,$$

added. Thus, this problem, which we call LPP$_1$, is:

$$\min_{x_1, x_2} z = -x_1 - x_2$$
$$\text{s.t. } 17x_1 + 32x_2 \leq 136$$
$$32x_1 + 15x_2 \leq 120$$
$$x_1, x_2 \geq 0$$
$$\boldsymbol{x_1 \leq 2}.$$

The second problem will have the constraint:

$$x_1 > \lceil x_1^0 \rceil = 3,$$

added. Thus, this problem, which we call LPP$_2$, is:

$$\min_{x_1, x_2} z = -x_1 - x_2$$
$$\text{s.t. } 17x_1 + 32x_2 \leq 136$$
$$32x_1 + 15x_2 \leq 120$$
$$x_1, x_2 \geq 0$$
$$\boldsymbol{x_1 \geq 3}.$$

We then update the set of problems to be solved as:

$$\Xi \leftarrow \Xi \cup \mathrm{LPP}_1 \cup \mathrm{LPP}_2 = \{\mathrm{LPP}_1, \mathrm{LPP}_2\}.$$

Figure 3.8 shows the two new linear optimization problems being added to the set Ξ. It also illustrates why we refer to the process of adding new optimization problems as 'branching.' What we are essentially doing in creating these new problems is adding some restrictions (the new constraint added) to the most recently solved problem. This results in the problems forming something of a tree, as more and more constraints are added to problems that are solved (this tree-like structure is further shown in Figures 3.9–3.16). LPP_0 is shaded in Figure 3.8 to illustrate that this problem is processed and no longer needs to be solved at this point. We use this convention throughout Figures 3.9–3.16 to indicate problems that have been processed.

Fig. 3.8 Initial Branching in the Branch-and-Bound Algorithm

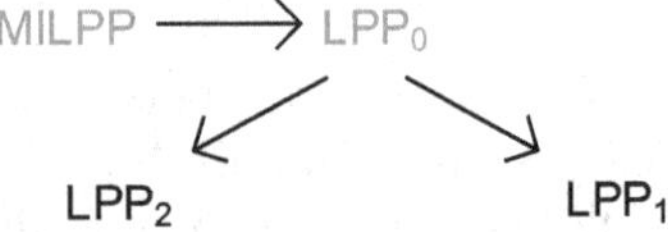

We next proceed to the Solving step and select one of the problems from Ξ to solve. In theory, we can select any of the problems in Ξ. As discussed in Section 3.5.5, there are different 'strategies' that can be employed to determine which problem in Ξ to solve. We choose to solve LPP_2 at this point, which has the optimal solution $\hat{x} = (3, 1.67)^\top$ and objective-function value $\hat{z} = -4.6$. We also update the set of problems remaining to be solved by removing LPP_2, which gives:

$$\Xi \leftarrow \{\mathrm{LPP}_1, \cancel{\mathrm{LPP}_2}\} = \{\mathrm{LPP}_1\}.$$

We now proceed to the Bound Updating and Branching step. The solution to LPP_2 falls into Case 3 of Step 3, thus we first check to see if we can update the lower bound, z^l. We find that $z^l \leq \hat{z} \leq z^u$, however because $z^u = +\infty$, we cannot update the lower bound. Next, because we have that $\hat{z} \leq z^u$, we know that we must add two new optimization problems to Ξ. To do this, we pick one of the variables that must be integer-valued in the original MILPP, but which has a non-integer value in $\hat{x}$. There is only one option for this, which is x_2. Thus, we create two new problems in which we add the constraints:

$$x_2 \leq \lfloor \hat{x}_2 \rfloor = 1,$$

and:

$$x_2 \geq \lceil \hat{x}_2 \rceil = 2,$$

to LPP_2. These two new problems, which we denote LPP_3 and LPP_4, are:

$$\min_{x_1, x_2} z = -x_1 - x_2$$

$$\text{s.t. } 17x_1 + 32x_2 \leq 136$$
$$32x_1 + 15x_2 \leq 120$$
$$x_1, x_2 \geq 0$$
$$x_1 \geq 3$$
$$\mathbf{x_2 \leq 1},$$

and:

$$\min_{x_1, x_2} z = -x_1 - x_2$$

$$\text{s.t. } 17x_1 + 32x_2 \leq 136$$
$$32x_1 + 15x_2 \leq 120$$
$$x_1, x_2 \geq 0$$
$$x_1 \geq 3$$
$$\mathbf{x_2 \geq 2},$$

respectively. The set of problems remaining to be solved is then updated:

$$\Xi \leftarrow \Xi \cup \text{LPP}_3 \cup \text{LPP}_4 = \{\text{LPP}_1, \text{LPP}_3, \text{LPP}_4\}.$$

Figure 3.9 shows the resulting tree structure of the problems when the two new LPPs are added to Ξ.

Fig. 3.9 First Bound Updating and Branching in the Branch-and-Bound Algorithm

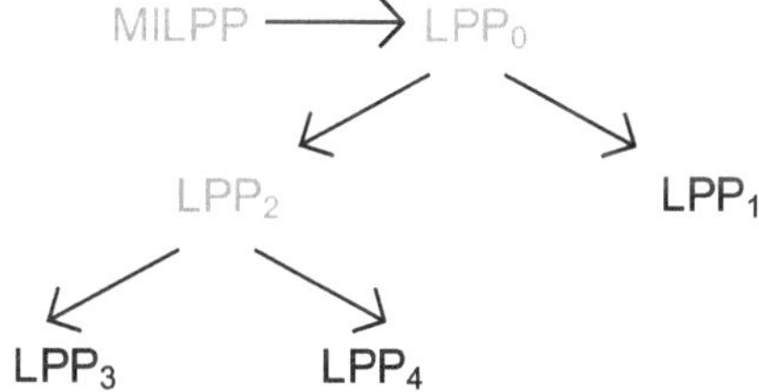

We next proceed to the Optimality Checking step. Because $\Xi \neq \emptyset$, the Branch-and-Bound Algorithm does *not* terminate and we instead return to Step 2.

In Step 2 we select a problem in Ξ to solve. We arbitrarily select LPP$_3$, which has optimal solution $\hat{x} = (3.281, 1)^\top$ and objective-function value $\hat{z} = -4.281$. We then remove LPP$_3$ from Ξ, giving:

$$\Xi \leftarrow \{\text{LPP}_1, \cancel{\text{LPP}_3}, \text{LPP}_4\} = \{\text{LPP}_1, \text{LPP}_4\},$$

and proceed to the Bound Updating and Branching step.

The solution to LPP_3 again falls into Case 3. We next find that because $z^l \leq \hat{z} \leq z^u$ but $z^u = +\infty$ we still cannot update the lower bound. Moreover, because $\hat{z} \leq z^u$ we must add two new optimization problems to Ξ. We choose x_1 as the variable on which to add constraints in these new problems. More specifically, the two new problems will consist of LPP_3 with the constraints:

$$x_1 \leq \lfloor \hat{x}_1 \rfloor = 3,$$

and:

$$x_1 \geq \lceil \hat{x}_1 \rceil = 4,$$

added. This gives the two new optimization problems:

$$\min_{x_1, x_2} z = -x_1 - x_2$$
$$\text{s.t. } 17x_1 + 32x_2 \leq 136$$
$$32x_1 + 15x_2 \leq 120$$
$$x_1, x_2 \geq 0$$
$$x_1 \geq 3$$
$$x_2 \leq 1$$
$$\mathbf{x_1 \leq 3,}$$

and:

$$\min_{x_1, x_2} z = -x_1 - x_2$$
$$\text{s.t. } 17x_1 + 32x_2 \leq 136$$
$$32x_1 + 15x_2 \leq 120$$
$$x_1, x_2 \geq 0$$
$$x_1 \geq 3$$
$$x_2 \leq 1$$
$$\mathbf{x_1 \geq 4,}$$

which we denote LPP_5 and LPP_6, respectively. We add these problems to Ξ, giving:

$$\Xi \leftarrow \Xi \cup \text{LPP}_5 \cup \text{LPP}_6 = \{\text{LPP}_1, \text{LPP}_4, \text{LPP}_5, \text{LPP}_6\}.$$

Figure 3.10 shows the tree-like structure of the problems in Ξ at the end of the second Bound Updating and Branching step.

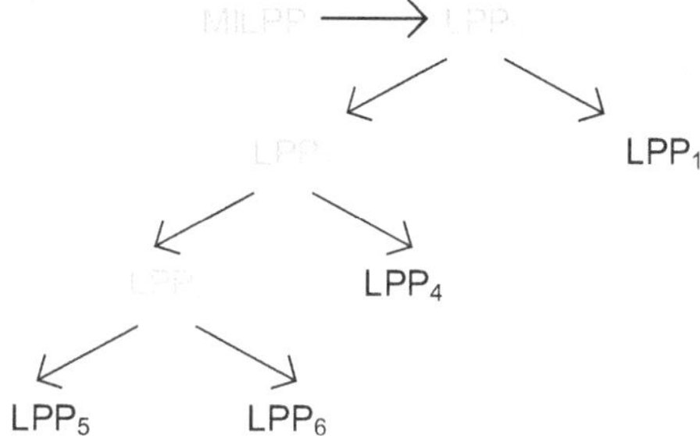

Fig. 3.10 Second Bound Updating and Branching in the Branch-and-Bound Algorithm

We then go to the Optimality Check step. Finding that $\Xi \neq \emptyset$, we return to Step 2 and pick a new problem in Ξ to solve.

In Step 2 we arbitrarily select LPP_5 to solve and find that it has as an optimal solution, $\hat{x} = (3, 1)^\top$, and corresponding objective-function value, $\hat{z} = -4$. We remove LPP_5 from Ξ, giving:

$$\Xi \leftarrow \{LPP_1, LPP_4, \cancel{LPP_5}, LPP_6\} = \{LPP_1, LPP_4, LPP_6\}.$$

Moving to the Bound Updating and Branching step, we find that $\hat{x}$ falls into Case 1, because it satisfies all of the integrality constraints of the original MILPP. We further have that $\hat{z} < z^u$. Thus, we can update the upper bound as:

$$z^u \leftarrow \hat{z} = -4,$$

and the best solution found thus far as:

$$x^b \leftarrow \hat{x} = (3, 1)^\top.$$

We also know that we do not have to add any other problems to Ξ and can proceed to the next step.

We go to the Optimality Check step and find that $\Xi \neq \emptyset$. Thus, we must return to Step 2 and select another problem in Ξ to solve. Figure 3.11 shows the new tree after Ξ is updated by removing LPP_5.

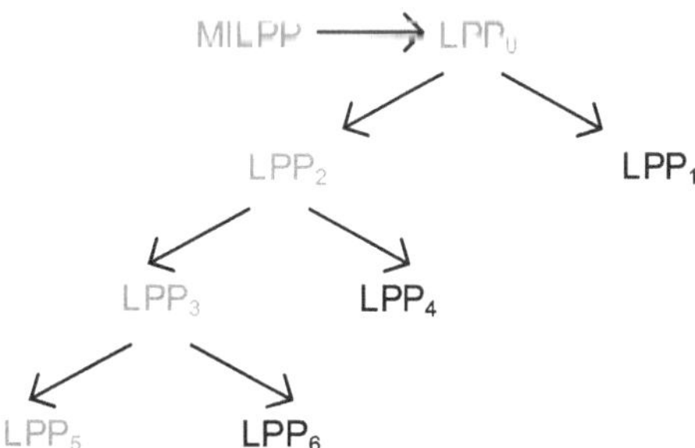

Fig. 3.11 Third Bound Updating and Branching in the Branch-and-Bound Algorithm

In Step 2 we next select LPP_6 to solve, but find that it is infeasible. We then update $\mathcal{E}$ to:

$$\mathcal{E} \leftarrow \{\text{LPP}_1, \text{LPP}_4, \cancel{\text{LPP}_6}\} = \{\text{LPP}_1, \text{LPP}_4\}.$$

When we proceed to the Bound Updating and Branching step there is nothing to be done, because LPP_6 is infeasible. Thus, we proceed to the Optimality Check step and because $\mathcal{E} \neq \emptyset$ we return to Step 2. Figure 3.12 shows the updated tree after LPP_6 is removed.

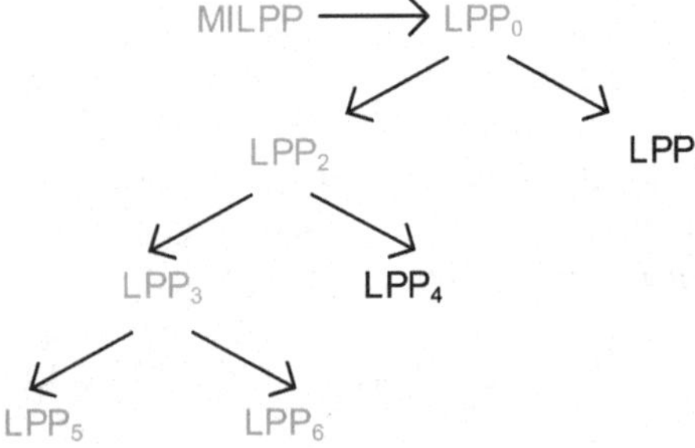

Fig. 3.12 Fourth Bound Updating and Branching in the Branch-and-Bound Algorithm

In Step 2 we next select LPP_4 to solve and find that this problem is also infeasible. We update $\mathcal{E}$ to:

$$\mathcal{E} \leftarrow \{\text{LPP}_1, \cancel{\text{LPP}_4}\} = \{\text{LPP}_1\}.$$

As in the previous iteration, there is nothing to be done in the Bound Updating and Branching step, because LPP_4 is infeasible but the Optimality Check step tells us to return to Step 2 because $\mathcal{E} \neq \emptyset$. Figure 3.13 shows the updated tree after LPP_6 is removed.

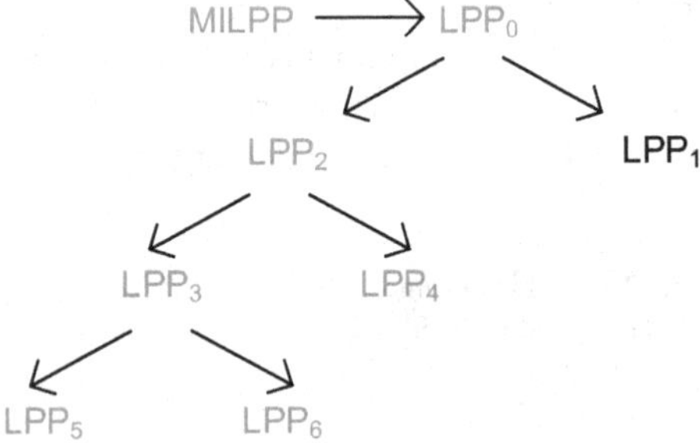

Fig. 3.13 Fifth Bound Updating and Branching in the Branch-and-Bound Algorithm

We now solve LPP_1, which has $\hat{x} = (2, 3.187)^\top$ as an optimal solution and $\hat{z} = -5.187$ as its corresponding optimal objective-function value. We update $\mathcal{E}$ to:

$$\mathcal{E} \leftarrow \{\cancel{\text{LPP}_1}\} = \{\}.$$

We next proceed to the Bound Updating and Branching step and find that $\hat{x}$ falls into Case 3. We next find that $z^l \leq \hat{z} \leq z^u$ and $z^u < +\infty$, thus we can update the lower bound to:

$$z^l \leftarrow \hat{z} = -5.187.$$

Furthermore, because we have that $\hat{z} < z^u$, we must add two new problems to Ξ. x_2 is the only variable with a non-integer value in $\hat{x}$ that must be integer-valued in the original MILPP. Thus, these problems are formed by adding the constraints:

$$x_2 \leq \lfloor \hat{x}_2 \rfloor = 3,$$

and:

$$x_2 \geq \lceil \hat{x}_2 \rceil = 4,$$

to LPP_1, giving:

$$\min_{x_1, x_2} z = -x_1 - x_2$$
$$\mathrm{s.t.}\ 17x_1 + 32x_2 \leq 136$$
$$32x_1 + 15x_2 \leq 120$$
$$x_1, x_2 \geq 0$$
$$x_1 \leq 2$$
$$\mathbf{x_2 \leq 3,}$$

and:

$$\min_{x_1, x_2} z = -x_1 - x_2$$
$$\mathrm{s.t.}\ 17x_1 + 32x_2 \leq 136$$
$$32x_1 + 15x_2 \leq 120$$
$$x_1, x_2 \geq 0$$
$$x_1 \leq 2$$
$$\mathbf{x_2 \geq 4,}$$

as our two new problems, which we denote as LPP_7 and LPP_8, respectively. Adding these to Ξ gives:

$$\Xi \leftarrow \Xi \cup \mathrm{LPP}_7 \cup \mathrm{LPP}_8 = \{\mathrm{LPP}_7, \mathrm{LPP}_8\},$$

and Figure 3.14 shows the updated tree. We next proceed to the Optimality Check. Because $\Xi \neq \emptyset$ we must return to Step 2.

In Step 2 we choose LPP_7 as the next problem to solve and find $\hat{x} = (2, 3)^\top$ to be its optimal solution and $\hat{z} = -5$ the corresponding objective-function value. We

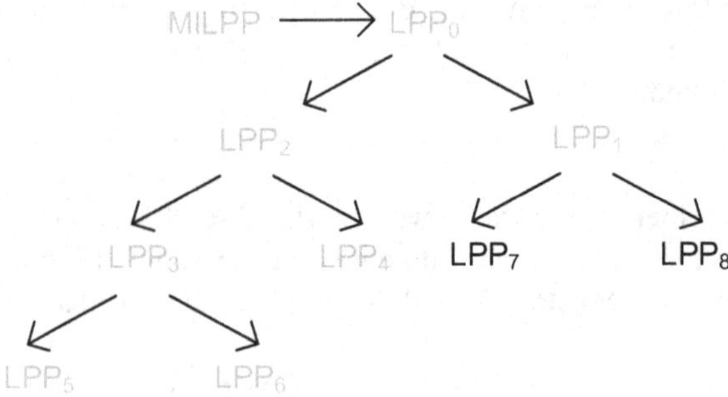

Fig. 3.14 Sixth Bound Updating and Branching in the Branch-and-Bound Algorithm

update Ξ, giving:

$$\Xi \leftarrow \{\cancel{\text{LPP}_7}, \text{LPP}_8\} = \{\text{LPP}_8\}.$$

In the Bound Updating and Branching step we find that $\hat{x}$ falls into Case 1, because $\hat{x}$ satisfies all of the integrality constraints of the original MILPP. We further have that $\hat{z} < z^u$. Thus, we update the upper bound:

$$z^u \leftarrow \hat{z} = -5,$$

and the best feasible solution found thus far:

$$x^b \leftarrow \hat{x} = (2, 3)^\top.$$

We do not have to add any problems to Ξ and thus proceed to the Optimality Check step. Because we still have $\Xi \neq \emptyset$ we return to Step 2. Figure 3.15 shows the new tree with LPP$_7$ removed.

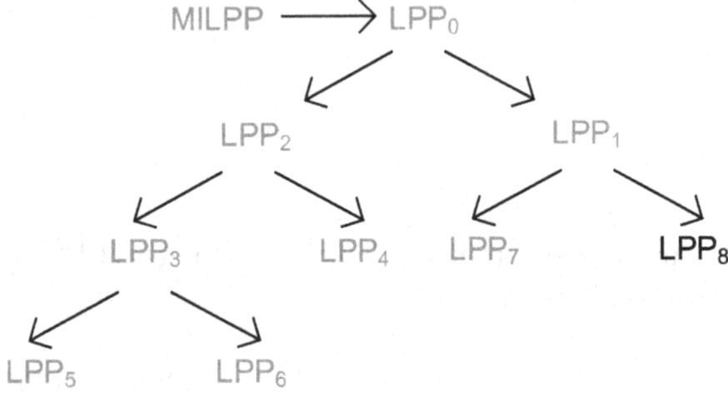

Fig. 3.15 Seventh Bound Updating and Branching in the Branch-and-Bound Algorithm

Returning to Step 2, we solve LPP$_8$ (as it is the only problem remaining in Ξ) and find that it has $\hat{x} = (0.471, 4)^\top$ as an optimal solution with objective-function value $\hat{z} = -4.471$. Removing LPP$_8$ from Ξ gives:

$$\Xi \leftarrow \{\mathbf{LPP_8}\} = \{\}.$$

In the Bound Updating and Branching step we find that $\hat{x}$ falls into Case 3. However, because $\hat{z} > z^u$ we do not update the upper bound nor do we add any more problems to Ξ. Instead, we proceed to the Optimality Check and because we have $\Xi = \emptyset$, we terminate the Branch-and-Bound Algorithm. Moreover, because we have found a solution that is feasible in the original MILPP, $x^b = (2, 3)^\top$, we report this as the optimal solution of the original MILPP. Figure 3.16 shows the final tree of problems after the Branch-and-Bound Algorithm terminates. $\square$

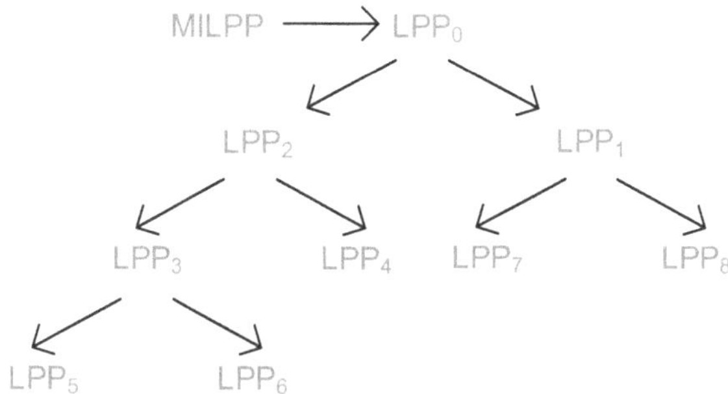

Fig. 3.16 Final branch-and-bound tree

3.5.3 Rationale Behind Branch-and-Bound Algorithm

The idea behind the Branch-and-Bound Algorithm is relatively straightforward. What the algorithm does is enumerates all of the different possible values that the integer variables can take in an MILPP. However, it does this in an intelligent fashion.

The heart of this enumeration is the Bound Updating and Branching step of the algorithm. Whenever a problem in Ξ gives a solution in which an integer variable has a non-integer value, the algorithm creates two new problems in which that variable is fixed to be on the 'two sides' of that non-integer value. That is the logic behind adding constraints of the form:

$$x_i \leq \lfloor \hat{x}_i \rfloor,$$

and:

$$x_i \geq \lceil \hat{x}_i \rceil,$$

in Step 3. As we note in the discussion of Steps 1 and 3 in Sections 3.5.2.2 and 3.5.2.4, the constraints that are used to generate the new problems do not discard any feasible solutions to the original MILPP. They only work to generate LPPs that eventually generate solutions that are feasible in the original MILPP.

The 'intelligence' comes into play because we do not always add new problems after solving a problem in Step 2. For instance, we do not add new problems in Cases 1 and 2 of the Bound Updating and Branching step. The reason we do not in Case 1 is that the most recently solved LPP has already given a solution that is feasible in the original MILPP. It is true that solving a new problem in which an additional constraint is added to this LPP may give another feasible solution. However, any such feasible solutions will give worse objective-function values than the solution already found (this is a consequence of the Relaxation-Optimality Property because the most recently solved LPP is a relaxation of any LPP that we generate by adding constraints).

We do not add any new problems after Case 2 because if the most recently solved LPP is infeasible, then any LPP we generate by adding more constraints is also guaranteed to be infeasible.

Finally, in Case 3 of the Bound Updating and Branching step we do not add new problems if the most recently solved LPP gives an objective-function value, $\hat{z}$, that is greater than the current upper bound, z^u. The reason behind this is that if we do add new problems, their optimal solutions (whether they be feasible in the original MILPP or not) will have objective-function values that are worse (higher) than $\hat{z}$. This is because the most recently solved LPP is a relaxation of these problems that would have added constraints. If $z^u < \hat{z}$ that means we already have a solution that is feasible in the original MILPP and which gives a better objective-function value than any feasible solutions from these problems could give.

The intelligent enumeration of the Branch-and-Bound Algorithm can be seen in the amount of work involved in solving the Photovoltaic Panel-Repair Problem in Example 3.1. The Branch-and-Bound Algorithm requires us to solve nine LPPs (when we count the first linear relaxation). As a result, we are implicitly examining nine possible solutions to the MILPP. Figure 3.1 shows that this problem has 15 feasible solutions. Because of its intelligence, we typically only have to examine a subset of solutions when solving a MILPP using the Branch-and-Bound Algorithm.

3.5.4 Branch-and-Bound Algorithm

We now give a more detailed outline of the Branch-and-Bound Algorithm. Lines 2–16 combine the Initialization and Initial Branching steps, which are discussed in Sections 3.5.2.1 and 3.5.2.2. First, Lines 2–4 initialize z^l, z^u, and $\mathcal{Z}$. Next the linear relaxation is solved in Line 5. Lines 6–9 terminate the algorithm if the linear relaxation shows the original MILPP to be infeasible or gives an optimal solution to the MILPP.

Branch-and-Bound Algorithm

1: **procedure** BRANCH AND BOUND
2:   $z^l \leftarrow -\infty$
3:   $z^u \leftarrow +\infty$
4:   $\Xi \leftarrow \emptyset$
5:   solve LPP_0 ▷ x^0, z^0 denote optimal solution and objective-function value
6:   **if** LPP_0 is infeasible **then**
7:       **stop**, original MILPP is infeasible
8:   **else if** x^0 satisfies integrality constraints of original MILPP **then**
9:       **stop**, x^0 is optimal in original MILPP
10:   **else**
11:       $z^l \leftarrow z^0$
12:       select a variable, x_i, to branch on
13:       generate LPP^-, which is LPP_0 with constraint $x_i \leq \lfloor x_i^0 \rfloor$ added
14:       generate LPP^+, which is LPP_0 with constraint $x_i \geq \lceil x_i^0 \rceil$ added
15:       $\Xi \leftarrow \Xi \cup LPP^- \cup LPP^+$
16:   **end if**
17:   **repeat**
18:       select a problem in Ξ ▷ denote the selected problem 'LPP'
19:       remove LPP from Ξ
20:       solve LPP   ▷ $\hat{x}, \hat{z}$ denote optimal solution and objective-function value
21:       **if** $\hat{x}$ satisfies integrality constraints of original MILPP **then**
22:           **if** $\hat{z} < z_u$ **then**
23:               $z^u \leftarrow \hat{z}$
24:               $x^b \leftarrow \hat{x}$
25:           **end if**
26:       **else if** LPP is feasible **then**
27:           **if** $z^l < \hat{z} \leq z^u$ and $z^u < +\infty$ **then**
28:               $z^l \leftarrow \hat{z}$
29:           **end if**
30:           **if** $\hat{z} < z^u$ **then**
31:               select a variable, x_i, to branch on
32:               generate LPP^-, which is LPP with constraint $x_i \leq \lfloor x_i^0 \rfloor$ added
33:               generate LPP^+, which is LPP with constraint $x_i \geq \lceil x_i^0 \rceil$ added
34:               $\Xi \leftarrow \Xi \cup LPP^- \cup LPP^+$
35:           **end if**
36:       **end if**
37:   **until** $\Xi = \emptyset$
38: **end procedure**

Otherwise, the lower bound is updated in Line 11. Note that if the linear relaxation is unbounded the value of z^l does not actually change, because $z^0 = -\infty$. In Line 12 we select one variable, which we denote x_i, to branch on. The variable to branch on can be any variable that has a non-integer value in x^0 but which is constrained to take on an integer value in the original MILPP. We then generate the two new LPPs, which have the same objective function and constraints as the linear relaxation, but each have one new constraint, which are:

$$x_i \leq \lfloor x_i^0 \rfloor,$$

and:

$$x_i \geq \lceil x_i^0 \rceil.$$

Lines 17–37 are the main iterative loop of the algorithm. We begin by selecting a problem, which we denote LPP, in Ξ, removing LPP from Ξ, and then solving LPP in Lines 18–20. We let $\hat{x}$ and $\hat{z}$ denote the optimal solution and objective-function value of LPP. Lines 21–36 are the Bound Updating and Branching process, which is outlined in Section 3.5.2.4. Lines 21–25 handle Case 1 of this process, where $\hat{x}$ satisfies the integrality constraints of the original MILPP. If so, we update the upper bound and the best solution found in Lines 23 and 24, so long as $\hat{z} < z^u$. We do not have to branch (*i.e.*, add new problems to Ξ) in this case.

Lines 26–36 handle Case 3 in Section 3.5.2.4. If LPP is feasible but $\hat{x}$ is not feasible in the original MILPP, we then update the lower bound in Line 28 if $z^l < \hat{z} \leq z^u$ and $z^u < +\infty$. Also, if $\hat{z} < z^u$ we select another variable to branch on and generate the two LPPs that are added to Ξ in Lines 31–34.

Note that Lines 17–37 do not explicitly discuss cases in which LPP is infeasible or unbounded. If LPP is infeasible, there is no bound updating to be done and no new LPPs to be generated and added to Ξ. Thus, we do not do anything after solving LPP in that iteration. If LPP is unbounded, then we generate new optimization problems that are added to Ξ in Lines 31–34.

Line 37 is the Optimality Check. This is because we continue the iterative loop until $\Xi = \emptyset$, which is the test conducted in Section 3.5.2.5.

3.5.5 Practical Considerations

There are three important practical issues to consider when applying the Branch-and-Bound Algorithm. The first involves which variable to branch on (*i.e.*, which variable to use when adding constraints to the new problems being added to Ξ) in the Initial Branching and the Bound Updating and Branching steps of the algorithm. One may be tempted to wonder which is the 'best' variable to branch on, in the sense of obtaining the optimal solution to the original MILPP as quickly as possible. Unfortunately, the answer to this question is usually highly problem-dependent and complex relationships between different variables can make it difficult to ascertain this *a priori*. Thus, no general rules are available. Most MILPP solvers employ heuristic rules, and the cost of a solver package is often tied to how much research and sophistication is involved in the rules implemented.

A second question is what order to process the problems in Ξ in Step 2 of the algorithm. There is, again, no general rule that applies to all problems because the efficiency of solving a MILPP is highly dependent on the structure of the objective function and constraints. Two heuristic rules that are commonly employed are known as **depth-first** and **breadth-first** strategies. A depth-first strategy, which we employ in Example 3.1, goes deep down the branches of the tree as quickly as possible. This is seen in the strategy that we employ because we solve LPP_2, followed by LPP_3 and LPP_5. A breadth-first strategy would, instead, stay at the top levels of the tree first before going deeper. Applying such a strategy to Example 3.1 would see us solve LPP_2 and LPP_1 before moving on the LPP_3, LPP_4, LPP_7, and LPP_8 and only then going on to LPP_5 and LPP_6.

The primary benefit of a depth-first strategy is that it quickly produces problems with many constraints that are either infeasible or give feasible solutions to the original MILPP. This allows us to tighten the upper and lower bounds and find feasible solutions relatively quickly. On the other hand, a breadth-first strategy allows us to solve LPPs that are very similar to each other (they only have a small number of constraints that differ from one another). This often allows us to solve the LPPs more quickly. That being said, there is no general rule as to which of the two strategies works most efficiently. Indeed, many solvers employ a combination of the two approaches.

A third issue, which is raised in Section 3.5.1, is that we often do not solve an MILPP to complete optimality. That means, we may terminate the Branch-and-Bound Algorithm before every problem in Ξ is solved. This process is normally governed by the upper and lower bounds. If these bounds are sufficiently close to one another, we may terminate the algorithm because we have a solution that is 'close enough' to optimal. For instance, in Example 3.1 we find the solution $x = (3, 1)^\top$, which is not optimal after, solving LPP_5. We also know, after solving LPP_5, that $z^l = -5.347$ and $z^u = -4$. Thus, we know that the solution we have is at most:

$$\left| \frac{z^u - z^l}{z^l} \right| = 0.25,$$

or 25% away from the optimal solution. In many cases, we may not be sufficiently comfortable to use the feasible solution that we have at hand. However, if we find a solution and know that it is at most 0.01% away from optimal, we may be happy with that. Indeed, by default most MILPP solvers do not solve MILPPs to complete optimality but stop once this so-called **optimality gap** is sufficiently small.

3.6   Solving Pure-Integer Linear Optimization Problems Using Cutting Planes

This section outlines a fundamentally different way of solving a mixed-integer linear optimization problem. Although the technique that we outline here can be applied to generic MILPPs, we focus our attention on pure-integer linear optimization problems. Interested readers may consult other more advanced texts [13] that discuss the generalization of this technique to generic MILPPs.

The idea of this algorithm is relatively straightforward and relies on some simple properties of MILPPs and LPPs. We first know that if the integrality constraints of a MILPP are relaxed, the resulting linear relaxation is an LPP. Thus, we know that an optimal solution to the linear relaxation will have the properties discussed in Section 2.3. More specifically, we argue in Section 2.3.1 that the feasible region of every LPP has at least one extreme point or corner that is optimal.

To see why this observation is important, examine the feasible region of the linear relaxation of the Photovoltaic Panel-Repair Problem, which is shown in Figure 3.17. We know from the properties of linear optimization problems that when we solve this linear relaxation, one of the four extreme points of the linear relaxation, which are highlighted as red squares in Figure 3.17, will be the solution given by the Simplex method. Unfortunately, only one of these extreme points:

$$(x_1, x_2) = (0, 0)^\top,$$

is a solution that is feasible in the original MILPP, and it is not an optimal solution to the MILPP.

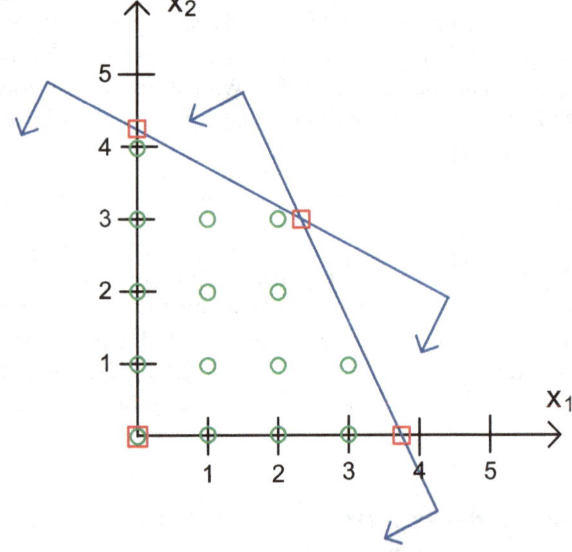

Fig. 3.17 Geometrical representation of the feasible region of the linear relaxation of the Photovoltaic Panel-Repair Problem

The idea that we explore in this section is to solve the MILPP by adding what are known as **cutting planes** [8]. A cutting plane is simply a constraint that cuts off solutions that are feasible in the linear relaxation but are not feasible in the original MILPP. Figure 3.18 shows the feasible region of the linear relaxation of the original MILPP when these cutting planes are added. Note that the additional constraints make the five extreme points of the feasible region (which are highlighted with red squares) coincide with points where x is integer-valued (and, thus, feasible in the original MILPP). The benefit of these cutting planes is that once we solve the linear relaxation, the Simplex method gives us one of the extreme points, which will have integer values for x, as an optimal solution. The important observation about cutting planes is that they change (and reduce) the feasible region of the linear relaxation. However, they do not change the feasible region of the original MILPP.

In practice, we do not have all of the cutting planes needed to solve a MILPP *a priori*. Instead, we generate the cutting planes in an iterative algorithm by solving the linear relaxation and finding constraints that cut off non-integer solutions. After a sufficient number of iterations, and adding a sufficient number of these constraints, we have a linear relaxation that gives an integer-valued solution that is feasible in the original MILPP.

We proceed in this section by first outlining how to generate cuts from a non-integer solution obtained from solving the linear relaxation of a pure-integer linear optimization problem. We then outline the iterative algorithm to generate cutting planes and demonstrate its use with an example.

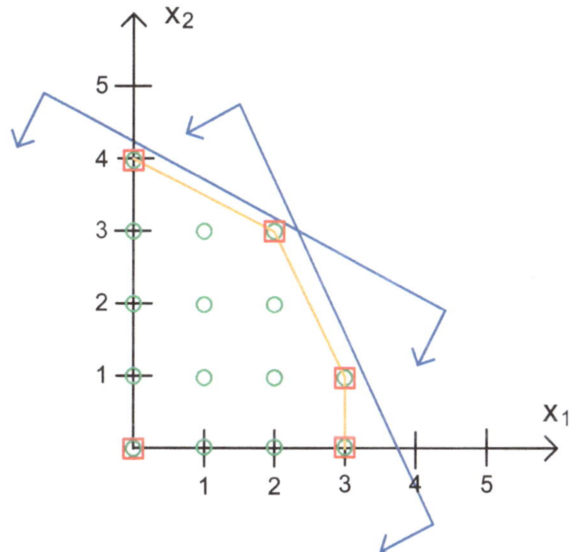

Fig. 3.18 Geometrical representation of the feasible region of the linear relaxation of the Photovoltaic Panel-Repair Problem with cutting planes added

3.6.1 Generating Cutting Planes

To derive a cutting plane, let us suppose that we solve the linear relaxation of a PILPP and obtain a solution, $\hat{x}$, which does not satisfy all of the integrality restrictions of the original PILPP. We further assume that the linear relaxation is converted to standard form (2.14)–(2.16):

$$\min_{x} c^\top x$$

$$\text{s.t. } Ax = b$$

$$x \geq 0,$$

which is introduced in Section 2.2.2.1. This form allows us to analyze the structural equality constraints by partitioning x into basic and non-basic variables (*cf.* Section 2.4.1 for further details). This partition allows us to write the equality constraints as:

$$[B \ N]\begin{pmatrix} x_B \\ x_N \end{pmatrix} = b, \tag{3.44}$$

where we partition A into:

$$A = [B \ N],$$

and the B submatrix is full-rank and we partition x into:

$$x = \begin{pmatrix} x_B \\ x_N \end{pmatrix}.$$

Equation (3.44) can be rewritten as:

$$Bx_B + Nx_N = b,$$

or, because B is full-rank, as:

$$x_B + B^{-1}Nx_N = B^{-1}b.$$

This can further be simplified as:

$$x_B + \tilde{N}x_N = \tilde{b}, \tag{3.45}$$

where:

$$\tilde{N} = B^{-1}N, \tag{3.46}$$

and:

$$\tilde{b} = B^{-1}b. \tag{3.47}$$

We next divide the values of $\tilde{N}$ and $\tilde{b}$ into their integer and non-integer components. More specifically, define:

$$\tilde{N}^I = \lfloor \tilde{N} \rfloor,$$

$$\tilde{N}^F = \tilde{N} - \tilde{N}^I,$$

$$\tilde{b}^I = \left\lfloor \tilde{b} \right\rfloor,$$

and:

$$\tilde{b}^F = \tilde{b} - \tilde{b}^I.$$

We can note a few properties of $\tilde{N}$, $\tilde{b}$, $\tilde{N}^I$, $\tilde{N}^F$, $\tilde{b}^I$, and $\tilde{b}^F$. First, we clearly have that:

$$\tilde{N} = \tilde{N}^I + \tilde{N}^F,$$

and:

$$\tilde{b} = \tilde{b}^I + \tilde{b}^F,$$

from the definitions of $\tilde{N}^I$, $\tilde{N}^F$, $\tilde{b}^I$, and $\tilde{b}^F$. Next, we know that $\tilde{N}^I$ and $\tilde{b}^I$ are integer-valued, because they are defined as the floors of $\tilde{N}$ and $\tilde{b}$, respectively. Finally, we have that $\tilde{N}^F$ and $\tilde{b}^F$ are non-integer-valued, non-negative, and less than 1, because they are defined as the difference between each of $\tilde{N}$ and $\tilde{b}$ and their floors.

Thus, we can rewrite (3.45) as:

$$x_B + (\tilde{N}^I + \tilde{N}^F)x_N = \tilde{b}^I + \tilde{b}^F,$$

or, by rearranging terms, as:

$$x_B + \tilde{N}^I x_N - \tilde{b}^I = \tilde{b}^F - \tilde{N}^F x_N. \tag{3.48}$$

We next consider a basic variable, which we denote $x_{B,i}$, which has a non-integer value in $\hat{x}$. Note that because the original PILPP requires all of the variables to be integer-valued, the value of $\hat{x}_{B,i}$ is not feasible in the PILPP. We can define the value of $x_{B,i}$ from (3.48) as:

$$x_{B,i} + \sum_{j=1}^{m} \tilde{N}_{i,j}^I x_{N,j} - \tilde{b}_i^I = \tilde{b}_i^F - \sum_{j=1}^{m} \tilde{N}_{i,j}^F x_{N,j}, \tag{3.49}$$

where m is the number of structural equality constraints when the linear relaxation is written in standard form.

Note that the left-hand side of (3.49) is, by definition, integer-valued in the original PILPP. To see why, note that all of the x's are restricted to be integer-valued in the original PILPP. Moreover, the coefficients $\tilde{N}_{i,j}^I$ and the constant $\tilde{b}_i^I$ are defined to be integer-valued as well. Thus, for (3.49) to hold, the right-hand side must be integer-valued as well.

Let us next examine the right-hand side of (3.49). We can first argue that $\tilde{b}_i^F > 0$. The reason for this is that we are focusing our attention on a basic variable that is not integer-valued in $\hat{x}$. Moreover, we know from the discussion in Section 2.4.2 that the values of basic variables are equal to $\tilde{b}$, because the non-basic variables are fixed equal to zero in the Simplex method. Thus, because $x_{B,i} = \tilde{b}_i$, if $\hat{x}_{B,i}$ is not integer-valued, then $\tilde{b}_i$ must have a strictly positive non-integer component, $\tilde{b}_i^F$. We further know that the second term on the right-hand side of (3.49):

$$\sum_{j=1}^{m} \tilde{N}_{i,j}^{F} x_{N,j},$$

is non-negative, because the coefficients, $\tilde{N}_{i,j}^{F}$, and the variables, $x_{N,j}$, are all non-negative. Because $\tilde{b}^F$ is defined as the non-integer component of $\tilde{b}$, it is by definition strictly less than 1. Thus, we can conclude that the right-hand side of (3.49) is an integer that is less than or equal to zero, or that:

$$\tilde{b}_i^F - \sum_{j=1}^{m} \tilde{N}_{i,j}^{F} x_{N,j} \leq 0,$$

which can also be written as:

$$\sum_{j=1}^{m} \tilde{N}_{i,j}^{F} x_{N,j} - \tilde{b}_i^F \geq 0. \tag{3.50}$$

Inequality (3.50) is the cutting plane that is generated by the solution, $\hat{x}$. This type of cutting plane is often referred to as a Gomory cut, as Gomory introduced this method of solving MILPPs [8]. Cuts of this form are used in the iterative algorithm that is outlined in the next section.

3.6.2   Outline of Cutting-Plane Algorithm

Building off of the derivation of the Gomory cut given in Section 3.6.1, we now outline the major steps of the Cutting-Plane Algorithm.

3.6.2.1   Step 0: Initialization

We begin the Cutting-Plane Algorithm by first solving the linear relaxation of the original PILPP. When we solve the linear relaxation, one of the following three outcomes is possible.

1. The linear relaxation may be infeasible. If so, then based on the Relaxation-Feasibility Property, the original PILPP is infeasible as well. As such, we terminate the Cutting-Plane Algorithm and report that the original PILPP is infeasible.
2. Solving the linear relaxation may give a solution that satisfies all of the integrality constraints of the original PILPP. If so, then based on the Relaxation-Optimality

Corollary, the optimal solution of the linear relaxation is also optimal in the original PILPP. Thus, we terminate the algorithm and report the solution found as being optimal.

3. If the solution to the linear relaxation does not fit into the first two cases (*i.e.*, we obtain a solution with non-integer values for some of the variables or the linear relaxation is unbounded), then we proceed with the Cutting-Plane Algorithm. Let $\hat{x}$ denote the optimal solution found from solving the linear relaxation.

3.6.2.2 Step 1: Cut Generation

Select a variable, which we denote x_i, that has a non-integer value in $\hat{x}$. In theory, any non-integer-valued variable can be used. In practice, it is often beneficial to select the one that has the largest non-integer component when $\tilde{b}$ is decomposed into $\tilde{b}^I$ and $\tilde{b}^F$. Generate cut (3.50) for the chosen variable.

3.6.2.3 Step 2: Solving

Add the cut generated in Step 1 to the most recently solved LPP and solve the resulting LPP.

3.6.2.4 Step 3: Optimality Check

When we solve the LPP in Step 2 there are three possible outcomes.

1. The LPP may be infeasible. If so, then the original PILPP is infeasible as well. As such, we terminate the Cutting-Plane Algorithm and report that the original PILPP is infeasible.
2. Solving the LPP may give a solution that satisfies all of the integrality constraints of the original PILPP. If so, then the optimal solution of the LPP is also optimal in the original PILPP. Thus, we terminate the algorithm and report the solution found as being optimal.
3. If the solution to the LPP does not fit into these two cases (*i.e.*, we obtain a solution with non-integer values for some of the variables or the LPP is unbounded) then we continue the algorithm by returning to Step 1. Let $\hat{x}$ denote the optimal solution found from solving the LPP.

Example 3.2 Consider the following variant of the Photovoltaic Panel-Repair Problem from Section 3.1.1. A type-A repair unit is now 10% more effective than a type-B unit. Moreover, each type-A unit has a mass of 2 kg and occupies 7 m^3 of space while a type-B unit has a mass of 1 kg and occupies 8 m^3. The shuttle can now carry at most 6 kg of repair units and has at most 28 m^3 of space available in its cargo bay. The payload specialists must determine how many units of each type to carry aboard

the spacecraft to maximize the effectiveness-weighted number of repair units sent to
the spacecraft.

To formulate this problem we let x_1 and x_2, respectively, denote the number of
type-A and -B repair units put into the spacecraft. The PILPP is then:

$$\min_{x_1,x_2} z = -\frac{11}{10}x_1 - x_2$$
$$\text{s.t. } 2x_1 + x_2 \leq 6$$
$$7x_1 + 8x_2 \leq 28$$
$$x_1, x_2 \geq 0$$
$$x_1, x_2 \in \mathbb{Z},$$

when the objective function is converted into a minimization.

To solve this problem using the Cutting-Planes Algorithm, we first convert the
linear relaxation:

$$\min_{x_1,x_2} z = -\frac{11}{10}x_1 - x_2$$
$$\text{s.t. } 2x_1 + x_2 \leq 6$$
$$7x_1 + 8x_2 \leq 28$$
$$x_1, x_2 \geq 0,$$

into standard form:

$$\min_{x_1,x_2,x_3,x_4} z = -\frac{11}{10}x_1 - x_2$$
$$\text{s.t. } 2x_1 + x_2 + x_3 = 6$$
$$7x_1 + 8x_2 + x_4 = 28$$
$$x_1, x_2, x_3, x_4 \geq 0,$$

by adding two new slack variables, x_3 and x_4. Solving this problem gives $\hat{x} = (20/9, 14/9, 0, 0)^\top$. Because this solution does not satisfy all of the integrality constraints of the original PILPP, we add a new cut.

To do so, we first recall that we have:

$$A = \begin{bmatrix} 2 & 1 & 1 & 0 \\ 7 & 8 & 0 & 1 \end{bmatrix},$$

and:

$$b = \begin{pmatrix} 6 \\ 28 \end{pmatrix}.$$

Because x_1 and x_2 are basic variables and x_3 and x_4 are non-basic variables, we know that the basis matrix will have the first two columns of A:

$$B = \begin{bmatrix} 2 & 1 \\ 7 & 8 \end{bmatrix},$$

and the N matrix will have the remaining columns:

$$N = \begin{bmatrix} 1 & 0 \\ 0 & 1 \end{bmatrix}.$$

Using (3.46) and (3.47) we have:

$$\tilde{N} = \begin{bmatrix} 8/9 & -1/9 \\ -7/9 & 2/9 \end{bmatrix},$$

and:

$$\tilde{b} = \begin{pmatrix} 20/9 \\ 14/9 \end{pmatrix}.$$

We can decompose these two into their integer and non-integer parts:

$$\tilde{N}^I = \begin{bmatrix} 0 & -1 \\ -1 & 0 \end{bmatrix},$$

$$\tilde{N}^F = \begin{bmatrix} 8/9 & 8/9 \\ 2/9 & 2/9 \end{bmatrix},$$

$$\tilde{b}^I = \begin{pmatrix} 2 \\ 1 \end{pmatrix}.$$

and:

$$\tilde{b}^F = \begin{pmatrix} 2/9 \\ 5/9 \end{pmatrix}.$$

We now select either of x_1 or x_2 to generate a cutting plane with. Because the non-integer components of $\hat{x}$ are given by $\tilde{b}^F$ and $\tilde{b}_2^F > \tilde{b}_1^F$, we generate a cut using x_2. From (3.50) this cut is given by:

$$\sum_{j=1}^{m} \tilde{N}_{2,j}^F x_{N,j} - \tilde{b}_2^F \geq 0,$$

or by:

$$\frac{2}{9} x_3 + \frac{2}{9} x_4 - \frac{5}{9} \geq 0,$$

when we substitute in the values of $\tilde{N}^F$ and $\tilde{b}^F$. We can further simplify this cut to:

$$2x_3 + 2x_4 - 5 \geq 0.$$

Adding this cut to the standard form of the linear relaxation of the PILPP gives:

$$\min_{x_1,x_2,x_3,x_4} z = -\frac{11}{10}x_1 - x_2$$
$$\text{s.t. } 2x_1 + x_2 + x_3 = 6$$
$$7x_1 + 8x_2 + x_4 = 28$$
$$2x_3 + 2x_4 - 5 \geq 0$$
$$x_1, x_2, x_3, x_4 \geq 0,$$

which we transform into the standard-form problem:

$$\min_{x_1,x_2,x_3,x_4,x_5} z = -\frac{11}{10}x_1 - x_2$$
$$\text{s.t. } 2x_1 + x_2 + x_3 = 6$$
$$7x_1 + 8x_2 + x_4 = 28$$
$$2x_3 + 2x_4 - x_5 = 5$$
$$x_1, x_2, x_3, x_4, x_5 \geq 0,$$

by adding the surplus variable, x_5, to the problem. Solving this problem gives $\hat{x} = (5/2, 1, 0, 5/2, 0)^\top$. This solution does not satisfy the integrality constraints of the original PILPP, because $\hat{x}_1 = 5/2$.

Thus, we must generate a new cutting plane. To do so, we first note that we now have:

$$A = \begin{bmatrix} 2 & 1 & 1 & 0 & 0 \\ 7 & 8 & 0 & 1 & 0 \\ 0 & 0 & 2 & 2 & -1 \end{bmatrix},$$

and:

$$b = \begin{pmatrix} 6 \\ 28 \\ 5 \end{pmatrix}.$$

Because x_1, x_2, and x_4 are basic variables we have:

$$B = \begin{bmatrix} 2 & 1 & 0 \\ 7 & 8 & 1 \\ 0 & 0 & 2 \end{bmatrix},$$

and:

$$N = \begin{bmatrix} 1 & 0 \\ 0 & 0 \\ 2 & -1 \end{bmatrix},$$

and from (3.46) and (3.47) we have:

$$\tilde{N} = \begin{bmatrix} 1 & -1/18 \\ -1 & 1/9 \\ 1 & -1/2 \end{bmatrix},$$

and:

$$\tilde{b} = \begin{pmatrix} 5/2 \\ 1 \\ 5/2 \end{pmatrix}.$$

Decomposing these into their integer and non-integer parts gives:

$$\tilde{N}^I = \begin{bmatrix} 1 & -1 \\ -1 & 0 \\ 1 & -1 \end{bmatrix},$$

$$\tilde{N}^F = \begin{bmatrix} 0 & 17/18 \\ 0 & 1/9 \\ 0 & 1/2 \end{bmatrix},$$

$$\tilde{b}^I = \begin{pmatrix} 1 \\ 1 \\ 1 \end{pmatrix}.$$

and:

$$\tilde{b}^F = \begin{pmatrix} 1/2 \\ 0 \\ 1/2 \end{pmatrix}.$$

Note that the value of $\hat{x}_4 = 5/2$ does not affect the feasibility of this solution in
the original PILPP, because the original PILPP only constraints x_1 and x_2 to be
integer-valued. However, if we examine the structural constraint:

$$7x_1 + 8x_2 + x_4 = 28,$$

it is clear that if x_4 takes on a non-integer value, then at least one of x_1 or x_2 will
also take on a non-integer value. Thus, we have the option of generating a cut using
either of x_1 or x_4. If we select x_4, the new cutting plane will be:

$$\frac{1}{2}x_5 - \frac{1}{2} \geq 0,$$

or:

$$x_5 - 1 \geq 0,$$

when simplified. Adding this constraint to the current LPP gives:

$$\min_{x_1,x_2,x_3,x_4,x_5} \quad z = -\frac{11}{10}x_1 - x_2$$

$$\text{s.t. } 2x_1 + x_2 + x_3 = 6$$
$$7x_1 + 8x_2 + x_4 = 28$$
$$2x_3 + 2x_4 - x_5 = 5$$
$$x_5 - 1 \geq 0$$
$$x_1, x_2, x_3, x_4, x_5 \geq 0,$$

which is:

$$\min_{x_1,x_2,x_3,x_4,x_5,x_6} \quad z = -\frac{11}{10}x_1 - x_2$$

$$\text{s.t. } 2x_1 + x_2 + x_3 = 6$$
$$7x_1 + 8x_2 + x_4 = 28$$
$$2x_3 + 2x_4 - x_5 = 5$$
$$x_5 - x_6 = 1$$
$$x_1, x_2, x_3, x_4, x_5, x_6 \geq 0,$$

in standard form.

Solving the new LPP gives $x = (23/9, 8/9, 0, 3, 1, 0)^\top$. Thus, we must generate a new cut. To do so, we note that we now have:

$$A = \begin{bmatrix} 2 & 1 & 1 & 0 & 0 & 0 \\ 7 & 8 & 0 & 1 & 0 & 0 \\ 0 & 0 & 2 & 2 & -1 & 0 \\ 0 & 0 & 0 & 0 & 1 & -1 \end{bmatrix},$$

$$b = \begin{pmatrix} 6 \\ 28 \\ 5 \\ 1 \end{pmatrix},$$

$$B = \begin{bmatrix} 2 & 1 & 0 & 0 \\ 7 & 8 & 1 & 0 \\ 0 & 0 & 2 & -1 \\ 0 & 0 & 0 & 1 \end{bmatrix},$$

and:

$$N = \begin{bmatrix} 1 & 0 \\ 0 & 0 \\ 2 & 0 \\ 0 & -1 \end{bmatrix}.$$

From (3.46) and (3.47) we also have:

$$\tilde{N} = \begin{bmatrix} 1 & -1/18 \\ -1 & 1/9 \\ 1 & -1/2 \\ 0 & -1 \end{bmatrix},$$

and:

$$\tilde{b} = \begin{pmatrix} 25/9 \\ 4/9 \\ 5 \\ 5 \end{pmatrix},$$

which are decomposed as:

$$\tilde{N}^I = \begin{bmatrix} 1 & -1 \\ -1 & 0 \\ 1 & -1 \\ 0 & -1 \end{bmatrix},$$

$$\tilde{N}^F = \begin{bmatrix} 0 & 17/18 \\ 0 & 1/9 \\ 0 & 1/2 \\ 0 & 0 \end{bmatrix},$$

$$\tilde{b}^I = \begin{pmatrix} 2 \\ 0 \\ 5 \\ 5 \end{pmatrix},$$

and:

$$\tilde{b}^F = \begin{pmatrix} 7/9 \\ 4/9 \\ 0 \\ 0 \end{pmatrix}.$$

Generating a cut with x_2 gives:

$$\frac{1}{9}x_6 - \frac{4}{9} \geq 0,$$

or:

$$x_6 - 4 \geq 0.$$

Adding this to our current LPP gives:

$$\min_{x_1,x_2,x_3,x_4,x_5,x_6} z = -\frac{11}{10}x_1 - x_2$$
$$\text{s.t. } 2x_1 + x_2 + x_3 = 6$$
$$7x_1 + 8x_2 + x_4 = 28$$
$$2x_3 + 2x_4 - x_5 = 5$$
$$x_5 - x_6 = 1$$
$$x_6 - 4 \geq 0$$
$$x_1, x_2, x_3, x_4, x_5, x_6 \geq 0,$$

which is:

$$\min_{x_1,x_2,x_3,x_4,x_5,x_6,x_7} z = -\frac{11}{10}x_1 - x_2$$
$$\text{s.t. } 2x_1 + x_2 + x_3 = 6$$
$$7x_1 + 8x_2 + x_4 = 28$$
$$2x_3 + 2x_4 - x_5 = 5$$
$$x_5 - x_6 = 1$$
$$x_6 - x_7 = 4$$
$$x_1, x_2, x_3, x_4, x_5, x_6, x_7 \geq 0,$$

in standard form. The optimal solution to this LPP is $x = (3, 0, 0, 7, 9, 4)^\top$, which satisfies all of the integrality constraints of the original PILPP. Thus, we terminate the Cutting-Plane Algorithm and report $(x_1, x_2) = (3, 0)$ as an optimal solution. $\square$

3.6.3 *Cutting-Plane Algorithm*

We now give a more detailed outline of the Cutting-Plane Algorithm. Lines 2–5 initialize the algorithm. We have three flags that indicate if the algorithm should terminate because the original PILPP is found to be infeasible or unbounded or if we find an optimal solution to the original PILPP. These three flags are initially set to zero in Lines 2–4. The algorithm also tracks a 'current LPP,' the constraints of which are updated as new cutting planes are generated. In Line 5, the current LPP is set equal to the linear relaxation of the PILPP.

Cutting-Plane Algorithm

1: **procedure** CUTTING PLANE
2:      $\tau^I \leftarrow 0$
3:      $\tau^U \leftarrow 0$
4:      $\tau^S \leftarrow 0$
5:      'current LPP' $\leftarrow$ 'linear relaxation of PILPP' $\triangleright$ $\hat{x}$ denotes optimal solution
6:      **repeat**
7:          solve current LPP
8:          **if** most recently solved LPP is infeasible **then**
9:              $\tau^I \leftarrow 1$
10:          **else if** $\hat{x}$ satisfies integrality constraints of original PILPP **then**
11:              **if** objective function is unbounded **then**
12:                  $\tau^U \leftarrow 1$
13:              **else**
14:                  $\tau^S \leftarrow 1$
15:              **end if**
16:          **else**
17:              determine B and N from final tableau of most recently solved LPP
18:              $\tilde{N} \leftarrow B^{-1}N$
19:              $\tilde{b} \leftarrow B^{-1}b$
20:              $\tilde{N}^I \leftarrow \lfloor \tilde{N} \rfloor$
21:              $\tilde{N}^F \leftarrow \tilde{N} - \tilde{N}^I$
22:              $\tilde{b}^I \leftarrow \lfloor \tilde{b} \rfloor$
23:              $\tilde{b}^F \leftarrow \tilde{b} - \tilde{b}^I$
24:              select a variable, x_i, to add a cut for
25:              add constraint $\sum_{j=1}^{m} \tilde{N}_{i,j}^F x_{N,j} - \tilde{b}_i^F \geq 0$ to most recently solved LPP
26:          **end if**
27:      **until** $\tau^I = 1$ or $\tau^U = 1$ or $\tau^S = 1$
28:      **if** $\tau^I = 1$ **then**
29:          original PILPP is infeasible
30:      **else if** $\tau^U = 1$ **then**
31:          original PILPP is unbounded
32:      **else**
33:          $\hat{x}$ is optimal in original PILPP
34:      **end if**
35: **end procedure**

Lines 6–27 are the main iterative loop of the algorithm. The current LPP is solved in Line 7 and one of four things happens depending on the solution found. If the current LPP is infeasible, then the flag τ^I is set equal to 1 in Lines 8 and 9 to indicate that the original PILPP is found to be infeasible. Next, if the optimal solution to the current LPP satisfies the integrality constraints of the original PILPP and the objective function is bounded, then we have found an optimal solution to the original PILPP and set the flag τ^S equal to 1 in Lines 13 and 14. Otherwise, if the objective function is unbounded, then the original PILPP is unbounded as well and the flag τ^U is set equal to 1 in Lines 11 and 12. Note that this case only applies if we have an integer-valued solution that gives an unbounded objective function. If the LPP solved in Line 7 is unbounded but we cannot guarantee that there is an integer-valued solution that is unbounded, this falls into the final case in which a new cut is added to the current problem. In the final case, in which the LPP is feasible but we do not find a solution that satisfies the integrality constraints of the original PILPP (or the LPP is unbounded but we cannot guarantee that there is an integer-valued solution that is unbounded), we add a new cut to the current problem in Lines 17–25.

This iterative process repeats until we terminate the algorithm in Line 27 for one of the three reasons (optimality, infeasibility, or unboundedness). Lines 28–34 determine what to report, based on which of τ^I, τ^U, or τ^S is equal to 1 when the algorithm terminates.

3.7 Final Remarks

Current techniques to efficiently solve large-scale MILPPs (*e.g.*, those implemented in CPLEX [10] or GUROBI [9]) combine the branch-and-bound algorithm with cutting-plane methods to reduce the feasible region of the LPPs that must be solved while also exploring the branch-and-bound tree. These hybrid techniques pay particular attention to adding cutting planes to the original linear relaxation of the MILPP before branching begins. The reason for this is that cutting planes added to the linear relaxation are carried through in all of the subsequent LPPs that are solved and tend to improve the quality of the solutions found. This, in turn, increases solution efficiency. Commercial MILPP solvers also incorporate numerous heuristics to determine which variable to branch on and which LPP in $\varXi$ to solve in each iteration of the Branch-and-Bound Algorithm. Bixby [2] provides an excellent overview of the evolution of MILPP solvers up to 2002. Commercial MILPP solvers, such as CPLEX [10] and GUROBI [9], can be easily accessed using mathematical programming languages [3, 7, 11, 14].

The two methods to solve MILPP that are outlined in this chapter are generic, in the sense that they can be applied to any generic MILPP. For very difficult and large-scale MILPPs, decomposition techniques are often employed. These techniques exploit the structure of a problem to determine intelligent ways in which to break the MILPP into smaller subproblems from which a good solution to the overall problem can be found. Another extension of the solution methods discussed here are to develop other types of cutting planes. These cutting planes often exploit the structure of the problem at hand. While Gomory cuts are guaranteed to eventually find an optimal solution to a MILPP, the number of cuts that may need to be added grows exponentially with the problem size. Other types of cuts may yield a solution more quickly than Gomory cuts alone can. Wolsey and Nemhauser [16], Bertsimas and Weismantel [1], and Rao [13] discuss these more advanced solution techniques, including problem decomposition, cutting planes, and hybrid methods. Castillo *et al.* [4] provide further discussion of modeling using mixed-integer linear optimization problems. We also refer interested readers to other relevant works on the topic of MILPPs [5, 12, 15].

Our discussion in this chapter focuses exclusively on mixed-integer linear optimization problems. It is a straightforward extension of the formulation techniques discussed in this chapter and in Chapter 4 to formulate mixed-integer nonlinear optimization problems. Indeed, such problems are gaining increased focused from the operations research community. However, the solution of such problems is still typically quite taxing and demanding compared to MILPPs. We refer interested readers to the work of Floudas [6], which provides an excellent introduction to the formulation and solution of mixed-integer nonlinear optimization problems.

3.8 GAMS Codes

This final section provides GAMS [3] codes for the main problems considered in this chapter. GAMS uses a variety of different solvers, among them CPLEX [10] and GUROBI [9], to actually solve MILPPs.

3.8.1 Photovoltaic Panel-Repair Problem

The Photovoltaic Panel-Repair Problem, which is introduced in Section 3.1.1, has the following GAMS formulation:

```
1  option OPTCR=0;
2  variable z;
3  integer variables x1, x2;
4  equations of, l1, l2;
5  of .. z =e= x1+x2;
6  l1 .. 17*x1+32*x2 =l= 136;
7  l2 .. 32*x1+15*x2 =l= 120;
8  model pv /all/;
9  solve pv using mip maximizing z;
```

Line 1 indicates that the solution tolerance should be 0. Otherwise, the solver may terminate once the optimality gap is sufficiently small but non zero, giving a near-optimal but not a fully optimal solution. Lines 2 and 3 are variable declarations, Line 4 gives names to the equations of the model, Line 5 defines the objective function, Lines 6 and 7 define the constraints, Line 8 defines the model, and Line 9 directs GAMS to solve it.

The GAMS output that provides information about the optimal solution is:

		LOWER	LEVEL	UPPER	MARGINAL
---- VAR	z	**-INF**	5.000	**+INF**	.
---- VAR	x1	.	2.000	100.000	1.000
---- VAR	x2	.	3.000	100.000	1.000

3.8.2 *Natural Gas-Storage Problem*

The Natural Gas-Storage Problem, which is introduced in Section 3.1.2, has the following GAMS formulation:

```
option OPTCR=0;
variable z;
positive variables y11, y12, y21, y22, y31, y32;
binary variables x1, x2, x3;
equations of, d1, d2, s1, s2, s3;
of .. z =e= y11+6*y12+2*y21+5*y22+3*y31+4*y32-8*x1-9*
    x2-7*x3;
d1 .. y11+y21+y31 =e= 10;
d2 .. y12+y22+y32 =e= 6;
s1 .. y11+y12 =l= 7*x1;
s2 .. y21+y22 =l= 8*x2;
s3 .. y31+y32 =l= 9*x3;
model gs /all/;
solve gs using mip maximizing z;
```

Line 1 indicates that the solution tolerance should be 0, Lines 2–4 are variable declarations, Line 5 gives names to the equations of the model, Line 6 defines the objective function, Lines 7–11 define the constraints, Line 12 defines the model, and Line 13 directs GAMS to solve it.

The GAMS output that provides information about the optimal solution is:

		LOWER	LEVEL	UPPER	MARGINAL
---- VAR	z	**-INF**	49.000	**+INF**	.
---- VAR	y11	.	1.000	**+INF**	.
---- VAR	y12	.	6.000	**+INF**	.
---- VAR	y21	.	.	**+INF**	.
---- VAR	y22	.	.	**+INF**	-2.000
---- VAR	y31	.	9.000	**+INF**	.
---- VAR	y32	.	.	**+INF**	-4.000
---- VAR	x1	.	1.000	1.000	-8.000
---- VAR	x2	.	.	1.000	-1.000
---- VAR	x3	.	1.000	1.000	11.000

3.8.3 *Electricity-Scheduling Problem*

The Electricity-Scheduling Problem, which is introduced in Section 3.1.3, has the
following GAMS formulation:

```
1   option OPTCR=0;
2   variable z;
3   positive variables p1, p2, p3;
4   binary variables x1, x2, x3;
5   equations of, b, l1u, l1d, l2u, l2d, l3u, l3d;
6   of   .. z =e= 2*p1+5*p2+1*p3 + 40*x1+50*x2+35*x3;
7   b    .. p1+p2+p3 =e= 50;
8   l1u  .. p1 =l= 20*x1;
9   l1d  .. p1 =g=  5*x1;
10  l2u  .. p2 =l= 40*x2;
11  l2d  .. p2 =g=  6*x2;
12  l3u  .. p3 =l= 35*x3;
13  l3d  .. p3 =g=  4*x3;
14  model es /all/;
15  solve es using mip minimizing z;
```

Line 1 indicates that the solution tolerance should be 0, Lines 2–4 are variable
declarations, Line 5 gives names to the equations of the model, Line 6 defines the
objective function, Lines 7–13 define the constraints, Line 14 defines the model, and
Line 15 directs GAMS to solve it.

The GAMS output that provides information about the optimal solution is:

```
1                          LOWER      LEVEL      UPPER      MARGINAL

3   ---- VAR z            -INF    140.000    +INF           .
4   ---- VAR p1             .       15.000    +INF           .
5   ---- VAR p2             .          .      +INF        3.000
6   ---- VAR p3             .       35.000    +INF           .
7   ---- VAR x1             .        1.000    1.000       40.000
8   ---- VAR x2             .          .      1.000       50.000
9   ---- VAR x3             .        1.000    1.000         EPS
```

3.8.4 *Oil-Transmission Problem*

The Oil-Transmission Problem, which is introduced in Section 3.1.4, has the following GAMS formulation:

```
1   option OPTCR=0;
2   variable z;
3   positive variables p1, p2;
4   variable f1, f2, f3;
5   binary variables x1, x2;
```

```
 6  equations of, b1, b2, b3, 112, 121, 113, 131, 123, 132
     ;
 7  of .. z =e= 2000*p1+3000*p2-50000*x1-55000*x2;
 8  b1 .. p1+p2 =e= 30;
 9  b2 .. p1-f1-f3 =e=  0;
10  b3 .. p2-f2+f3 =e=  0;
11  112 .. f3 =l=  10;
12  121 .. f3 =g= -10;
13  113 .. f1 =l=  12 + 11*x1;
14  131 .. f1 =g= -12 - 11*x1;
15  123 .. f2 =l=  11 + 12*x2;
16  132 .. f2 =g= -11 - 12*x2;
17  model ot /all/;
18  solve ot using MIP maximizing z;
```

Line 1 indicates that the solution tolerance should be 0, Lines 2–5 are variable declarations, Line 6 gives names to the equations of the model, Line 7 defines the objective function, Lines 8–16 define the constraints, Line 17 defines the model, and Line 18 directs GAMS to solve it.

The GAMS output that provides information about the optimal solution is:

```
 1                        LOWER       LEVEL       UPPER      MARGINAL
 3  ---- VAR z           -INF     35000.000      +INF           .
 4  ---- VAR p1             .             .       +INF    -1000.000
 5  ---- VAR p2             .        30.000       +INF           .
 6  ---- VAR f1          -INF        10.000       +INF           .
 7  ---- VAR f2          -INF        20.000       +INF           .
 8  ---- VAR f3          -INF       -10.000       +INF           .
 9  ---- VAR x1             .             .       1.000    -5.000E+4
10  ---- VAR x2             .         1.000       1.000    -5.500E+4
```

3.8.5 Charging-Station Problem

The Charging-Station Problem, which is introduced in Section 3.1.5, has the following GAMS formulation:

```
 1  option OPTCR=0;
 2  variable z;
 3  binary variables x1, x2, x3;
 4  equations of, n1, n2, n3, n4;
 5  of .. z =e= 10*x1+12*x2+13*x3;
 6  n1 .. 1*x1 + 0*x2 + 1*x3 =g= 1;
 7  n2 .. 0*x1 + 1*x2 + 0*x3 =g= 1;
 8  n3 .. 1*x1 + 1*x2 + 0*x3 =g= 1;
 9  n4 .. 0*x1 + 0*x2 + 1*x3 =g= 1;
10  model cs /all/;
11  solve cs using mip minimizing z;
```

Line 1 indicates that the solution tolerance should be 0, Lines 2 and 3 are variable declarations, Line 4 gives names to the equations of the model, Line 5 defines the objective function, Lines 6–9 define the constraints, Line 10 defines the model, and Line 11 directs GAMS to solve it.

The GAMS output that provides information about the optimal solution is:

		LOWER	LEVEL	UPPER	MARGINAL
3	---- VAR z	**-INF**	25.000	**+INF**	.
4	---- VAR x1	.	.	1.000	10.000
5	---- VAR x2	.	1.000	1.000	12.000
6	---- VAR x3	.	1.000	1.000	13.000

3.8.6 Wind Farm-Maintenance Problem

The Wind Farm-Maintenance Problem, which is introduced in Section 3.1.6, has the following GAMS formulation:

```
option OPTCR=0;
variable z;
binary variables x11,x12,x13,x21,x22,x23,x31,x32,x33;
equations of, f1, f2, f3, t1, t2, t3;
of .. z =e= 10*x11+12*x12+14*x13+9*x21+8*x22+15*x23
           +10*x31+5*x32+15*x33;
f1 .. x11 + x12 + x13 =e= 1;
f2 .. x21 + x22 + x23 =e= 1;
f3 .. x31 + x32 + x33 =e= 1;
t1 .. x11 + x21 + x31 =e= 1;
t2 .. x12 + x22 + x32 =e= 1;
t3 .. x13 + x23 + x33 =e= 1;
model wf /all/;
solve wf using mip minimizing z;
```

Line 1 indicates that the solution tolerance should be 0, Lines 2 and 3 are variable declarations, Line 4 gives names to the equations of the model, Line 5 defines the objective function, Lines 6–11 define the constraints, Line 12 defines the model, and Line 13 directs GAMS to solve it.

The GAMS output that provides information about the optimal solution is:

		LOWER	LEVEL	UPPER	MARGINAL
3	---- VAR z	**-INF**	28.000	**+INF**	.
4	---- VAR x11	.	.	1.000	10.000
5	---- VAR x12	.	.	1.000	12.000
6	---- VAR x13	.	1.000	1.000	14.000
7	---- VAR x21	.	1.000	1.000	9.000
8	---- VAR x22	.	.	1.000	8.000
9	---- VAR x23	.	.	1.000	15.000
10	---- VAR x31	.	.	1.000	10.000

```
11    ----  VAR  x32             .          1.000       1.000        5.000
12    ----  VAR  x33             .            .         1.000       15.000
```

3.9 Exercises

3.1 Steve owns two warehouses containing 120 and 100 photovoltaic panels, respectively. He uses these two warehouses to serve customers in three markets, which have demands for 40, 70, and 50 units, respectively, which must be met exactly. Per-unit transportation costs between each warehouse and each market are given in Table 3.8. Formulate an MILPP to determine how many panels Steve should ship from each warehouse to each market to minimize total transportation cost.

Table 3.8 Per-unit transportation costs [$] for Exercise 3.1

	Warehouse 1	Warehouse 2
Market 1	14	12
Market 2	13	10
Market 3	11	11

3.2 An electricity producer may use some combination of electricity-generating units that are powered by either wind or natural gas. The capacity of each unit (either wind- or natural gas-powered) is 10 kW. The cost of building a wind-powered unit is $2500 and its operating cost is $6/kWh. The cost of building a natural gas-powered unit is $2000 and its operating cost is $80/kWh.

If the demand to be supplied is 100 kW and the producer may not build more than four wind-powered units, how many units of each type should the electricity producer build to minimize the sum of building and generation costs?

3.3 Design, formulate, and solve an assignment problem similar to the Wind Farm-Maintenance Problem, which is introduced in Section 3.1.6.

3.4 Design, formulate, and solve a knapsack problem similar to the Photovoltaic Panel-Repair Problem, which is introduced in Section 3.1.1.

3.5 Design, formulate, and solve a set-covering problem similar to the Charging-Station Problem, which is introduced in Section 3.1.5.

3.6 Linearize the cost function:

$$\text{cost} = \begin{cases} 0, & \text{if } x = 0, \\ 5 + x, & \text{if } 0 < x \leq 10, \end{cases}$$

using integer variables.

3.7 Linearize the piecewise-linear cost function:

$$\text{cost} = \begin{cases} \frac{1}{2}x, & \text{if } 0 \le x \le 1, \\ \frac{1}{2} + \frac{2}{3}(x-1), & \text{if } 1 < x \le 2, \\ \frac{7}{6} + \frac{1}{3}(x-2), & \text{if } 2 < x \le 3, \end{cases}$$

using integer variables.

3.8 Design and formulate an instance of an Alternative Constraint, as outlined in Section 3.3.4.

3.9 Consider the MILPP:

$$\min_{x_1,x_2} z = -x_1 - x_2$$

$$\text{s.t. } 2x_1 + x_2 \le 13$$

$$x_1 + 2x_2 \le 12$$

$$x_1, x_2 \ge 0$$

$$x_1, x_2 \in \mathbb{Z}.$$

The linear relaxation of this problem has the optimal solution $x^0 = (14/3, 11/3)^\top$. If the Branch-and-Bound Algorithm is to be applied where x_1 is the branching variable, determine the new LPPs that would be added to $\varXi$.

3.10 Generate a cutting plane for the MILPP in Exercise 3.9.

3.11 Solve the Photovoltaic Panel-Repair Problem using the Cutting-Plane Algorithm.

3.12 Solve the Natural Gas-Storage Problem using the Branch-and-Bound Algorithm.

References

1. Bertsimas D, Weismantel R (2005) Optimization over integers. Dynamic Ideas, Belmont
2. Bixby RE (2002) Solving real-world linear programs: a decade and more of progress. Oper Res 50:3–15
3. Brook A, Kendrick D, Meeraus A (1988) GAMS – a user's guide. ACM, New York. www.gams.com
4. Castillo E, Conejo AJ, Pedregal P, García R, Alguacil N (2002) Building and solving mathematical programming models in engineering and science. Pure and applied mathematics series, Wiley, New York
5. Conforti M, Cornuejols G, Zambelli G (2016) Integer programming. Springer, New York
6. Floudas CA (1995) Nonlinear and mixed-integer optimization: fundamentals and applications. Oxford University Press, Oxford

7. Fourer R, Gay DM, Kernighan BW (2002) AMPL: a modeling language for mathematical programming, 2nd edn. Cengage Learning, Boston. www.ampl.com
8. Gomory RE (1960) An algorithm for the mixed integer problem. RAND Corporation, Santa Monica
9. Gurobi Optimization, Inc. (2010) Gurobi optimizer reference manual, Version 3.0. Houston, Texas. www.gurobi.com
10. IBM ILOG CPLEX Optimization Studio (2016) CPLEX user's manual, Version 12 Release 7. IBM Corp. www.cplex.com
11. Lubin M, Dunning I (2015) Computing in operations research using Julia. INFORMS J Comput 27(2):238–248. www.juliaopt.org
12. Papadimitriou CH, Steiglitz K (1998) Combinatorial optimization: algorithms and complexity. Dover Publications, Mineola
13. Rao SS (2009) Engineering optimization: theory and practice, 4th edn. Wiley, Hoboken
14. Roelofs M, Bisschop J (2016) AIMMS the language reference, AIMMS B.V., Haarlem, The Netherlands. www.aimms.com
15. Wolsey LA (1998) Integer programming. Wiley interscience series in discrete mathematics and optimization, Wiley, New York
16. Wolsey LA, Nemhauser GL (1999) Integer and combinatorial optimization. Wiley interscience series in discrete mathematics and optimization, Wiley, New York

Chapter 4
Nonlinear Optimization

Chapters 2 and 3 are devoted to problems with a linear structure. More specifically, the problems studied there all have an objective function and constraints that are linear in the decision variables. Although linear optimization problems are very common and can model a variety of real-world problems, we are sometimes faced with modeling a system that includes important nonlinearities. When either the constraints or objective function of an optimization problem are nonlinear in the decision variables, we say that we are faced with a nonlinear optimization problem or nonlinear programming problem (NLPP).

In this chapter, we begin by first introducing some nonlinear optimization problems, then discuss methods to solve NLPPs. The example nonlinear optimization problems that we introduce draw on a wide swath of problem domains spanning finance, design, planning, and energy systems. We then discuss an analytical approach to solving NLPPs. This method uses what are called **optimality conditions**—properties that an optimal set of decision variables has to exhibit. Optimality conditions can be likened to the technique to solve an LPP that draws on the Strong-Duality Property, which is discussed in Section 2.7.4. In Chapter 5 we introduce another approach to solving NLPPs, **iterative solution algorithms**, which are techniques used by software packages to solve NLPPs. These iterative algorithms can be likened to using the Simplex method to solve linear optimization problems or the branch-and-bound or cutting-plane methods to solve mixed-integer linear problems.

4.1 Motivating Examples

In this section, we present a variety of nonlinear optimization problems, which are motivated by a mixture of geometric, mechanical, and electrical systems.

© Springer International Publishing AG 2017

R. Sioshansi and A.J. Conejo, *Optimization in Engineering*,

Springer Optimization and Its Applications 120, DOI 10.1007/978-3-319-56769-3_4

4.1.1   Geometric Examples

4.1.1.1   Packing-Box Problem

A company must determine the dimensions of a cardboard box to maximize its volume. The box can use at most $60\,\text{cm}^2$ of cardboard. For structural reasons, the bottom and top faces of the box must be of triple weight (*i.e.*, three pieces of cardboard).

There are three decision variables in this problem, h, w, and d, which are the height, width, and depth of the box in cm, respectively (see Figure 4.1).

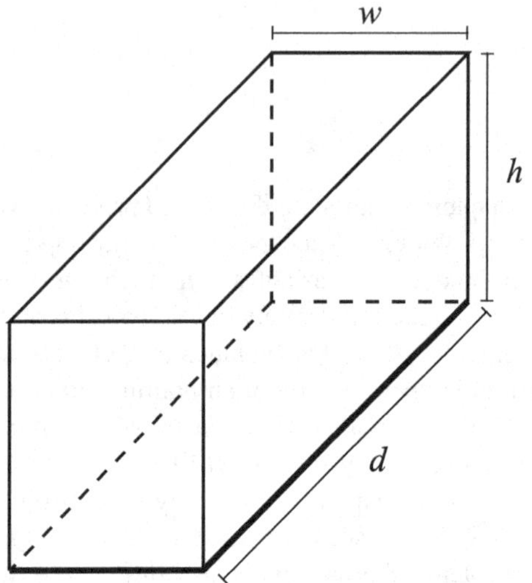

Fig. 4.1 Illustration of the Packing-Box Problem

The objective is to maximize the volume of the box:

$$\max_{h,w,d}\ hwd.$$

There are two types of constraints. First, we must ensure that the box uses no more than $60\,\text{cm}^2$ of cardboard, noting that the top and bottom of the box are of triple weight:

$$2wh + 2dh + 6wd \le 60.$$

The second type of constraint ensures that the dimensions of the box are non-negative, as negative dimensions are physically impossible:

$$w, h, d \ge 0.$$

Putting all of this together, the NLPP can be written as:

$$\max_{h,w,d} \; hwd$$

$$\text{s.t. } 2wh + 2dh + 6wd \leq 60$$

$$w, h, d \geq 0.$$

4.1.1.2 Awning Problem

A box, which is h m high and w m wide, is placed against the side of a building (see Figure 4.2). The building owner would like to construct an awning of minimum length that completely covers the box.

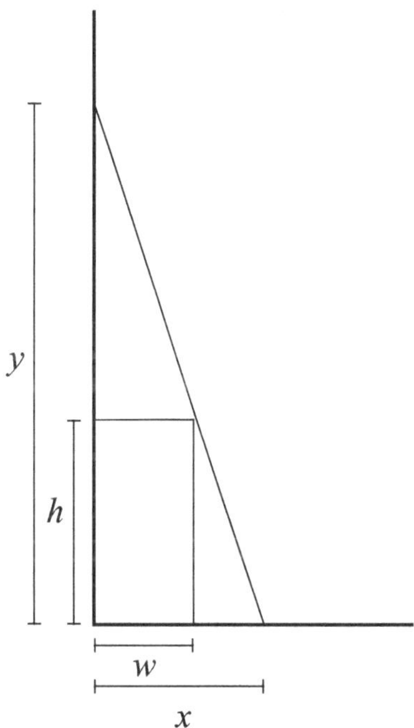

Fig. 4.2 Illustration of the Awning Problem

There are two decision variables in this problem. The first, x, measures the distance between the building wall and the point at which the awning is anchored to the ground, in meters. The second, y, measures the height of the anchor point of the awning to the building wall, also in meters.

The objective is to minimize the length of the awning. From the Pythagorean theorem this is:

$$\min_{x,y} \; \sqrt{x^2 + y^2}.$$

We have two types of constraints in this problem. The first ensures that the upper-right corner of the box is below the awning (which ensures that the box is wholly contained by the awning). To derive this constraint, we compute the height of the awning w m away from the wall as:

$$y - \frac{y}{x}w.$$

To ensure that the upper-right corner of the box is below the awning, this point must be at least h m high, giving our first constraint:

$$y - \frac{y}{x}w \geq h.$$

We must also ensure that the distances between the anchor points of the awning and the building and ground are non-negative:

$$x, y \geq 0.$$

Thus, our NLPP is:

$$\min_{x,y} \sqrt{x^2 + y^2}$$

$$\text{s.t. } y - \frac{y}{x}w \geq h$$

$$x, y \geq 0.$$

4.1.1.3   Facility-Location Problem

A retailer must decide where to place a single distribution center to service N retail locations in a region (see Figure 4.3). Retail location n is at coordinates (x_n, y_n). Each week V_n trucks leave the distribution center carrying goods to retail location n, and then return to the distribution center. These trucks can all travel on straight paths from the distribution center to the retail location. The company would like to determine where to place the distribution center to minimize the total distance that all of the trucks must travel each week.

There are two variables in this problem, a and b, which denote the coordinates of the distribution center.

The objective is to minimize the total distance traveled by all of the trucks. As illustrated in Figure 4.3, each truck that serves retail location n must travel a distance of:

$$\sqrt{(x_n - a)^2 + (y_n - b)^2},$$

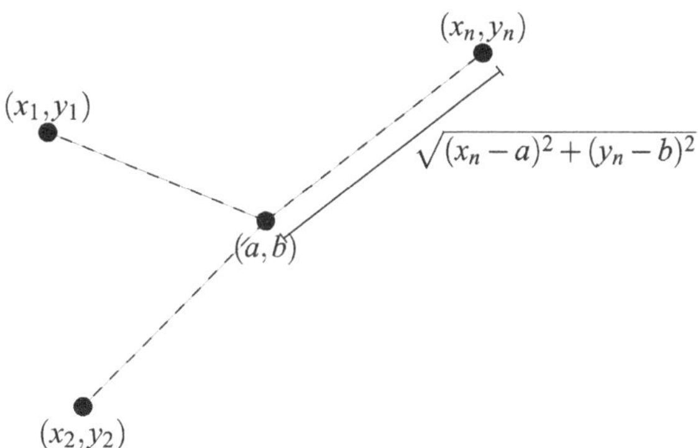

Fig. 4.3 Illustration of the Facility-Location Problem

to get there from the distribution center. Because V_n trucks must travel to retail location n and they make a roundtrip, the total distance covered by all of the trucks to retail location n is:

$$2V_n\sqrt{(x_n - a)^2 + (y_n - b)^2}.$$

The objective is to minimize this total distance, or:

$$\min_{a,b} \sum_{n=1}^{N} 2V_n\sqrt{(x_n - a)^2 + (y_n - b)^2}.$$

Because the distribution center can be placed anywhere, this problem has no constraints.

Thus, our NLPP is simply:

$$\min_{a,b} \sum_{n=1}^{N} 2V_n\sqrt{(x_n - a)^2 + (y_n - b)^2}.$$

4.1.1.4 Cylinder Problem

A brewery would like to design a cylindrical vat (see Figure 4.4), in which to brew beer. The material used to construct the top of the vat costs $\$c_1$ per m^2. The material cost of the side and bottom of the vat is proportional to the volume of the vat. This is because the side and bottom must be reinforced to hold the volume of liquid that is brewed in the vat. This material cost is $\$c_2 V$ per m^2, where V is the volume of the vat. The value of the beer that is produced in the vat during its usable life is proportional to the volume of the vat, and is given by $\$N$ per m^3. The brewery would like to design the vat to maximize the net profit earned over its usable life.

This NLPP has two decision variables, h and r, which measure the height and radius, respectively, of the vat in meters.

The objective is to maximize the value of the beer brewed less the cost of building the vat. The value of the beer produced by the vat is given by:

$$N\pi r^2 h.$$

The cost of top of the vat is:

$$c_1 \pi r^2.$$

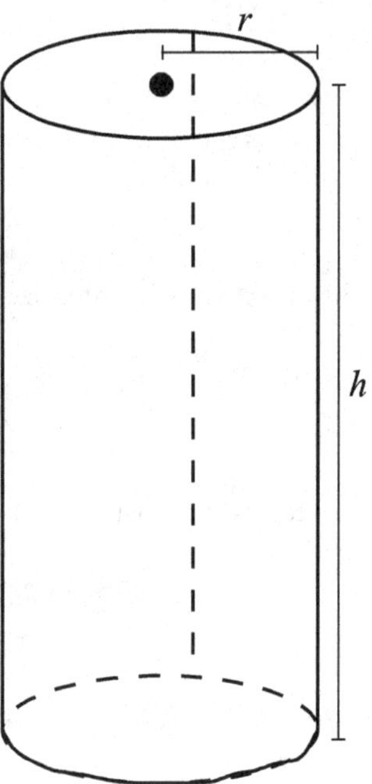

Fig. 4.4 Illustration of the Cylinder Problem

The per-unit material cost (in \$/m^3) of the side and bottom of the vat is given by:

$$c_2 \pi r^2 h.$$

Thus the bottom of the vat costs:

$$c_2 \pi r^2 h \pi r^2,$$

and the side costs:

$$c_2 \pi r^2 h 2\pi r h.$$

Thus, the objective, which is to maximize the value of the beer produced by the vat less its material costs, is:

$$\max_{h,r} \; N\pi r^2 h - c_1 \pi r^2 - c_2 \pi r^2 h \cdot (\pi r^2 + 2\pi rh).$$

There is one type of constraint, which ensures that the vat has non-negative dimensions:

$$r, h \geq 0.$$

Putting all of this together, the NLPP can be written as:

$$\max_{h,r} \; N\pi r^2 h - c_1 \pi r^2 - c_2 \pi r^2 h \cdot (\pi r^2 + 2\pi rh)$$

$$\text{s.t. } r, h \geq 0.$$

4.1.2 Mechanical Examples

4.1.2.1 Machining-Speed Problem

A company produces widgets that need to be machined. Each widget needs to spend t_p minutes being prepared before being machined. The time that each widget spends being machined depends on the machine speed. This machining time is given by:

$$\frac{\lambda}{v},$$

where λ is a constant and v is the machine speed. The tool used to machine the widgets needs to be replaced periodically, due to wear. Each time the tool is replaced the machine is out of service (and widgets cannot be produced) for t_c minutes. The amount of time it takes for the tool to be worn down depends on the machine speed, and is given by:

$$\left(\frac{C}{v}\right)^{1/n},$$

where C and n are constants. The cost of replacement tools is negligible. Each widget sells for a price of $\$p$ and uses $\$m$ worth of raw materials. Moreover, every minute of production time incurs $\$h$ of overhead costs. The company would like to determine the optimal machine speed that maximizes the average per-minute profit of the widget-machining process.

This problem has a single decision variable, v, which is the machine speed.

To derive the objective, we first note that each widget produced earns $\$(p - m)$ of revenue less material costs. Dividing this by the number of widgets produced per minute gives an expression for net revenue per minute. Producing each widget takes three steps. The first is preparation, which takes t_p minutes. Each widget must also be machined, which takes:

$$\frac{\lambda}{v},$$

minutes. Finally, each widget produced reduces the usable life of the machine tool by the fraction:

$$\frac{\lambda/v}{(C/v)^{1/n}}.$$

Thus, each widget contributes:

$$t_c \frac{\lambda/v}{(C/v)^{1/n}},$$

minutes toward tool-replacement time. Therefore, each widget takes a total of:

$$t_p + \frac{\lambda}{v} + t_c \frac{\lambda/v}{(C/v)^{1/n}},$$

minutes of time to produce. The objective, which is to maximize average per-minute profits of the widget-machining process, is, thus, given by:

$$\max_v \ \frac{p - m}{t_p + \frac{\lambda}{v} + t_c \frac{\lambda/v}{(C/v)^{1/n}}} - h,$$

where we also subtract the overhead costs.

The only constraint on this problem is that the machine speed needs to be non-negative:

$$v \geq 0.$$

Thus, our NLPP is:

$$\max_v \ \frac{p - m}{t_p + \frac{\lambda}{v} + t_c \frac{\lambda/v}{(C/v)^{1/n}}} - h$$

$$\text{s.t. } v \geq 0.$$

4.1.2.2 Hanging-Chain Problem

A chain consisting of N links, each of which is 10 cm in length and 50 g in mass, hangs between two points a distance L cm apart, as shown in Figure 4.5. The chain will naturally hang in a configuration that minimizes its potential energy. Formulate a nonlinear optimization problem to determine the configuration of the chain links when it is hanging.

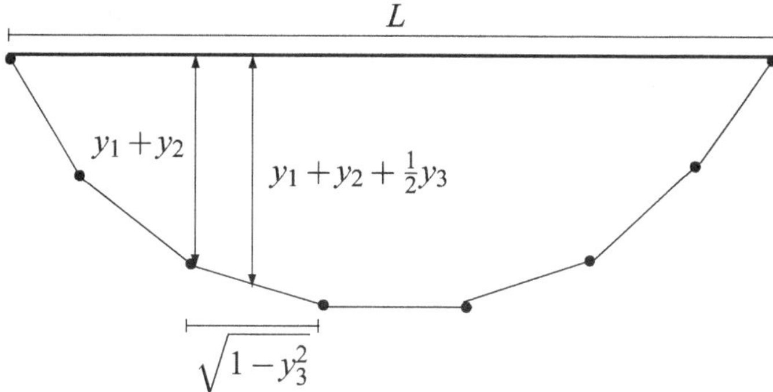

Fig. 4.5 Illustration of the Hanging-Chain Problem

To formulate this problem, we define variables, $y_1, y_2, \ldots, y_N$. We let y_n measure the vertical displacement of the right end of the nth chain link from the right end of the $(n - 1)$th link in cm (see Figure 4.5).

The objective of the NLPP is to minimize the potential energy of the chain. This is, in turn, the sum of the potential energies of each chain link. The potential energy of chain link n is given by the product of its mass, the gravitational constant, and the vertical displacement of its midpoint from the ground. The displacement of the midpoint nth link is given by:

$$y_1 + y_2 + \cdots + y_{n-1} + \frac{1}{2}y_n.$$

This is illustrated for the third link in Figure 4.5. Thus, the objective of the NLPP is:

$$\min_{y_1,\ldots,y_N} 50g \left[\frac{1}{2}y_1 + \left(y_1 + \frac{1}{2}y_2 \right) + \cdots + \left(y_1 + \cdots + y_{N-1} + \frac{1}{2}y_N \right) \right]$$

where g is the gravitational constant. This objective can be further simplified to:

$$\min_{y} 50g \sum_{n=1}^{N} \left(N - n + \frac{1}{2} \right) y_n,$$

by adding and simplifying terms.

This problem has two constraints. The first is to ensure that the vertical displacements of the chain links all sum to zero, meaning that its two anchor points are at the same height:

$$\sum_{n=1}^{N} y_n = 0.$$

The other constraint ensures that the horizontal displacements of the chain links sum to L, which ensures that the two anchor points of the chain are L cm apart. Because we know that chain link n has a length of 10 cm and a vertical displacement (relative to the $(n-1)$th link) of y_n, we can use the Pythagorean theorem to compute its horizontal displacement (relative to the $(n-1)$th link) as:

$$\sqrt{10 - y_n^2}.$$

Because this constraint requires that these horizontal displacements sum to L it can be written as:

$$\sum_{n=1}^{N} \sqrt{10 - y_n^2} = L.$$

Putting all of this together, the NLPP can be written as:

$$\min_{y} \; 50g \sum_{n=1}^{N} \left(N - n + \frac{1}{2} \right) y_n$$

$$\text{s.t.} \; \sum_{n=1}^{N} y_n = 0$$

$$\sum_{n=1}^{N} \sqrt{10 - y_n^2} = L.$$

4.1.3 Planning Examples

4.1.3.1 Return-Maximization Problem

An investor has a sum of money to invest in N different assets. Asset n generates a rate of return, which is given by r_n. Define:

$$\bar{r}_n = \mathbb{E}[r_n],$$

as the expected rate of return of asset n. The risk of the portfolio is measured by the covariance between the returns of the different assets. Let:

$$\sigma_{n,m} = \sum_{n=1}^{N} \sum_{m=1}^{N} \mathbb{E}[(r_n - \mathbb{E}[r_n])(r_m - \mathbb{E}[r_m])],$$

be the covariance between the returns of assets n and m. By convention the covariance between the return of asset n and itself, $\sigma_{n,n}$, is equal to σ_n^2, which is the variance of the return of asset n. The investor would like to determine how to allocate the sum of money to maximize the expected return on investment while limiting the variance of the portfolio to be no more than $\bar{s}$.

To formulate this problem, we define N decision variables, $w_1, w_2, \ldots, w_N$, where w_n represents the fraction of the money available that is invested in asset n.

The objective function is to maximize the expected return on investment:

$$\max_{w_1,\ldots,w_N} \sum_{n=1}^{N} \mathbb{E}[r_n w_n] = \sum_{n=1}^{N} \mathbb{E}[r_n] w_n = \sum_{n=1}^{N} \bar{r}_n w_n.$$

There are three types of constraints. First, we must ensure that the portfolio variance is no greater than $\bar{s}$. We can compute the variance of the portfolio as:

$$\sum_{n=1}^{N} \sum_{m=1}^{N} \mathbb{E}[(r_n w_n - \mathbb{E}[r_n w_n])(r_m w_m - \mathbb{E}[r_m w_m])].$$

Factoring the w's out of the expectations gives:

$$\sum_{n=1}^{N} \sum_{m=1}^{N} \mathbb{E}[(r_n - \mathbb{E}[r_n])(r_m - \mathbb{E}[r_m])] w_n w_m = \sum_{n=1}^{N} \sum_{m=1}^{N} \sigma_{n,m} w_n w_m,$$

where the term on the right-hand side of the equality follows from the definition of the σ's. Thus, our portfolio-variance constraint is:

$$\sum_{n=1}^{N} \sum_{m=1}^{N} \sigma_{n,m} w_n w_m \leq \bar{s}.$$

Next, we must ensure that all of the money is invested, meaning that the w's sum to one:

$$\sum_{n=1}^{N} w_n = 1.$$

We, finally, must ensure that the amounts invested are non-negative:

$$w_n \geq 0, \forall\, n = 1, \ldots, N.$$

Taking these together, our NLPP is:

$$\max_{w} \sum_{n=1}^{N} \bar{r}_n w_n$$

$$\text{s.t.} \sum_{n=1}^{N} \sum_{m=1}^{N} \sigma_{n,m} w_n w_m \leq \bar{s}$$

$$\sum_{n=1}^{N} w_n = 1$$

$$w_n \geq 0, \forall\, n = 1, \ldots, N.$$

4.1.3.2 Variance-Minimization Problem

Consider the basic setup in the Return-Maximization Problem, which is introduced in Section 4.1.3.1. The investor now wishes to select a portfolio that has minimal variance while achieving at least $\bar{R}$ as the expected rate of return.

We retain the same decision variables in this problem that we define in the Return-Maximization Problem, which is introduced in Section 4.1.3.1. Specifically, define N decision variables, $w_1, w_2, \ldots, w_N$, where w_n represents the fraction of the money available that is invested in asset n.

The objective is to minimize portfolio variance. Using the expression for portfolio variance that is derived in Section 4.1.3.1, the objective function is:

$$\min_{w_1, \ldots, w_N} \sum_{n=1}^{N} \sum_{m=1}^{N} \sigma_{n,m} w_n w_m.$$

We again have three types of constraints. The first ensures that the portfolio achieves the desired minimum expected return. We can, again, use the expression in Section 4.1.3.1 for expected portfolio return to write this constraint as:

$$\sum_{n=1}^{N} \bar{r}_n w_n \geq \bar{R}.$$

We must also ensure that the full sum of money is invested:

$$\sum_{n=1}^{N} w_n = 1,$$

and that the amounts invested are non-negative:

$$w_n \geq 0, \forall\, n = 1, \ldots, N.$$

Taking these together, our NLPP is:

$$\min_{w} \sum_{n=1}^{N} \sum_{m=1}^{N} \sigma_{n,m} w_n w_m$$

$$\text{s.t.} \sum_{n=1}^{N} \bar{r}_n w_n \geq \bar{R}$$

$$\sum_{n=1}^{N} w_n = 1$$

$$w_n \geq 0, \forall\, n = 1, \ldots, N.$$

4.1.3.3 Inventory-Planning Problem

A store needs to plan its inventory of three different sizes of T-shirts—small, medium, and large—for the next season. Small T-shirts cost \$1 each, medium ones \$2 each, and large ones \$4 each. Small T-shirts sell for \$10 each, medium ones for \$12, and large ones for \$13. Although the store does not know the demand for the three T-shirt sizes, it forecasts that the demands during the next season for the three are independent and uniformly distributed between 0 and 3000. T-shirts that are unsold at the end of the season must be discarded, which is costless. The store wants to make its ordering decision to maximize its expected profits from T-shirt sales.

We define three decision variables for this problem, x_s, x_m, and x_l, which denote the number of small, medium, and large T-shirts ordered, respectively.

The objective is to maximize expected profits. To compute revenues from small T-shirt sales, we let a_s and D_s denote the number of small T-shirts sold and demanded, respectively. We can then express the expected revenues from small T-shirt sales, as a function of x_s as:

$$10\mathbb{E}[a_s] = 10(\mathbb{E}[a_s|D_s > x_s]\text{Prob}\{D_s > x_s\} + \mathbb{E}[a_s|D_s \leq x_s]\text{Prob}\{D_s \leq x_s\}).$$

We know that if $D_s > x_s$, then the number of shirts sold is $a_s = x_s$. Furthermore, because D_s is uniformly distributed, we know that this occurs with probability:

$$1 - \frac{x_s}{3000}.$$

Otherwise, if $D_s \leq x_s$, then the number of shirts sold is $a_s = D_s$. We compute the expected sales quantity in this case as:

$$\int_0^{x_s} \frac{1}{3000} D\, dD = \frac{1}{2} \cdot \frac{1}{3000} x_s^2,$$

because we know the uniformly distributed demand has a density function of $1/3000$. Combining these observations, we can compute the expected revenues from small T-shirt sales as:

$$10\mathbb{E}[a_s] = 10(\mathbb{E}[a_s|D_s > x_s]\mathrm{Prob}\{D_s > x_s\} + \mathbb{E}[a_s|D_s \leq x_s]\mathrm{Prob}\{D_s \leq x_s\})$$
$$= 10\left[x_s \cdot \left(1 - \frac{x_s}{3000}\right) + \frac{1}{6000} x_s^2\right]$$
$$= 10\left(x_s - \frac{x_s^2}{6000}\right).$$

Because the expected revenues from medium and large T-shirt sales have a similar form, the total expected revenue is given by:

$$10\left(x_s - \frac{x_s^2}{6000}\right) + 12\left(x_m - \frac{x_m^2}{6000}\right) + 13\left(x_l - \frac{x_l^2}{6000}\right).$$

Thus, the objective, which is to maximize expected profits, is given by:

$$\max_{x_s, x_m, x_l} 10\left(x_s - \frac{x_s^2}{6000}\right) + 12\left(x_m - \frac{x_m^2}{6000}\right) + 13\left(x_l - \frac{x_l^2}{6000}\right) - x_s - 2x_m - 4x_l.$$

The only constraints are to ensure that the number of shirts ordered are non-negative and less than 3000 (because there is a maximum possible demand of 3000 units for each type of shirt):

$$0 \leq x_s, x_l, x_m \leq 3000.$$

Thus, our NLPP is:

$$\max_x 10\left(x_s - \frac{x_s^2}{6000}\right) + 12\left(x_m - \frac{x_m^2}{6000}\right) + 13\left(x_l - \frac{x_l^2}{6000}\right) - x_s - 2x_m - 4x_l$$

s.t. $0 \leq x_s, x_l, x_m \leq 3000.$

4.1.4 Energy-Related Examples

4.1.4.1 Economic-Dispatch Problem

An electricity transmission network consists of N nodes. There is a generator, which can produce energy, attached to each node. The cost of producing q_n MW from node n over the next hour is given by:

$$c_n(q_n) = a_{n,0} + a_{n,1}q_n + a_{n,2}q_n^2,$$

and the generator at node n must produce at least Q_n^- but no more than Q_n^+ MW. Each node also has demand for energy, with D_n MW of demand over the next hour at node n. The nodes in the network are connected to each other by transmission lines and we let Ω_n denote the set of nodes that are directly connected to node n by a transmission line. For any node n and $m \in \Omega_n$, the power flow over the line directly connecting nodes n and m is equal to:

$$f_{n,m} = Y_{n,m} \sin(\theta_n - \theta_m),$$

where $Y_{n,m}$ is the (constant) electrical admittance of the line connecting nodes n and m, and θ_n and θ_m are the phase angles of nodes n and m, respectively. We have that $Y_{n,m} = Y_{m,n}$ and we use the sign convention that $f_{n,m} = -f_{m,n}$ is positive if there is a net power flow from node n to m. There is a limit on how much power can flow along each transmission line. Let $L_{n,m} = L_{m,n} > 0$ denote the flow limit on the line directly connecting node n to node m.

The demand at node n can either be satisfied by the generator at node n or by power imported from other nodes. We assume that the network is 'lossless,' meaning that the total power produced at the N nodes must equal the total demand among the N nodes. The goal of the power system operator is to determine how much to produce from each generator and how to set the power flows and phase angles to serve customers' demands at minimum total cost.

There are three types of variables in this problem. The first are the production levels at each node, which we denote by $q_1, \ldots, q_N$, letting q_n be the MWh of energy produced at node n. The second are the phase angles at each of the nodes, which we denote by $\theta_1, \ldots, \theta_N$, with θ_n representing the phase angle of node n. The third is the flow on the lines, which we denote by $f_{n,m}$ for all $n - 1, \ldots, N$ and all $m \in \Omega_n$. We let $f_{n,m}$ denote the flow, in MWh, on the line connecting nodes n and m. As noted before, we use the sign convention that if $f_{n,m} > 0$ this means that there is a net flow from node n to node m and $f_{n,m} < 0$ implies a net flow from node m to node n.

The objective is to minimize the total cost of producing energy to serve the system's demand:

$$\min_{q,\theta,f} \sum_{n=1}^{N} c_n(q) = \sum_{n=1}^{N} \left[a_{n,0} + a_{n,1}q_n + a_{n,2}q_n^2 \right].$$

There are five set of constraints in the problem. The first set ensures that the local demand at each node is satisfied by either local supply or energy imported from other nodes:

$$D_n = q_n + \sum_{m \in \Omega_n} f_{m,n}, \ \forall \, n = 1, \ldots, N.$$

Next, we need to ensure that the total amount generated equals the amount demanded. This constraint arises because the network is assumed to be lossless. Otherwise, without this constraint, energy could either be generated and not consumed anywhere in the network or it could be consumed without having been produced anywhere in the network. This constraint is written as:

$$\sum_{n=1}^{N} D_n = \sum_{n=1}^{N} q_n.$$

We also have equalities that define the flow on each line in terms of the phase angles at the end of the line:

$$f_{n,m} = Y_{n,m} \sin(\theta_n - \theta_m), \ \forall \, n = 1, \ldots, N, m \in \Omega_n.$$

We must also ensure that the flows do not violate their limits:

$$f_{n,m} \le L_{n,m}, \ \forall \, n = 1, \ldots, N, m \in \Omega_n,$$

and that the power generated at each node is between its bounds:

$$Q_n^- \le q_n \le Q_n^+, \ \forall \, n = 1, \ldots, N.$$

Thus, our NLPP is:

$$\min_{q,\theta,f} \ \sum_{n=1}^{N} \left[a_{n,0} + a_{n,1} q_n + a_{n,2} q_n^2 \right]$$

$$\text{s.t. } D_n = q_n + \sum_{m \in \Omega_n} f_{m,n}, \ \forall \, n = 1, \ldots, N$$

$$\sum_{n=1}^{N} D_n = \sum_{n=1}^{N} q_n$$

$$f_{n,m} = Y_{n,m} \sin(\theta_n - \theta_m), \ \forall \, n = 1, \ldots, N, m \in \Omega_n$$

$$f_{n,m} \le L_{n,m}, \ \forall \, n = 1, \ldots, N, m \in \Omega_n$$

$$Q_n^- \le q_n \le Q_n^+, \ \forall \, n = 1, \ldots, N.$$

One interesting property of this optimization problem, which is worth noting, is that there are decision variables that do not appear in the objective function (specifically, the θ's and f's). Despite this, θ and f must be listed as decision variables. If they are not, that would imply that their values are fixed, meaning (in the context of this problem) that the system operator does not have the ability to decide the flows on the transmission lines or the phase angles at the nodes. This would be an overly restrictive problem formulation, given the problem description.

4.2 Types of Nonlinear Optimization Problems

In the following sections of this chapter, we concern ourselves with three broad classes of successively more difficult nonlinear optimization problems. Moreover, when analyzing these problems it is always helpful to put them into a standard form. By doing so, we are able to apply the same generic tools to solve these classes of NLPPs. These standard forms are akin to the standard and canonical forms of LPPs, which are introduced in Section 2.2.

We now introduce these three types of optimization problems and refer back to the motivating problems in Section 4.1 to give examples of each. We also use these examples to demonstrate how problems can be converted to these three standard forms.

4.2.1 Unconstrained Nonlinear Optimization Problems

An **unconstrained nonlinear optimization problem** has an objective function that is being minimized, but does not have any constraints on what values the decision variables can take. An unconstrained nonlinear optimization problem can be generically written as:

$$\min_{x \in \mathbb{R}^n} \; f(x),$$

where $f(x) : \mathbb{R}^n \to \mathbb{R}$ is the objective function.

Among the motivating examples given in Section 4.1, the Facility-Location Problem in Section 4.1.1.3 is an example of an unconstrained problem. The objective function of the Facility-Location Problem that is given in Section 4.1.1.3 is already a minimization. Thus, no further work is needed to convert this problem into a standard-form unconstrained problem. As discussed in Section 2.2.2.1, a maximization problem can be converted to minimization problem by multiplying the objective function through by -1.

4.2.2 Equality-Constrained Nonlinear Optimization Problems

An **equality-constrained nonlinear optimization problem** has an objective function that is being minimized and a set of m equality constraints that have zeros on their right-hand sides. An equality-constrained nonlinear optimization problem can be generically written as:

$$\min_{x \in \mathbb{R}^n} \; f(x)$$

$$\text{s.t. } h_1(x) = 0$$

$$h_2(x) = 0$$

$$\vdots$$

$$h_m(x) = 0,$$

where $f(x)$ is the objective function and $h_1(x), h_2(x), \ldots, h_m(x)$ are the m equality-constraint functions. The objective and constraint functions map the n-dimensional vector, x, to scalar values.

Among the examples that are given in Section 4.1, the Hanging-Chain Problem in Section 4.1.2.2 is an example of an equality-constrained problem. The objective function that is given in Section 4.1.2.2 is already a minimization, thus it does not have to be manipulated to put it into the standard form for an equality-constrained problem. We can convert the constraints into standard form by subtracting all of the terms from one side of the equality. This gives the following standard-form NLPP for the Hanging-Chain Problem:

$$\min_{y \in \mathbb{R}^N} f(y) = 50g \sum_{n=1}^{N} \left(N - n + \frac{1}{2} \right) y_n$$

$$\text{s.t. } h_1(y) = \sum_{n=1}^{N} y_n = 0$$

$$h_2(y) = \sum_{n=1}^{N} \sqrt{1 - y_n^2} - L = 0.$$

4.2.3 Equality- and Inequality-Constrained Nonlinear Optimization Problems

An **equality- and inequality-constrained nonlinear optimization problem** has an objective function that is being minimized, a set of m equality constraints that have zeros on their right-hand sides, and a set of r less-than-or-equal-to constraints that have zeros on their right-hand sides. An equality- and inequality-constrained problem can be generically written as:

$$\min_{x\in\mathbb{R}^n} \ f(x)$$

$$\text{s.t. } h_1(x) = 0$$
$$h_2(x) = 0$$
$$\vdots$$
$$h_m(x) = 0$$
$$g_1(x) \le 0$$
$$g_2(x) \le 0$$
$$\vdots$$
$$g_r(x) \le 0,$$

where $f(x)$ is the objective function, $h_1(x), h_2(x), \ldots, h_m(x)$ are the m equality-constraint functions, and $g_1(x), g_2(x), \ldots, g_r(x)$ are the r inequality-constraint functions. The objective and constraint functions all map the n-dimensional vector, x, to scalar values. Note that one could have a problem with only inequality constraints, i.e., $m = 0$, meaning that there are no equality constraints.

All of the other motivating examples given in Section 4.1, that are not categorized as being unconstrained or equality constrained, are examples of equality- and inequality-constrained problems. To demonstrate how a problem can converted to the generic form, take as an example the Return-Maximization Problem, which is given in Section 4.1.3.1. We convert the maximization to a minimization by multiplying the objective function through by -1. The constraints are similarly manipulated to yield the standard form, giving the following standard-form NLPP for the Return-Maximization Problem:

$$\min_{w\in\mathbb{R}^N} \ f(w) = -\sum_{n=1}^{N} \bar{r}_n w_i$$

$$\text{s.t. } h_1(w) = \sum_{n=1}^{N} w_n - 1 = 0$$

$$g_1(w) = \sum_{n=1}^{N}\sum_{m=1}^{N} \sigma_{n,m} w_n w_m - \bar{s} \le 0$$

$$g_2(w) = -w_1 \le 0$$

$$\vdots$$

$$g_{N+1}(w) = -w_N \le 0.$$

4.3 Global and Local Minima

When we solve an optimization problem, we want to find a feasible solution that makes the objective as small as possible among all feasible solutions. In other words, we are searching for what is known as a **global minimum**. The difficulty that we encounter is that for most problems, we can only find what are known as **local minima**. These concepts are both defined in the following subsections. As we discuss below, one can find a global minimum by exhaustively searching for all local minima and picking the one that gives the best objective-function value. This is the approach we must take to solve most nonlinear optimization problems.

4.3.1 Global Minima

Given a nonlinear optimization problem, a feasible solution, x^*, is a **global minimum** if $f(x^*) \leq f(x)$ for all other feasible values of x.

Figure 4.6 illustrates the global minimum of a parabolic objective function. Because there are no restriction on what values of x may be chosen, the example in Figure 4.6 is an unconstrained problem. Clearly, x^* gives the smallest objective value and satisfies the definition of a global minimum.

Fig. 4.6 The global minimum of a parabolic objective function

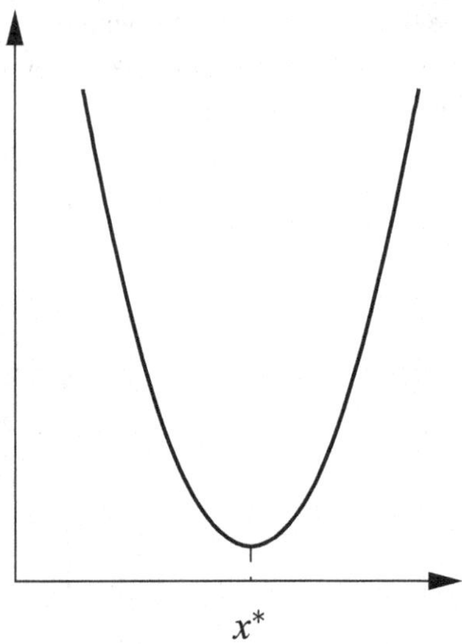

Figure 4.7 demonstrates the effect of adding the inequality constraint, $x \geq \hat{x}$, to the optimization problem that is illustrated in Figure 4.6. Although x^* still gives the smallest objective-function value, it is no longer feasible. Thus, it does not satisfy the definition of a global minimum. One can tell from visual inspection that $\hat{x}$ is in fact the global minimum of the parabolic objective function when the constraint is added.

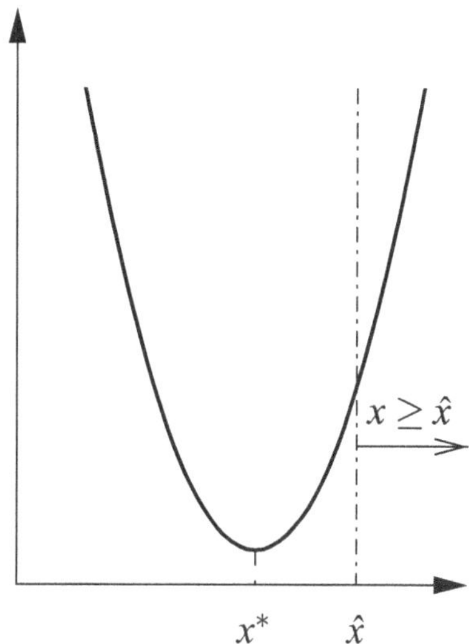

Fig. 4.7 The global minimum of a parabolic objective function with an inequality constraint

It is also important to note that as with a linear or mixed-integer linear optimization problem, a nonlinear problem can have multiple global minima. This is illustrated in Figure 4.8, where we see that all values of x between x^1 and x^2 are global minima.

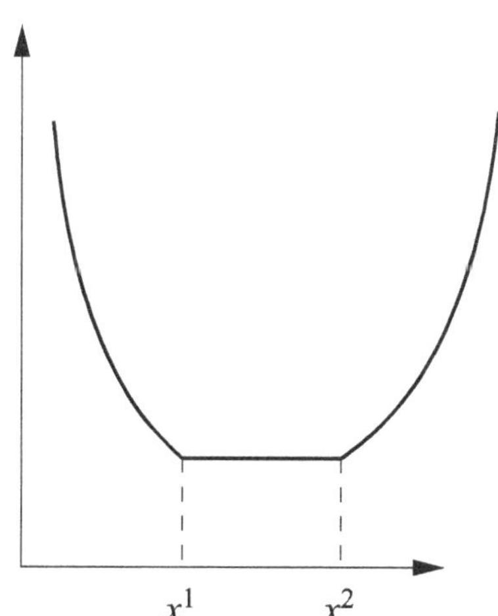

Fig. 4.8 An objective function that has multiple (indeed, an infinite number of) global minima

4.3.2 Local Minima

Although we are typically looking for a global minimum when solving an NLPP, for most problems the best we can do is find **local minima**. This is because the methods used to find minima use local information (*i.e.*, derivatives). As discussed below, we find a global minimum of most NLPPs by exhaustively searching for all local minima and choosing the one that gives the smallest objective-function value.

We now define a local minimum and then illustrate the concept with some examples.

Given a nonlinear optimization problem, a feasible solution, x^*, is a **local minimum** if $f(x^*) \leq f(x)$ for all other feasible values of x that are close to x^*.

Figure 4.9 shows an objective function with four local minima, labeled x^1, x^2, x^3, and x^4. Each of these points satisfies the definition of a local minimum, because other feasible points that are close to them give the same or higher objective-function values. Among the four local minima, one of them, x^4 is also a global minimum. This illustrates the way that we normally go about finding a global minimum. We exhaustively search for all local minima and then choose the one that gives the smallest objective-function value. We must also pay attention to ensure that the problem is not unbounded—if it is then the problem does not have a global minimum (even though it may have local minima).

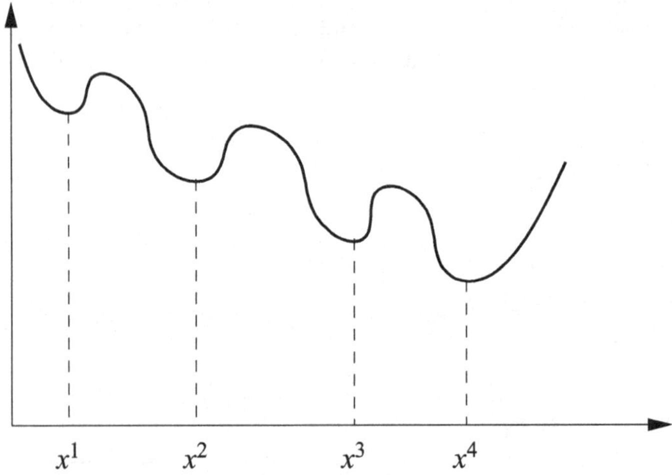

Fig. 4.9 Local and global minima of an objective function

Figure 4.10 demonstrates how adding a constraint affects the definition of a local minimum. Here we have the same objective function as in Figure 4.9, but have added the constraint $x \geq \hat{x}$. As in Figure 4.9, x^3 and x^4 are still local minima and x^4 is the global minimum. However, x^1 and x^2 are not local minima, because they are no longer feasible. Moreover, $\hat{x}$ is a local minimum. To see why, we first note that values of x that are close to $\hat{x}$ but to its left give smaller objective-function values than $\hat{x}$ does. However, these points to the left of $\hat{x}$ are not considered in the definition of a local minimum, because we only consider *feasible* points that are close to $\hat{x}$. If we restrict attention to feasible points that are close to $\hat{x}$ (*i.e.*, points to the right of $\hat{x}$) then we see that $\hat{x}$ does indeed satisfy the definition of a local minimum.

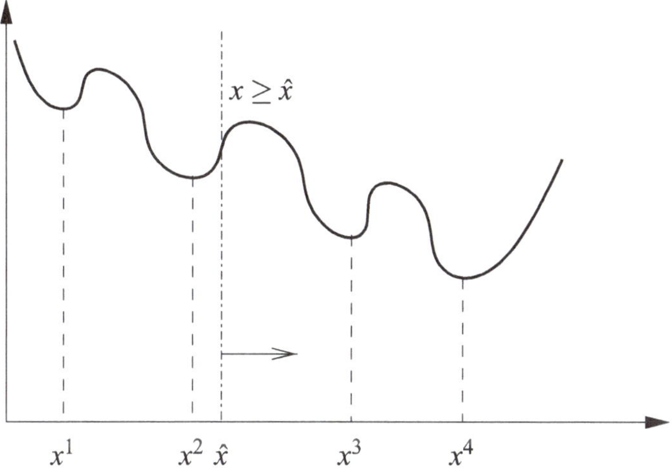

Fig. 4.10 Local and global minima of an objective function with an inequality constraint

The objective function that is shown in Figure 4.8 also demonstrates why the weak inequality in the definition of a local minimum is important. All of the points between x^1 and x^2 in this figure (which we argue in Section 4.3.1 are global minima) are also local minima.

4.4 Convex Nonlinear Optimization Problems

We note in Section 4.3 that for most optimization problems, the best we can do is find local minima. Thus, in practice, finding global minima can be tedious because it requires us to search for all local minima and pick the one that gives the smallest objective-function value. There is one special class of optimization problems, which are called **convex optimization problems**, for which it is easier to find a global minimum. A convex optimization problem has the property that any local minimum is guaranteed to be a global minimum. Thus, finding a global minimum of a convex optimization problem is relatively easy, because we are done as soon as we find a local minimum.

In this section, we first define what a convex optimization problem is and then discuss ways to test whether a problem has the needed convexity property. We finally show the result that any local minimum of a convex optimization problem is guaranteed to be a global minimum.

4.4.1 Convex Optimization Problems

To define a convex optimization problem, we consider a more generic form of an optimization problem than those given in Section 4.2. Here we write a generic optimization problem as:

$$\min_{x}\ f(x)$$

$$\text{s.t. } x \in X.$$

As before, $f(x)$ is the objective function that we seek to minimize. The set $X \subseteq \mathbb{R}^n$ represents the feasible region of the problem. In the case of an unconstrained nonlinear optimization problem we would have $X = \mathbb{R}^n$. If we have a problem with a mixture of equality and inequality constraints, we define X as:

$$X = \{x \in \mathbb{R}^n : h_1(x) = 0, \ldots, h_m(x) = 0, g_1(x) \le 0, \ldots, g_r(x) \le 0\},$$

the set of decision-variable vectors, x, that simultaneously satisfy all of the constraints. Using this more generic form of an optimization problem, we now define a convex optimization problem.

> An optimization problem of the form:
>
> $$\min_{x}\ f(x)$$
>
> $$\text{s.t. } x \in X \subseteq \mathbb{R}^n,$$
>
> is a **convex optimization problem** if the set X is convex and $f(x)$ is a convex function on the set X.

Determining if a problem is convex boils down to determining two things: (i) is the feasible region convex and (ii) is the objective function convex on the feasible region? We discuss methods that can be used to answer these two questions in the following sections.

4.4.2 Determining if a Feasible Region is Convex

Recall the following definition of a convex set from Section B.1.

A set $X \subseteq \mathbb{R}^n$ is said to be a **convex set** if for any two points x^1 and $x^2 \in X$ and for any value of $\alpha \in [0, 1]$ we have that:

$$\alpha x^1 + (1 - \alpha)x^2 \in X.$$

One way to test whether a set is convex is to use the definition directly, as demonstrated in the following example.

Example 4.1 Consider the feasible region of the Packing-Box Problem that is introduced in Section 4.1.1.1. The feasible region of this problem is:

$$X = \{(w \, h \, d) : 2wh + 2dh + 6wd \leq 60, w \geq 0, h \geq 0, d \geq 0\}.$$

Note that the points:

$$(w \, h \, d) = (1 \, 1 \, 29/4),$$

and:

$$(\hat{w} \, \hat{h} \, \hat{d}) = (29/4 \, 1 \, 1),$$

are both in the feasible region, X. However, if we take the midpoint of these two feasible points:

$$(\tilde{w} \, \tilde{h} \, \tilde{d}) = \frac{1}{2}(w \, h \, d) + \frac{1}{2}(\hat{w} \, \hat{h} \, \hat{d}) = (33/8 \, 1 \, 33/8),$$

we see that this point is infeasible because:

$$2\tilde{w}\tilde{h} + 2\tilde{d}\tilde{h} + 6\tilde{w}\tilde{d} \approx 118.59.$$

Thus, the set X is not convex and the Packing-Box Problem is not a convex optimization problem. $\square$

In practice, the definition of a convex set can be cumbersome to work with. For this reason, we use the following three properties, which can make it easier to show that a feasible set is convex.

4.4.2.1 Linear Constraints

The first property involves equality and inequality constraints that are linear in the decision variables. We can show graphically and through a simple proof that linear constraints always define convex feasible regions.

> **Linear-Constraint Property:** Any equality or inequality constraint that is linear in the decision variables defines a convex feasible set.

Before proving this result, Figure 4.11 graphically demonstrates this result. As the figure shows, a linear equality constraint defines a straight line in two dimensions, which is a convex set. We also know from the discussion in Section 2.1.1 that a linear equality constraint defines a hyperplane in three or more dimensions. All of these sets are convex. The figure also demonstrates the convexity of feasible sets defined by linear inequalities. In two dimensions, a linear inequality defines the space on one side of a line (also known as a halfspace). In higher dimensions, halfspaces generalize to the space on one side of a plane or hyperplane. These are also convex sets, as seen in Figure 4.11.

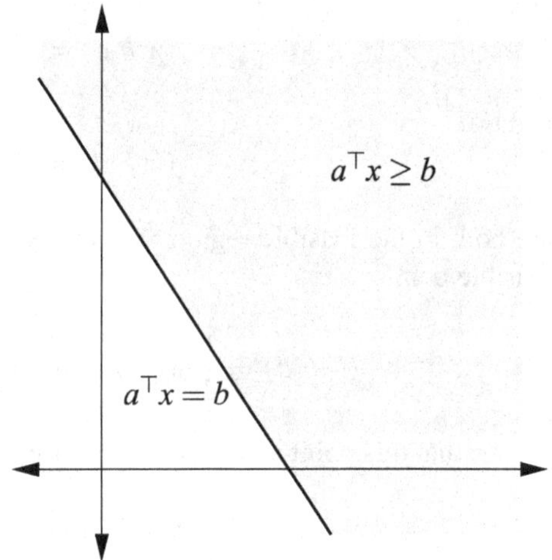

Fig. 4.11 Illustration of the feasible region defined by linear equality and inequality constraints in two dimensions

We now prove the Linear-Constraint Property.

Consider a linear equality constraint of the form:

$$\begin{pmatrix} a_1 & a_2 & \cdots & a_n \end{pmatrix} \begin{pmatrix} x_1 \\ x_2 \\ \vdots \\ x_n \end{pmatrix} = a^\top x = b,$$

where b is a scalar. Suppose that there are two points, x^1 and x^2, which are feasible. This means that:

$$a^\top x^1 = b, \tag{4.1}$$

and:

$$a^\top x^2 = b. \tag{4.2}$$

Take any value of $\alpha \in [0, 1]$ and consider the convex combination of x^1 and x^2:

$$\alpha x^1 + (1 - \alpha)x^2.$$

If we multiply this point by a we have:

$$a^\top[\alpha x^1 + (1-\alpha)x^2] = a^\top \alpha x^1 + a^\top(1-\alpha)x^2 = \alpha a^\top x^1 + (1-\alpha)a^\top x^2. \tag{4.3}$$

Substituting (4.1) and (4.2) into (4.3) gives:

$$a^\top[\alpha x^1 + (1 - \alpha)x^2] = \alpha b + (1 - \alpha)b = b,$$

or:

$$a^\top[\alpha x^1 + (1 - \alpha)x^2] = b.$$

Thus, $\alpha x^1 + (1 - \alpha)x^2$ is feasible and the feasible set defined by the linear equality constraint is convex.

To show the same result for the case of a linear inequality constraint, consider a linear inequality constraint of the form:

$$a^\top x \geq b.$$

Suppose that there are two points, x^1 and x^2, which are feasible in this inequality constraint. This means that:

$$a^\top x^1 \geq b, \tag{4.4}$$

and:

$$a^\top x^2 \geq b. \tag{4.5}$$

Again, take any value of $\alpha \in [0, 1]$ and consider the convex combination of x^1 and x^2, $\alpha x^1 + (1 - \alpha)x^2$. If we multiply this point by a we have:

$$a^\top[\alpha x^1 + (1 - \alpha)x^2] = \alpha a^\top x^1 + (1 - \alpha)a^\top x^2. \tag{4.6}$$

Substituting (4.4) and (4.5) into (4.6) and noting that $\alpha \geq 0$ and $1 - \alpha \geq 0$ gives:

$$a^\top [\alpha x^1 + (1 - \alpha)x^2] \geq \alpha b + (1 - \alpha)b = b,$$

or:

$$a^\top [\alpha x^1 + (1 - \alpha)x^2] \geq b.$$

Thus, $\alpha x^1 + (1 - \alpha)x^2$ is feasible and the feasible set defined by the linear inequality constraint is convex.

The following example demonstrates how this convexity result involving linear constraints can be used.

Example 4.2 Consider the constraints of the Variance-Minimization Problem that is introduced in Section 4.1.3.2, which are:

$$\sum_{n=1}^{N} \bar{r}_n w_n \geq \bar{R},$$

$$\sum_{n=1}^{N} w_n = 1,$$

and:

$$w_n \geq 0, \forall\, n = 1, \ldots, N.$$

Recall, also, that the decision-variable vector in this problem is w. Each of these constraints is linear in the decision variables, meaning that each constraint individually defines a convex feasible region.

We do not, yet, know if the whole feasible region of the problem is convex. We only know that each constraint on its own gives a convex set of points that satisfies it. We discuss the Intersection-of-Convex-Sets Property in Section 4.4.2.3, which allows us to draw the stronger conclusion that this problem does indeed have a convex feasible region. □

4.4.2.2　Convex-Inequality Constraints

The second property that we use to test whether an optimization problem has a convex feasible region is that less-than-or-equal-to constraints that have a convex function on the left-hand side define convex feasible regions. This property is stated as follows.

Convex-Inequality Property: Any inequality constraint of the form:

$$g(x) \leq 0,$$

which has a convex function, $g(x)$, on the left-hand side defines a convex feasible set.

To show this property, recall the following definition of a convex function from Section B.2.

Given a convex set, $X \subseteq \mathbb{R}^n$, a function defined on X is said to be a **convex function** on X if for any two points, x^1 and $x^2 \in X$, and for any value of $\alpha \in [0, 1]$ we have that:

$$\alpha f(x^1) + (1 - \alpha)f(x^2) \geq f(\alpha x^1 + (1 - \alpha)x^2).$$

Figures 4.12 and 4.13 illustrate the Convex-Inequality Property graphically. Figure 4.12 shows the feasible set, $g(x) \leq 0$, where $g(x)$ is a convex parabolic function while Figure 4.13 shows the case of a convex absolute value function. It is important to note that the Convex-Inequality Property only yields a one-sided implication—the feasible set defined by a constraint of the form $g(x) \leq 0$ where $g(x)$ is a convex function gives a convex feasible set. However, it may be the case that a constraint of the form $g(x) \leq 0$ where $g(x)$ is non-convex gives a convex feasible

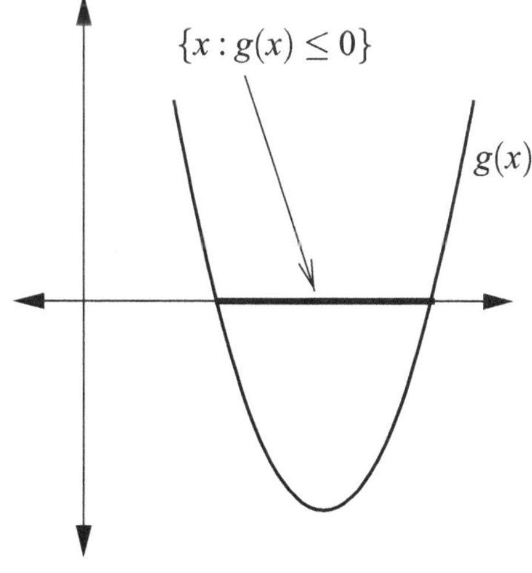

Fig. 4.12 A convex parabola, which defines a convex feasible region when on the left-hand side of a less-than-or-equal-to constraint

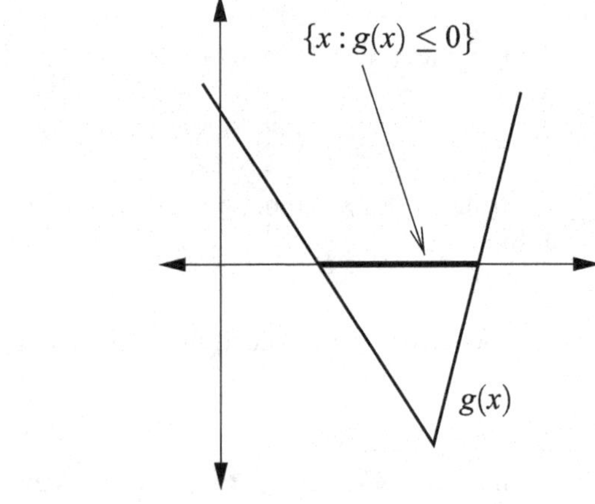

Fig. 4.13 A convex absolute value function, which defines a convex feasible region when on the left-hand side of a less-than-or-equal-to constraint

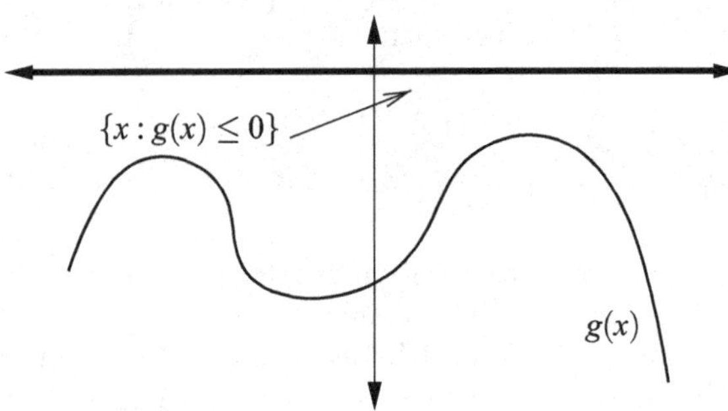

Fig. 4.14 A non-convex function, which defines a convex feasible region when on the left-hand side of a less-than-or-equal to constraint

set. Figure 4.14 demonstrates this by showing a non-convex function, $g(x)$, which gives a convex feasible region when on the left-hand side of a less-than-or-equal-to zero constraint, because the feasible region is all $x \in \mathbb{R}$.

We now give a proof of the Convex-Inequality Property.

Consider an inequality constraint of the form:

$$g(x) \leq 0,$$

where $g(x)$ is a convex function. Suppose that there are two feasible points, x^1 and x^2. This means that:

$$g(x^1) \leq 0, \tag{4.7}$$

and:

$$g(x^2) \leq 0. \tag{4.8}$$

Take $\alpha \in [0, 1]$ and consider the convex combination of x^1 and x^2, $\alpha x^1 + (1 - \alpha)x^2$. If we plug this point into the constraint we know that:

$$g(\alpha x^1 + (1 - \alpha)x^2) \leq \alpha g(x^1) + (1 - \alpha)g(x^2), \qquad (4.9)$$

because $g(x)$ is a convex function. Substituting (4.7) and (4.8) into (4.9) and noting that $\alpha \geq 0$ and $1 - \alpha \geq 0$ gives:

$$g(\alpha x^1 + (1 - \alpha)x^2) \leq \alpha 0 + (1 - \alpha)0 = 0,$$

or:

$$g(\alpha x^1 + (1 - \alpha)x^2) \leq 0,$$

meaning that the point $\alpha x^1 + (1-\alpha)x^2$ is feasible and the constraint, $g(x) \leq 0$, defines a convex feasible set.

Figure 4.15 illustrates the intuition behind the Convex-Inequality Property. If two points, x^1 and x^2, are feasible, this means that $g(x^1) \leq 0$ and $g(x^2) \leq 0$. Because $g(x)$ is convex, we know that at any point between x^1 and x^2 the function $g(x)$ is below the secant line connecting $g(x^1)$ and $g(x^2)$ (*cf.* Section B.2 for further details). However, because both $g(x^1)$ and $g(x^2)$ are less than or equal to zero, the secant line is also less than or equal to zero. Hence, the function is also less than or equal to zero at any point between x^1 and x^2.

Fig. 4.15 Illustration of the Convex-Inequality Property

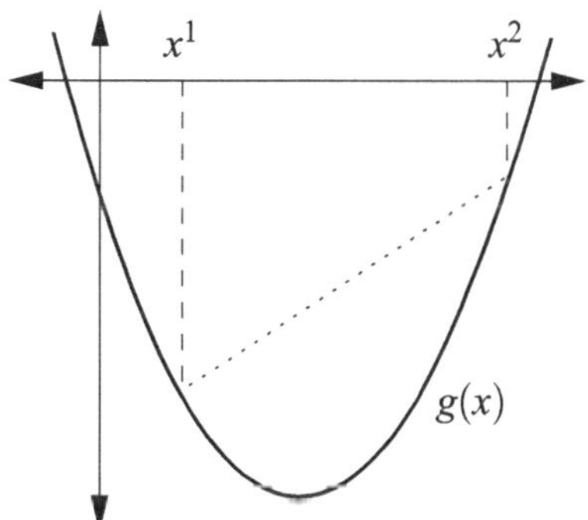

4.4.2.3 Intersection of Convex Sets

The third property that is useful to show that an optimization problem has a convex feasible region is that the intersection of any number of convex sets is convex. This is useful because the feasible region of an optimization problem is defined as the intersection of the feasible sets defined by each constraint. If each constraint individually defines a convex set, then the feasible region of the overall problem is convex as well. To more clearly illustrate this, consider the standard-form equality- and inequality-constrained NLPP:

$$\min_{x \in \mathbb{R}^n} f(x)$$

$$\text{s.t. } h_1(x) = 0$$
$$h_2(x) = 0$$
$$\vdots$$
$$h_m(x) = 0$$
$$g_1(x) \leq 0$$
$$g_2(x) \leq 0$$
$$\vdots$$
$$g_r(x) \leq 0.$$

The overall feasible region of this problem is:

$$X = \{x \in \mathbb{R}^n : h_1(x) = 0, \ldots, h_m(x) = 0, g_1(x) \leq 0, \ldots, g_r(x) \leq 0\}.$$

Another way to view this feasible region is to define the feasible region defined by each constraint individually as:

$$X_1 = \{x \in \mathbb{R}^n : h_1(x) = 0\},$$

$$\vdots$$

$$X_m = \{x \in \mathbb{R}^n : h_m(x) = 0\},$$

$$X_{m+1} = \{x \in \mathbb{R}^n : g_1(x) \leq 0\},$$

$$\vdots$$

and:

$$X_{m+r} = \{x \in \mathbb{R}^n : g_r(x) \leq 0\}.$$

We can then define X as the intersection of all of these individual sets:

$$X = X_1 \cap \cdots \cap X_m \cap X_{m+1} \cap \cdots \cap X_{m+r}.$$

If each of the sets, $X_1, \cdots, X_{m+r}$, is convex, then their intersection, X, is convex as well, meaning that the problem has a convex feasible region.

We now state and prove the result regarding the intersection of convex sets.

Intersection-of-Convex-Sets Property: Let $X_1, \ldots, X_k \subseteq \mathbb{R}^n$ be a collection of convex sets. Their intersection:

$$X = X_1 \cap \cdots \cap X_k,$$

is convex.

We show this by contradiction. Suppose that the statement is not true, meaning that the set X is not convex. This means that there exist points, $x^1, x^2 \in X$ and a value of $\alpha \in [0, 1]$ for which $\alpha x^1 + (1 - \alpha)x^2 \notin X$. If $x^1 \in X$, then x^1 must be in each of $X_1, \ldots, X_k$, because X is defined as the intersection of these sets. Likewise, if $x^2 \in X$ it must be in each of $X_1, \ldots, X_k$. Because each of $X_1, X_2, \ldots, X_k$ is convex, then $\alpha x^1 + (1 - \alpha)x^2$ must be in each of $X_1, \ldots, X_k$. However, if $\alpha x^1 + (1 - \alpha)x^2$ is in each of $X_1, \ldots, X_k$ then it must be in X, because X is defined as the intersection of $X_1, \ldots, X_k$. This gives a contradiction, showing that X must be a convex set.

Figure 4.16 illustrates the idea underlying the proof of the Intersection-of-Convex-Sets Property graphically for the case of the intersection of two convex sets, X^1 and X^2, in $\mathbb{R}^2$. The points x^1 and x^2 are both contained in the sets X_1 and X_2, thus they are both contained in X, which is the intersection of X_1 and X_2. Moreover, if we draw a line segment connecting x_1 and x_2 we know that this line segment must be contained in the set X_1, because X_1 is convex. The line segment must also be contained in the

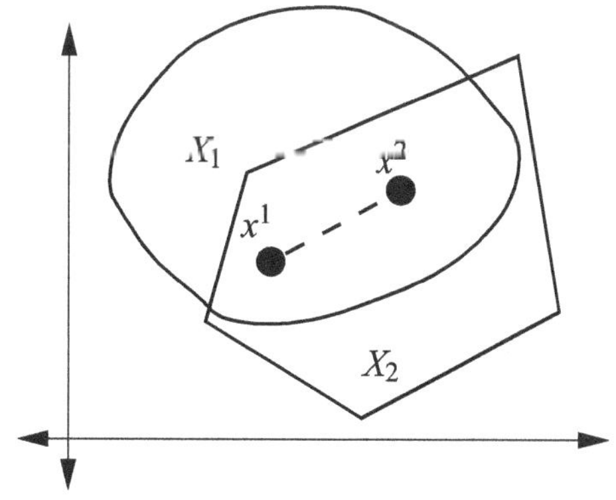

Fig. 4.16 Illustration of the Intersection-of-Convex-Sets Property

set X_2 for the same reason. Because X is defined as the collection of points that is common to both X_1 and X_2, this line segment must also be contained in X, showing that X is a convex set.

The following example demonstrates the use of the Intersection-of-Convex-Sets Property to show that an optimization problem has a convex feasible region.

Example 4.3 Consider the following optimization problem:

$$\min_{x} \; f(x)$$

$$\text{s.t. } g_1(x) = x_1^2 + x_2^2 - 4 \le 0$$

$$g_2(x) = -x_1 + x_2 + 1 \le 0,$$

where the objective, $f(x)$, is an arbitrary function. To show that the feasible region of this problem is convex, consider the first constraint. Note that we have:

$$\nabla^2 g_1(x) = \begin{bmatrix} 2 & 0 \\ 0 & 2 \end{bmatrix},$$

which is a positive-definite matrix (*cf.* Section A.2 for the definition of positive-definite matrices), meaning that $g_1(x)$ is a convex function (*cf.* Section B.2 for the use of the Hessian matrix as a means of testing whether a function is convex). Because $g_1(x)$ is on the left-hand side of a less-than-or-equal-to constraint, it defines a convex set. The second constraint is linear in x_1 and x_2, thus we know that it defines a convex set. The overall feasible region of the problem is defined by the intersection of the feasible regions defined by each constraint, each of which is convex. Thus, the overall feasible region of the problem is convex. Figure 4.17 illustrates the feasible region defined by each constraint and the overall feasible region of the problem.   □

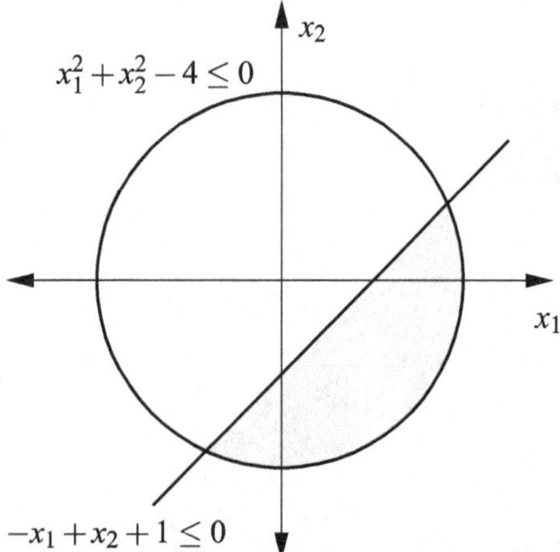

Fig. 4.17 Feasible region of optimization problem in Example 4.3

As another example of using the intersection property, recall that in Example 4.2 we note that each constraint of the Variance-Minimization Problem, which is introduced in Section 4.1.3.2, is linear. Thus, each constraint individually defines a convex set. Because the feasible region of the overall problem is defined as the intersection of these sets, the problem's overall feasible region is convex.

4.4.3 Determining if an Objective Function is Convex

The definition of a convex function is given in Section B.2. Although it is possible to show that an objective function is convex using this basic definition, it is typically easier to test whether a function is convex by determining whether its Hessian matrix is positive semidefinite. If it is the function is convex, otherwise the function is not (*cf.* Section B.2 for further details). We demonstrate this with the following example.

Example 4.4 Recall the Packing-Box Problem that is introduced in Section 4.1.1.1. This problem is formulated as:

$$\max_{h,w,d} \ hwd$$

$$\text{s.t. } 2wh + 2dh + 6wd \leq 60$$

$$w \geq 0$$

$$h \geq 0$$

$$d \geq 0.$$

Converting this problem to standard form it becomes:

$$\min_{h,w,d} \ f(h, w, d) = -hwd$$

$$\text{s.t. } 2wh + 2dh + 6wd - 60 \leq 0$$

$$- w \leq 0$$

$$- h \leq 0$$

$$- d \leq 0.$$

The Hessian of the objective function is:

$$\nabla^2 f(h, w, d) = \begin{bmatrix} 0 & -d & -w \\ -d & 0 & -h \\ -w & -h & 0 \end{bmatrix},$$

which is not positive semidefinite. Thus, the objective function of this problem is not convex and as such this is not a convex optimization problem. □

It is important to note that the definition of a convex optimization problem only requires the objective function to be convex on the feasible region. The following example demonstrates why this is important.

Example 4.5 Consider the unconstrained single-variable optimization problem:

$$\min_{x} \; f(x) = \sin(x).$$

We have:

$$\nabla^2 f(x) = -\sin(x),$$

which we know varies in sign for different values of x. Thus, this unconstrained optimization problem is not convex.

Suppose we add bound constraints and the problem becomes:

$$\min_{x} \; f(x) = \sin(x)$$

$$\text{s.t. } \pi \le x \le 2\pi.$$

The Hessian of the objective function remains the same. In this case, however, we only require the Hessian to be positive semidefinite over the feasible region (*i.e.*, for values of $x \in [\pi, 2\pi]$). Substituting these values of x into the Hessian gives us non-negative values. Thus, we have a convex optimization problem when the constraints are added. □

4.4.4 *Global Minima of Convex Optimization Problems*

Having defined a convex optimization problem and discussed how to determine if an optimization problem is convex, we now turn to proving the following important Global-Minimum-of-Convex-Problem Property.

Global-Minimum-of-Convex-Problem Property: Consider an optimization problem of the form:

$$\min_{x} \; f(x)$$

$$\text{s.t. } x \in X \subseteq \mathbb{R}^n,$$

where X is a convex set and $f(x)$ is a convex function on the set X. Any local minimum of this problem is also a global minimum.

We prove this by contradiction. To do so, suppose that this is not true. That means that there is a point, x^*, which is a local minimum but not a global minimum. Thus, there is another point, $\hat{x} \in X$, which is a global minimum and we have that $f(\hat{x}) < f(x^*)$.

Now, consider convex combinations of x^* and $\hat{x}$, $\alpha x^* + (1 - \alpha)\hat{x}$, where $\alpha \in [0, 1]$. Because X is convex, we know such points are feasible in the problem (as this is what it means for the set, X, to be convex). We also know, because $f(x)$ is convex, that:

$$f(\alpha x^* + (1 - \alpha)\hat{x}) \leq \alpha f(x^*) + (1 - \alpha)f(\hat{x}).$$

Because $\hat{x}$ is a global minimum but x^* is not, we also know that:

$$\alpha f(x^*) + (1 - \alpha)f(\hat{x}) < \alpha f(x^*) + (1 - \alpha)f(x^*) = f(x^*).$$

Combining these two inequalities gives:

$$f(\alpha x^* + (1 - \alpha)\hat{x}) < f(x^*).$$

If we let α get close to 1, then $\alpha x^* + (1 - \alpha)\hat{x}$ gets close to x^*. The last inequality says that these points (obtained for different values of α close to 1), which are feasible and close to x^*, give objective-function values that are better than that of x^*. This contradicts x^* being a local minimum. Thus, it is not possible for a convex optimization problem to have a local minimum that is not also a global minimum.

4.5   Optimality Conditions for Nonlinear Optimization Problems

One method of finding local minima of nonlinear optimization problems is by analyzing what are known as **optimality conditions**. There are two varieties of optimality conditions that we use—necessary and sufficient conditions. **Necessary conditions** are conditions that a solution must satisfy to be a local minimum. A solution that satisfies a necessary condition could possibly be a local minimum. Conversely, a vector of decision-variable values that does not satisfy a necessary condition *cannot* be a local minimum. **Sufficient conditions** are conditions that guarantee that a solution is a local minimum. However, a solution that does not satisfy a sufficient condition *cannot* be ruled out as a possible local minimum.

With the exception of convex optimization problems, we do not typically have conditions that are *both* necessary and sufficient. This means that the best we can typically do is to use necessary conditions to find solution that *might* be local minima. Any of those solutions that also satisfy sufficient conditions are guaranteed to be local minima. However, if there are solutions that only satisfy the necessary but not the sufficient conditions, they may or may not be local minima. There is no way to definitively guarantee one way or another.

We have different sets of optimality conditions for the three types of nonlinear optimization problems that are introduced in Section 4.2—unconstrained, equality-constrained, and equality- and inequality-constrained problems. We examine each of these three problem types in turn.

4.5.1 Unconstrained Nonlinear Optimization Problems

An unconstrained nonlinear optimization problem has the general form:

$$\min_{x \in \mathbb{R}^n} \; f(x),$$

where $f(x)$ is the objective being minimized and there are no constraints on the decision variables.

4.5.1.1 First-Order Necessary Condition for Unconstrained Nonlinear Optimization Problems

We begin by stating and proving what is known as the first-order necessary condition (FONC) for a local minimum. We also demonstrate its use and limitations through some examples

First-Order Necessary Condition for Unconstrained Nonlinear Optimization Problems: Consider an unconstrained nonlinear optimization problem of the form:

$$\min_{x \in \mathbb{R}^n} \; f(x).$$

If x^* is a local minimum, then $\nabla f(x^*) = 0$.

Suppose x^* is a local minimum. This means that any point close to x^* must give an objective-function value that is no less than that given by x^*. Consider points, $x^* + d$, that are close to x^* (meaning that $||d||$ is close to zero). We can use a first-order Taylor approximation (*cf.* Appendix A) to compute the objective-function value at such a point as:

$$f(x^* + d) \approx f(x^*) + d^\top \nabla f(x^*).$$

Because x^* is a local minimum, we must have $f(x^*+d) \geq f(x^*)$. Substituting the Taylor approximation in for $f(x^* + d)$ this can be written as:

$$f(x^*) + d^\top \nabla f(x^*) \geq f(x^*).$$

Subtracting $f(x^*)$ from both sides, this becomes:

$$d^\top \nabla f(x^*) \geq 0. \tag{4.10}$$

This inequality must also apply for the point x^*-d. That is, $f(x^*-d) \geq f(x^*)$. If we substitute the Taylor approximation for $f(x^* - d)$ into this inequality, we have:

$$d^\top \nabla f(x^*) \leq 0. \tag{4.11}$$

Combining (4.10) and (4.11) implies that we must have:

$$d^\top \nabla f(x^*) = 0.$$

Finally, note that we must have $d^\top \nabla f(x^*) = 0$ for any choice of d (so long as $||d||$ is close to zero). The only way that this holds is if $\nabla f(x^*) = 0$.

Figure 4.18 graphically illustrates the idea behind the FONC. If x^* is a local minimum, then there is a small neighborhood of points, which is represented by the area within the dotted circle, on which x^* provides the best objective-function value. If we examine the Taylor approximation of $f(x^*+d)$ and $f(x^*-d)$, where d is chosen so $x^* + d$ and $x^* - d$ are within this dotted circle, this implies that $d^\top \nabla f(x^*) = 0$. Because this must hold for any choice of d within the dotted circle (*i.e.*, moving in any direction away from x^*) this implies that we must have $\nabla f(x^*) = 0$.

Fig. 4.18 Illustration of the FONC for an unconstrained nonlinear optimization problem

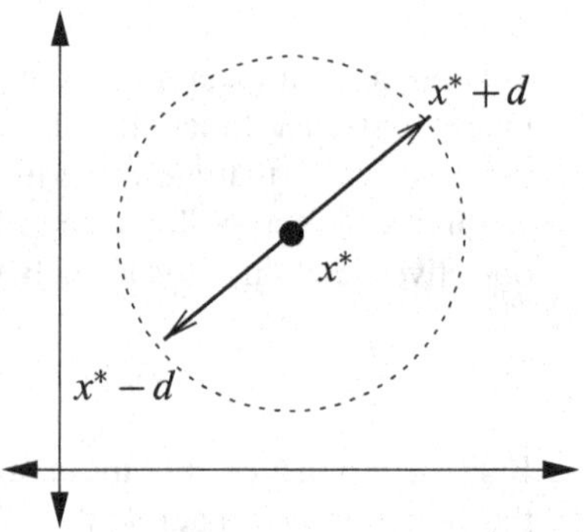

Points that satisfy the FONC (*i.e.*, points that have a gradient equal to zero) are also known as **stationary points**. We now demonstrate the use of the FONC with the following examples.

Example 4.6 Consider the unconstrained problem:

$$\min_{x}\ f(x) = (x_1 - 3)^2 + (x_2 + 4)^2.$$

To find stationary points, we set the gradient of $f(x)$ equal to zero, which gives:

$$\nabla f(x) = \begin{pmatrix} 2(x_1 - 3) \\ 2(x_2 + 4) \end{pmatrix} = \begin{pmatrix} 0 \\ 0 \end{pmatrix},$$

or:

$$x^* = \begin{pmatrix} 3 \\ -4 \end{pmatrix}.$$

It is easy to confirm that this point is in fact a local and global minimum, because the objective function is bounded below by zero. It we plug x^* into $f(x)$ we see that this point gives an objective-function value of zero, thus we have a local and global minimum. $\qquad\square$

It is important to note from this example that the FONC always results in a system of n equations, where n is the number of decision variables in the optimization problem. This is because the gradient is computed with respect to the decision variables, giving one first-order partial derivative for each decision variable. The fact that the number of equations is equal to the number of variables does not necessarily mean that the FONC has a unique solution. There could be multiple solutions or no solution, as demonstrated in the following examples.

Example 4.7 Consider the unconstrained problem:

$$\min_{x}\ f(x) = x_1 - 2x_2.$$

The gradient of this objective function is:

$$\nabla f(x) = \begin{pmatrix} 1 \\ -2 \end{pmatrix},$$

which cannot be made to equal zero. Because this problem does not have any stationary points, it cannot have any local minima. In fact, this problem is unbounded. To see this, note that:

$$\lim_{x_1 \to -\infty, x_2 \to +\infty} f(x) = -\infty. \qquad \square$$

Example 4.8 Consider the unconstrained problem:

$$\min_x \; f(x) = x^3 - x^2 - 4x - 6.$$

To find stationary points, we set the gradient of $f(x)$ equal to zero, which gives:

$$\nabla f(x) = 3x^2 - 2x - 4 = 0,$$

or:

$$x^* \in \left\{ \frac{2 - \sqrt{52}}{6}, \frac{2 + \sqrt{52}}{6} \right\}.$$

Both of these are stationary points and thus candidate local minima, based on the FONC.

Figure 4.19 shows the objective function of this problem. Based on visual inspection, it is clear that only one of these two stationary points, $x^* = (2+\sqrt{52})/6$, is a local minimum, whereas $(2 - \sqrt{52})/6$ is a local maximum. Moreover, visual inspection shows us that this objective function is also unbounded. Although $x^* = (2+\sqrt{52})/6$ is a local minimum, this particular problem does not have a global minimum because:

$$\lim_{x \to -\infty} f(x) = -\infty. \qquad \square$$

Example 4.8 illustrates an important limitation of the FONC. Although this condition eliminates non-stationary points that cannot be local minima, it does not distinguish between local minima and local maxima. This is also apparent in Figure 4.9— the local minima that are highlighted in this figure are all stationary points where the gradient is zero. However, there are three local maxima, which are also stationary points. The following example further demonstrates this limitation of the FONC.

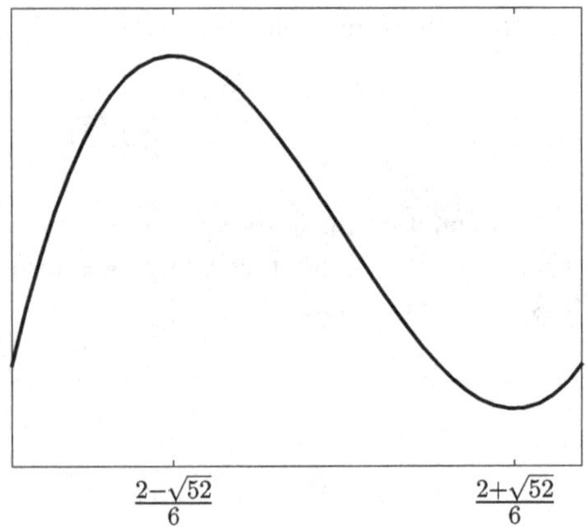

Fig. 4.19 Objective function
in Example 4.8

Example 4.9 Consider the unconstrained problem:

$$\max_{x} \ f(x),$$

where the objective is an arbitrary function. To solve this problem, we convert it to
a minimization of the form:

$$\min_{x} \ -f(x),$$

and search for stationary points:

$$\nabla(-f(x)) = -\nabla f(x) = 0.$$

Multiplying through by -1, the FONC can be written as:

$$\nabla f(x) = 0,$$

which also is the FONC for the following problem:

$$\min_{x} \ f(x).$$

This means that the FONCs for finding local minima and local maxima of $f(x)$ are
the same and the FONC cannot distinguish between the two.   □

4.5.1.2 Second-Order Necessary Condition for Unconstrained Nonlinear Optimization Problems

This limitation of the FONC—that it cannot distinguish between local minima and
maxima—motivates the second-order necessary condition (SONC) for a local minimum, which we now introduce.

Second-Order Necessary Condition for Unconstrained Nonlinear Optimization Problems: Consider an unconstrained nonlinear optimization problem of the form:

$$\min_{x \in \mathbb{R}^n} \ f(x).$$

If x^* is a local minimum, then $\nabla^2 f(x^*)$ is positive semidefinite.

Suppose x^* is a local minimum. Consider points, $x^* + d$, that are close to x^* (meaning that $||d||$ is close to zero). We can use a second-order Taylor approximation to compute the objective-function value at such a point as:

$$f(x^* + d) \approx f(x^*) + d^\top \nabla f(x^*) + \frac{1}{2} d^\top \nabla^2 f(x^*) d.$$

If x^* is a local minimum and d is sufficiently small in magnitude we must have:

$$f(x^* + d) \approx f(x^*) + d^\top \nabla f(x^*) + \frac{1}{2} d^\top \nabla^2 f(x^*) d \geq f(x^*),$$

or:

$$d^\top \nabla f(x^*) + \frac{1}{2} d^\top \nabla^2 f(x^*) d \geq 0. \tag{4.12}$$

We further know from the FONC that if x^* is a local minimum then we must have $\nabla f(x^*) = 0$, thus (4.12) becomes:

$$d^\top \nabla^2 f(x^*) d \geq 0,$$

when we multiply it through by 2. Because this inequality must hold for any choice of d (so long as $||d||$ is close to zero), this implies that $\nabla^2 f(x^*)$ is positive semidefinite (*cf.* Section A.2 for the definition of a positive semidefinite matrix).

The SONC follows very naturally from the FONC, by analyzing the second-order Taylor approximation of $f(x)$ at points close to a local minimum. The benefit of the SONC is that it can, in many instances, differentiate between local minima and maxima, as demonstrated in the following examples.

Example 4.10 Consider the following unconstrained optimization problem, which is introduced in Example 4.6:

$$\min_{x} \ f(x) = (x_1 - 3)^2 + (x_2 + 4)^2.$$

We know from Example 4.6 that:

$$x^* = \begin{pmatrix} 3 \\ -4 \end{pmatrix},$$

is the only stationary point. We also conclude in Example 4.6 that this point is a local and global minimum, by showing that the objective function attains its lower bound at this point. We can further compute the Hessian of the objective, which is:

$$\nabla^2 f(x) = \begin{bmatrix} 2 & 0 \\ 0 & 2 \end{bmatrix},$$

and is positive semidefinite, confirming that the stationary point found satisfies the SONC.   □

Example 4.11 Consider the following unconstrained optimization problem, which is introduced in Example 4.8:

$$\min_x \; f(x) = x^3 - x^2 - 4x - 6.$$

We know from Example 4.8 that this problem has two stationary points:

$$x^* \in \left\{ \frac{2 - \sqrt{52}}{6}, \frac{2 + \sqrt{52}}{6} \right\}.$$

The Hessian of this objective function is:

$$\nabla^2 f(x) = 6x - 2.$$

If we substitute the two stationary points into this Hessian we see that:

$$\nabla^2 f((2 - \sqrt{52})/6) = -\sqrt{52} < 0,$$

and:

$$\nabla^2 f((2 + \sqrt{52})/6) = \sqrt{52} > 0.$$

The Hessian is not positive semidefinite at $(2 - \sqrt{52})/6$, thus this point cannot be a local minimum, which is confirmed graphically in Figure 4.19. The Hessian is positive semidefinite at $(2 + \sqrt{52})/6$, thus this point remains a candidate local minimum (because it satisfies both the FONC and SONC). We can graphically confirm in Figure 4.19 that this point is indeed a local minimum.   □

Example 4.12 Consider the unconstrained optimization problem:

$$\min_x \ f(x) = \sin(x_1 + x_2).$$

The FONC is:

$$\nabla f(x) = \begin{pmatrix} \cos(x_1 + x_2) \\ \cos(x_1 + x_2) \end{pmatrix} = \begin{pmatrix} 0 \\ 0 \end{pmatrix},$$

which gives stationary points of the form:

$$x^* = \begin{pmatrix} x_1 \\ x_2 \end{pmatrix},$$

where:

$$x_1 + x_2 \in \left\{ \cdots, -\frac{3\pi}{2}, -\frac{\pi}{2}, \frac{\pi}{2}, \frac{3\pi}{2}, \cdots \right\}.$$

The Hessian of the objective function is:

$$\nabla^2 f(x) = \begin{bmatrix} -\sin(x_1 + x_2) & -\sin(x_1 + x_2) \\ -\sin(x_1 + x_2) & -\sin(x_1 + x_2) \end{bmatrix}.$$

The determinants of the principal minors of this matrix are $-\sin(x_1 + x_2)$, $-\sin(x_1 + x_2)$, and 0. Thus, values of x^* for which $-\sin(x_1 + x_2) \geq 0$ satisfy the SONC. Substituting the stationary points found above into $-\sin(x_1 + x_2)$ gives points of the form:

$$x^* = \begin{pmatrix} x_1 \\ x_2 \end{pmatrix},$$

where:

$$x_1 + x_2 \in \left\{ \cdots, -\frac{5\pi}{2}, -\frac{\pi}{2}, \frac{3\pi}{2}, \frac{7\pi}{2}, \cdots \right\},$$

that satisfy both the FONC and SONC and are candidate local minima. We further know that the sine function is bounded between -1 and 1. Because these values of x that satisfy both the FONC and SONC make the objective function attain its lower bound of -1, we know that these points are indeed global and local minima. $\square$

Although the SONC can typically distinguish between local minima and maxima, points that satisfy the FONC and SONC are not necessarily local minima, as demonstrated in the following example.

Example 4.13 Consider the unconstrained optimization problem:

$$\min_{x} \ f(x) = x^3.$$

The FONC is $\nabla f(x) = 3x^2 = 0$, which gives $x^* = 0$ as the unique stationary point. We further have that $\nabla^2 f(x) = 6x$ and substituting $x^* = 0$ into the Hessian gives a value of zero, meaning that it is positive semidefinite. Thus, this stationary point also satisfies the SONC. However, visual inspection of the objective function, which is shown in Figure 4.20, shows that this point is not a local minimum. Instead, it is a point of inflection or a saddle point.   □

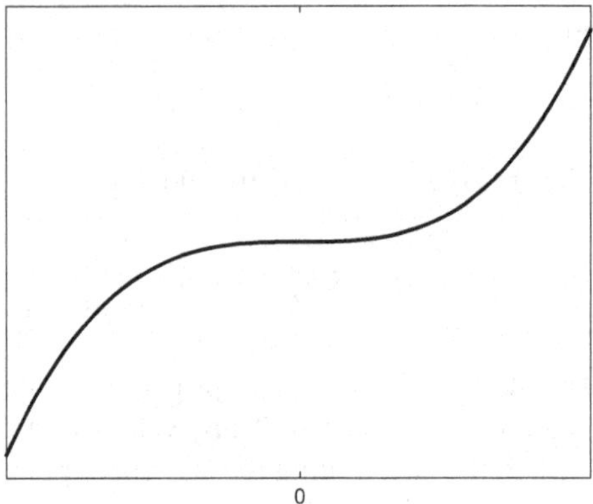

Fig. 4.20 Objective function in Example 4.13

The stationary point found in Example 4.13 is an example of a **saddle point**. A saddle point is a stationary point that is neither a local minimum nor maximum. Stationary points that have an indefinite Hessian matrix (*i.e.*, the Hessian is neither positive nor negative semidefinite) are guaranteed to be saddle points. However, Example 4.13 demonstrates that stationary points where the Hessian is positive or negative semidefinite can be saddle points as well.

This limitation of the FONC and the SONC—that they cannot eliminate saddle points—motivates our study of second-order sufficient conditions (SOSCs). A point that satisfies the SOSC is guaranteed to be a local minimum. The limitation of the SOSC, however, is that it is not typically a necessary condition. This means that there could be local minima that do not satisfy the SOSC. Thus, points that satisfy the FONC and SONC but do not satisfy the SOSC are still candidates for being local minima. The one exception to this is a convex optimization problem, in which case we have conditions that are both necessary and sufficient. We begin with a general SOSC that can be applied to any problem. We then consider the special case of convex optimization problems.

4.5.1.3 General Second-Order Sufficient Condition for Unconstrained Nonlinear Optimization Problems

We now state and demonstrate the general SOSC that can be applied to any problem.

General Second-Order Sufficient Condition for Unconstrained Nonlinear Optimization Problems: Consider an unconstrained nonlinear optimization problem of the form:

$$\min_{x \in \mathbb{R}^n} \; f(x).$$

If x^* satisfies $\nabla f(x^*) = 0$ and $\nabla^2 f(x^*)$ is positive definite, then x^* is a local minimum.

Let λ be the smallest eigenvalue of $\nabla^2 f(x^*)$. If we consider a point $x^* + d$ that is close to x^* (*i.e.*, by choosing a d with $||d||$ that is close to zero), the objective-function value at this point is approximately:

$$f(x^* + d) \approx f(x^*) + d^\top \nabla f(x^*) + \frac{1}{2} d^\top \nabla^2 f(x^*) d.$$

Because x^* is assumed to be stationary, we can rewrite this as:

$$f(x^* + d) - f(x^*) \approx \frac{1}{2} d^\top \nabla^2 f(x^*) d. \tag{4.13}$$

Using the Quadratic-Form Bound that is discussed in Section A.1, we further have that:

$$\frac{1}{2} d^\top \nabla^2 f(x^*) d \geq \frac{1}{2} \lambda ||d||^2 > 0, \tag{4.14}$$

where the last inequality follows from $\nabla^2 f(x^*)$ being positive definite, meaning that $\lambda > 0$. Combining (4.13) and (4.14) gives:

$$f(x^* + d) - f(x^*) > 0,$$

meaning that x^* gives an objective-function value that is strictly better than any point that is close to it. Thus, x^* is a local minimum.

The SOSC follows a similar line of reasoning to the SONC. By analyzing a second-order Taylor approximation of the objective function at points close to x^*, we can show that x^* being stationary and the Hessian being positive definite is sufficient for

x^* to be a local minimum. We demonstrate the use of the SOSC with the following examples.

Example 4.14 Consider the following unconstrained optimization problem, which is introduced in Example 4.6:

$$\min_{x} \ f(x) = (x_1 - 3)^2 + (x_2 + 4)^2.$$

This problem has a single stationary point:

$$x^* = \begin{pmatrix} 3 \\ -4 \end{pmatrix},$$

and the Hessian of the objective function is positive definite at this point. Because this point satisfies the SOSC, it is guaranteed to be a local minimum. □

Example 4.15 Consider the following unconstrained optimization problem, which is introduced in Example 4.8:

$$\min_{x} \ f(x) = x^3 - x^2 - 4x - 6.$$

This problem has two stationary points:

$$x^* \in \left\{ \frac{2 - \sqrt{52}}{6}, \frac{2 + \sqrt{52}}{6} \right\},$$

and the Hessian of the objective is positive definite at $(2 + \sqrt{52})/6$. Because this point satisfies the SOSC, it is guaranteed to be a local minimum, as confirmed in Figure 4.19. □

As discussed above, a limitation of the SOSC is that it is not a necessary condition for a point to be a local minimum. This means that there can be points that are local minima, yet do not satisfy the SOSC. We demonstrate this with the following example.

Example 4.16 Consider the following unconstrained optimization problem, which is introduced in Example 4.12:

$$\min_{x} \ f(x) = \sin(x_1 + x_2).$$

Points of the form:

$$x^* = \begin{pmatrix} x_1 \\ x_2 \end{pmatrix},$$

where:

$$x_1 + x_2 \in \left\{ \cdots, -\frac{5\pi}{2}, -\frac{\pi}{2}, \frac{3\pi}{2}, \frac{7\pi}{2}, \cdots \right\},$$

satisfy both the FONC and SONC. However, the determinant of the second leading principal minor of the Hessian of the objective function is zero at these points. Thus, the Hessian is only positive semidefinite (as opposed to positive definite) at these points. As such, these points do not satisfy the SOSC and are not guaranteed to be local minima on the basis of that optimality condition. However, we argue in Example 4.12 that because the objective function attains its lower bound of -1 at these points, they must be local and global minima. Thus, this problem has local minima that do not satisfy the SOSC. $\hfill\square$

4.5.1.4 Second-Order Sufficient Condition for Convex Unconstrained Optimization Problems

In the case of a convex optimization problem, it is much easier to guarantee that a point is a global minimum. This is because the FONC alone is also sufficient for a point to be a global minimum. This means that once we find a stationary point of a convex unconstrained optimization problem, we have a global minimum.

Sufficient Condition for Convex Unconstrained Optimization Problems:
Consider an unconstrained nonlinear optimization problem of the form:

$$\min_{x \in \mathbb{R}^n} \ f(x).$$

If the objective function, $f(x)$, is convex then the FONC is sufficient for a point to be a global minimum.

A differentiable convex function has the property that the tangent line to the function at any point, x^*, lies below the function (*cf.* Section B.2 for further details). Mathematically, this means that:

$$f(x) \geq f(x^*) + \nabla f(x^*)^\mathsf{T}(x - x^*), \ \forall x, x^*.$$

If we pick an x^* that is a stationary point, then this inequality becomes:

$$f(x) \geq f(x^*), \ \forall x,$$

which is the definition of a global minimum.

This proposition follows very simply from the properties of a convex function. The basic definition of a convex function is that every secant line connecting two points on the function be above the function. Another definition for a convex function that is once continuously differentiable is that every tangent line to the function is below the function. This gives the inequality in the proof above. If we examine the tangent line at a stationary point, such as that shown in Figure 4.21, the tangent is a horizontal line. Because the function must be above this horizontal line, that means it cannot attain a value that is lower than the value given by the stationary point, which is exactly the definition of a global minimum.

We demonstrate the use of this property with the following example.

Example 4.17 Consider the unconstrained nonlinear optimization problem:

$$\min_{x} \; f(x) = x_1^2 + x_2^2 + 2x_1 x_2.$$

The FONC is:

$$\nabla f(x) = \begin{pmatrix} 2x_1 + 2x_2 \\ 2x_1 + 2x_2 \end{pmatrix} = \begin{pmatrix} 0 \\ 0 \end{pmatrix},$$

which gives stationary points of the form:

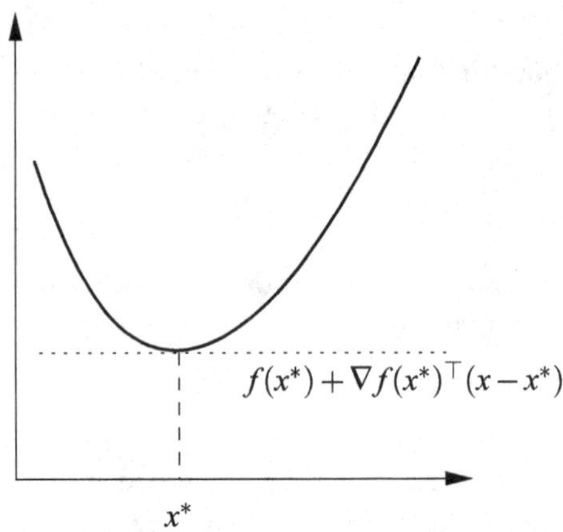

Fig. 4.21 Illustration of proof of sufficiency of FONC for a convex unconstrained optimization problem

$$x^* = \begin{pmatrix} x \\ -x \end{pmatrix}.$$

We further have that:

$$\nabla^2 f(x) = \begin{bmatrix} 2 & 2 \\ 2 & 2 \end{bmatrix},$$

which is positive semidefinite for all x. This means that the objective function is convex and this is a convex optimization problem. Thus, the stationary points we have found are global minima, because the FONC is sufficient for finding global minima of a convex optimization problem.   □

Finally, it is worth noting that the stationary points that are found in Example 4.17 do not satisfy the general SOSC, because the Hessian of the objective is not positive definite. Nevertheless, we can conclude that the stationary points found are global minima, because the problem is convex. This, again, demonstrates the important limitation of the SOSC, which is that it is generally only a sufficient and not necessary condition for local minima.

4.5.2   Equality-Constrained Nonlinear Optimization Problems

We now examine optimality conditions for equality-constrained nonlinear optimization problems. As with the unconstrained case, we begin by first discussing and demonstrating the use of an FONC for a local minimum. Although the SONC and SOSC for unconstrained problems can be generalized to the constrained case, these conditions are complicated. Thus, we restrict our attention to discussing sufficient conditions for the special case of a convex equality-constrained problem. More general cases of equality-constrained problems are analyzed by Bertsekas [2] and Luenberger and Ye [7].

The FONC that we discuss also has an important technical requirement, known as **regularity**. We omit this regularity requirement from the statement of the FONC and instead defer discussing regularity to Section 4.5.2.2. This is because in many cases the regularity condition is satisfied or does not affect the optimality conditions. However, we provide an example in Section 4.5.2.2 that shows how a local minimum can fail to satisfy the FONC if the regularity condition is not satisfied.

First-Order Necessary Condition for Equality-Constrained Nonlinear Optimization Problems: Consider an equality-constrained nonlinear optimization problem of the form:

$$\min_{x \in \mathbb{R}^n} \ f(x)$$

$$\text{s.t. } h_1(x) = 0$$

$$h_2(x) = 0$$

$$\vdots$$

$$h_m(x) = 0.$$

If x^* is a local minimum of this problem, then there exist m **Lagrange multipliers**, $\lambda_1^*, \lambda_2^*, \ldots, \lambda_m^*$, such that:

$$\nabla f(x^*) + \sum_{i=1}^{m} \lambda_i^* \nabla h_i(x^*) = 0.$$

The FONC for an equality-constrained problem require us to solve not only for values of the decisions variables in the original problem (*i.e.*, for x) but also for an additional set of m Lagrange multipliers. Note that the number of Lagrange multipliers is always equal to the number of equality constraints in the original problem. We demonstrate the use of the FONC and Lagrange multipliers to find candidate local minima in the following example.

Example 4.18 Consider the equality-constrained problem:

$$\min_{x} \ f(x) = 4x_1^2 + 3x_2^2 + 2x_1x_2 + 4x_1 + 6x_2 + 3$$

$$\text{s.t. } h_1(x) = x_1 - 2x_2 - 1 = 0$$

$$h_2(x) = x_1^2 + x_2^2 - 1 = 0.$$

To apply the FONC, we define two Lagrange multipliers, λ_1 and λ_2, which are associated with the two constraints. The FONC is then:

$$\nabla f(x^*) + \sum_{i=1}^{2} \lambda_i^* \nabla h_i(x^*) = 0,$$

or:

$$\begin{pmatrix} 8x_1^* + 2x_2^* + 4 \\ 6x_2^* + 2x_1^* + 6 \end{pmatrix} + \lambda_1^* \begin{pmatrix} 1 \\ -2 \end{pmatrix} + \lambda_2^* \begin{pmatrix} 2x_1^* \\ 2x_2^* \end{pmatrix} = \begin{pmatrix} 0 \\ 0 \end{pmatrix}.$$

Note that this is a system of two equations with four unknowns—the two original decision variables (x_1 and x_2) and the two Lagrange multipliers (λ_1 and λ_2). We do have two additional conditions that x must satisfy, which are the original constraints of the problem. If we add these two constraints, we now have the following system of four equations with four unknowns:

$$8x_1^* + 2x_2^* + 4 + \lambda_1^* + 2\lambda_2^* x_1^* = 0$$

$$6x_2^* + 2x_1^* + 6 - 2\lambda_1^* + 2\lambda_2^* x_2^* = 0$$

$$x_1^* - 2x_2^* - 1 = 0$$

$$x_1^{*2} + x_2^{*2} - 1 = 0.$$

This system of equations has two solutions:

$$(x_1^*, x_2^*, \lambda_1^*, \lambda_2^*) = (1, 0, 4, -8),$$

and:

$$(x_1^*, x_2^*, \lambda_1^*, \lambda_2^*) = (-3/5, -4/5, 24/25, -6/5).$$

Because these are the only two values of x and λ that satisfy the constraints of the problem and the FONC, the candidate values of x that can be local minima are:

$$\begin{pmatrix} x_1^* \\ x_2^* \end{pmatrix} = \begin{pmatrix} 1 \\ 0 \end{pmatrix},$$

and:

$$\begin{pmatrix} x_1^* \\ x_2^* \end{pmatrix} = \begin{pmatrix} -3/5 \\ -4/5 \end{pmatrix}.$$

We know that this problem is bounded, because the feasible region is bounded and the objective function does not asymptote. Thus, we know one of these two candidate points must be a global minimum. If we substitute these values into the objective function we have:

$$f\begin{pmatrix} 1 \\ 0 \end{pmatrix} = 11,$$

and:

$$f\begin{pmatrix} -3/5 \\ -4/5 \end{pmatrix} = \frac{3}{25}.$$

Because it gives a smaller objective-function value, we know that:

$$\begin{pmatrix} x_1^* \\ x_2^* \end{pmatrix} = \begin{pmatrix} -3/5 \\ -4/5 \end{pmatrix},$$

is the global minimum of this problem. $\square$

This example illustrates an important property of the FONC. When we add the original constraints of the problem, the number of equations we have is always equal to the number of unknowns that we solve for. This is because we have $n + m$ unknowns—n decision variables from the original problem and an additional m Lagrange multipliers (one for each constraint). We also have $n + m$ equations. There are n equations that come directly from the FONC, $i.e.$, the:

$$\nabla f(x^*) + \sum_{i=1}^{m} \lambda_i^* \nabla h_i(x^*) = 0.$$

This is because the gradient vectors have one partial derivative for each of the n original problem variables. We also have an additional m equations that come from the original constraints of the problem.

Note, however, that just as in the unconstrained case, having the same number of equations as unknowns does not imply that there is necessarily a unique solution to the FONC. We could have multiple solutions, as we have in Example 4.18, no solution (which could occur if the problem is infeasible or unbounded), or a unique solution.

Just as in the unconstrained case, the FONC give us candidate solutions that could be local minima. Moreover, points that do not satisfy the FONC *cannot* be local minima. Thus, the FONC typically eliminate many possible points from further consideration. Nevertheless, the FONC cannot necessarily distinguish between local minima, local maxima, and saddle points. The SONC and SOSC for unconstrained problems can be generalized to the equality-constrained case. However the most general second-order conditions are beyond the level of this book. Interested readers are referred to more advanced texts that cover these topics [2, 7]. We, instead, focus on a sufficient condition for the special case of a convex equality-constrained problem, which we now state.

Sufficient Condition for Convex Equality-Constrained Nonlinear Optimization Problems: Consider an equality-constrained nonlinear optimization problem of the form:

$$\min_{x \in \mathbb{R}^n} \ f(x)$$

$$\text{s.t. } h_1(x) = 0$$

$$h_2(x) = 0$$

$$\vdots$$

$$h_m(x) = 0.$$

If the constraint functions, $h_1(x), h_2(x), \ldots, h_m(x)$, are all linear in x and the objective, $f(x)$, is convex on the feasible region then the FONC is sufficient for a point to be a global minimum.

This result follows because an optimization with linear equality constraints and a convex objective function is a convex optimization problem. Convex optimization problems have the property that the FONC is sufficient for a point to be a local minimum [2]. Moreover, we know that any local minimum of a convex problem is a global minimum (*cf.* the Global-Minimum-of-Convex-Problem Property that is discussed in Section 4.4.4). Taken together, these properties give the sufficiency result. We now demonstrate the use of this property with the following example.

Example 4.19 Consider the equality-constrained problem:

$$\min_{x,y,z} f(x) = (x-3)^2 - 10 + (2y-4)^2 - 14 + (z-6)^2 - 6$$

$$\text{s.t. } h_1(x) = x + y + z - 10 = 0.$$

To solve this problem we introduce one Lagrange multiplier, λ_1. The FONC and the original constraint of the problem give us the following system of equations:

$$2(x-3) + \lambda_1 = 0$$

$$4(2y-4) + \lambda_1 = 0$$

$$2(z-6) + \lambda_1 = 0$$

$$x + y + z - 10 = 0.$$

The one solution to this system of equations is:

$$(x^*\ y^*\ z^*\ \lambda_1^*) = (23/9\ 17/9\ 50/9\ 8/9).$$

The constraint of this problem is linear and the Hessian of the objective function is:

$$\nabla^2 f(x, y, z) = \begin{bmatrix} 2 & 0 & 0 \\ 0 & 8 & 0 \\ 0 & 0 & 2 \end{bmatrix},$$

which is positive definite, meaning that the objective function is convex. Thus, the solution to the FONC is guaranteed to be a global minimum. $\square$

It is important to stress that this is only a *sufficient condition*. The problem given in Example 4.18 does not satisfy this condition, because the second constraint is not linear. Nevertheless, we are able to find a global minimum of the problem in that example by appealing to the fact that the problem is bounded and, thus, it must have a well defined global minimum which is also a local minimum. Because the FONC only gives us two candidate points that could be local minima, we know that the one that gives the smallest objective-function value is a global minimum.

4.5.2.1   Geometric Interpretation of the First-Order Necessary Condition for Equality-Constrained Nonlinear Optimization Problems

A general mathematical proof of the FONC for equality-constrained problems is beyond the level of this book (interested readers are referred to more advanced texts [2] for such a proof). We can, however, provide a geometric interpretation of the

FONC for equality-constrained problems. The FONC can be rewritten as:

$$\nabla f(x^*) = -\sum_{i=1}^{m} \lambda_i^* \nabla h_i(x^*),$$

which says that the gradient of the objective function at a local minimum must be a linear combination of the gradients of the constraint functions.

To understand why this is so, take a simple case of a problem with a single equality constraint and suppose that x^* is a local minimum of the problem. If so, we know that $h_1(x^*) = 0$ (i.e., x^* is feasible in the equality constraint). Now, consider directions, d, in which to move away from x^*. We know that a point, $x^* + d$, is feasible if and only if $h_1(x^* + d) = 0$. If we suppose that $||d||$ is close to zero, we can estimate the value of the constraint function at this point using a first-order Taylor approximation as:

$$h_1(x^* + d) \approx h_1(x^*) + d^\top \nabla h_1(x^*).$$

Because x^* is feasible, we have that $h_1(x^*) = 0$ and the Taylor approximation simplifies to:

$$h_1(x^* + d) \approx d^\top \nabla h_1(x^*).$$

Thus, $x^* + d$ is feasible so long as $d^\top \nabla h_1(x^*) = 0$. Put another way, the only directions in which we can feasibly move away from x^* are directions that are perpendicular to the gradient of the constraint function.

Let us now consider what effect moving in a feasible direction, d, away from x^* would have on the objective-function value. Again, assuming that $||d||$ is close to zero, we can estimate the objective-function value at this point using a first-order Taylor approximation as:

$$f(x^* + d) \approx f(x^*) + d^\top \nabla f(x^*).$$

Examining this Taylor approximation tells us that there are three possible things that can happen to the objective function if we move away from x^*. One is that $d^\top \nabla f(x^*) < 0$, meaning that the objective gets better. Clearly, this cannot happen if x^* is a local minimum, because that contradicts the definition of a local minimum. Along the same lines, if $d^\top \nabla f(x^*) > 0$ then we could feasibly move in the direction $-d$ and improve the objective function. This is because $-d^\top \nabla h_1(x^*) = 0$, meaning that this is a feasible direction in which to move away from x^*, and $-d^\top \nabla f(x^*) < 0$, meaning that the objective function improves. Clearly this cannot happen either. The third possibility is that $d^\top \nabla f(x^*) = 0$, meaning that the objective remains the same. This is the only possibility that satisfies the requirement of x^* being a local minimum.

In other words, if x^* is a local minimum, then we want to ensure that the only directions that we can feasibly move in are perpendicular to the gradient of the objective function. The FONC ensures that this is true, because it forces the gradient of the objective function to be a multiple of the gradient of the constraint. That way if we have a feasible direction, d, which has the property that $d^\top \nabla h_1(x^*) = 0$ then we also have that:

$$d^\top \nabla f(x^*) = -\lambda_1 d^\top \nabla h_1(x^*) = 0,$$

where the first equality comes from the FONC. With more than one constraint, we have to ensure that directions that are feasible to move in with respect to all of the constraints do not give an objective function improvement, which the FONC does.

Figure 4.22 graphically illustrates the FONC for a two-variable single-constraint problem. The figure shows the contour plot of the objective function and a local minimum, x^*, which gives an objective-function value of 0. The gradient of the constraint function points downward, thus the only directions that we can feasibly move away from x^* (based on the first-order Taylor approximation) is given by the dashed horizontal line. However, looking at the objective function gradient at this point, we see that these feasible directions we can move in give no change in the objective-function value.

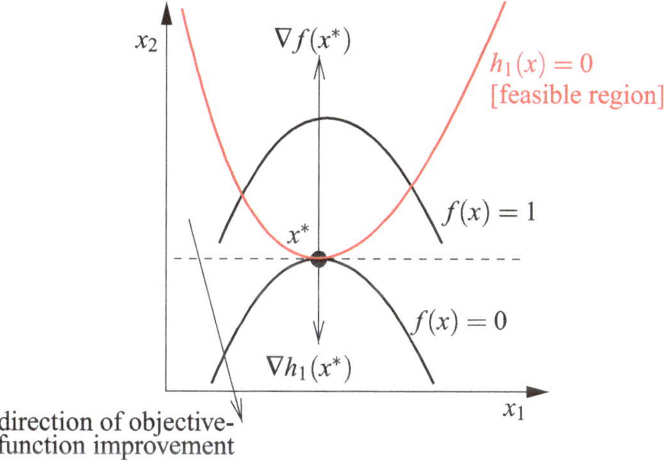

Fig. 4.22 Illustration of the FONC for an equality-constrained problem

Figure 4.23 demonstrates why a point that violates the FONC cannot be a local minimum. In this case, $\nabla f(x^*)$ is not a multiple of $\nabla h_1(x^*)$. Thus, if we move away from x^* in the direction d, which is feasible based on the first-order Taylor approximation of the constraint function, the objective function decreases. This violates the definition of a local minimum.

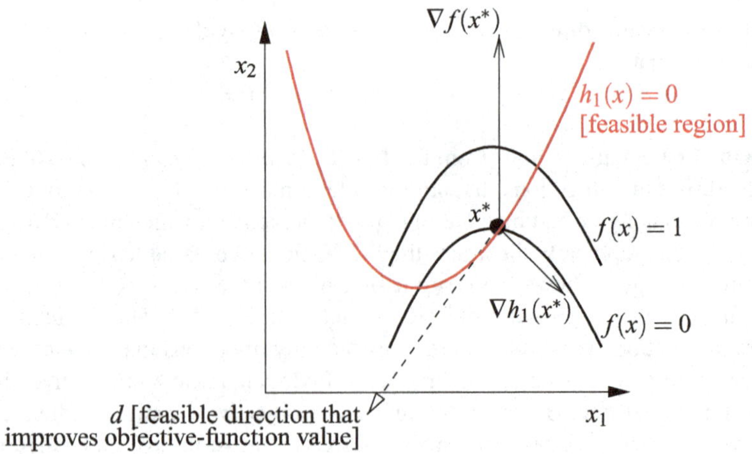

Fig. 4.23 Illustration of a point for which the FONC for an equality-constrained problem fails

4.5.2.2 An Added Wrinkle—Regularity

The FONC for equality-constrained problems has one additional technical requirement, which is known as regularity. A point is said to be **regular** if the gradients of the constraint functions at that point are all linearly independent. As the following example demonstrates, problems can have local minima that do not satisfy the regularity requirement, in which case they may not satisfy the FONC.

Example 4.20 Consider the equality-constrained problem:

$$\min_{x} \ f(x) = 2x_1 + 2x_2$$

$$\text{s.t. } h_1(x) = (x_1 - 1)^2 + x_2^2 - 1 = 0$$

$$h_2(x) = (x_1 + 1)^2 + x_2^2 - 1 = 0.$$

To apply the FONC to this problem, we define two Lagrange multipliers, λ_1 and λ_2, associated with the two constraints. The FONC and constraints of the problem are:

$$2 + 2\lambda_1(x_1 - 1) + 2\lambda_2(x_1 + 1) = 0$$

$$2 + 2\lambda_1 x_2 + 2\lambda_2 x_2 = 0$$

$$(x_1 - 1)^2 + x_2^2 - 1 = 0$$

$$(x_1 + 1)^2 + x_2^2 - 1 = 0.$$

Simultaneously, solving the two constraints gives $(x_1, x_2) = (0, 0)$. Substituting these values into the FONC gives:

$$2 - 2\lambda_1 + 2\lambda_2 = 0$$

$$2 = 0,$$

which clearly has no solution.

However, $(x_1, x_2) = (0, 0)$ is the only feasible solution in the constraints, thus it is by definition a global and local minimum. This means that this problem has a local minimum that does not satisfy the FONC. $\square$

Figure 4.24 illustrates why the FONC fails in Example 4.20. The figure shows the feasible regions defined by each of the two constraints and x^*, which is the unique feasible solution. Because x^* is the only feasible solution, it is by definition a local and global minimum. However, at this point the gradients of the constraints are the two horizontal vectors shown in the figure, which are not linearly independent. Because the gradient of the objective function is not horizontal, it is impossible to write it as a linear combination of the constraint gradients. This is a consequence of the problem in Example 4.20 having a local minimum that violates the regularity assumption.

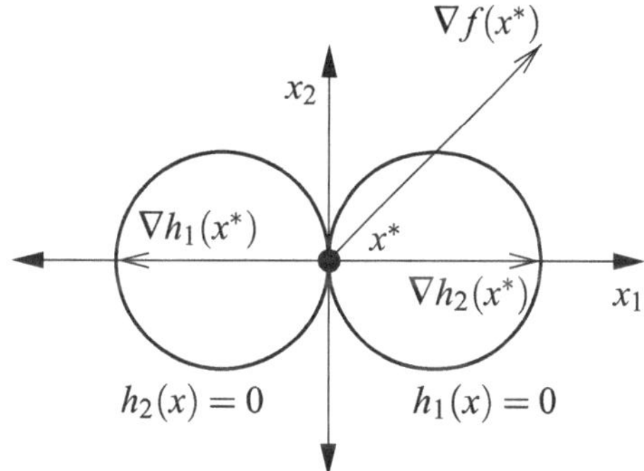

Fig. 4.24 Illustration of why the FONC fails for the equality-constrained problem in Example 4.20

It is worth noting that a point that violates the regularity assumption can still satisfy the FONC. For instance, if the objective function of the problem in Example 4.20 is changed to $f(x) = 2x_1$, the gradient of the objective function becomes:

$$\nabla f(x^*) = \begin{pmatrix} 2 \\ 0 \end{pmatrix}.$$

This gradient is a horizontal vector and can be written as a linear combination of the constraint function gradients. Interested readers are referred to more advanced texts [2], which discuss two important aspects of this regularity issue. One is a more general version of the FONC that does not require regularity. The other is what is known as constraint-qualification conditions. A problem that satisfies these constraint-qualification conditions are guaranteed to have local minima that satisfy the FONC.

4.5.3 Equality- and Inequality-Constrained Nonlinear Optimization Problems

As in the equality-constrained case, we examine equality- and inequality-constrained problems by using a FONC. We then discuss sufficient conditions for the special case of a convex equality- and inequality-constrained problem. Interested readers are referred to more advanced texts [2, 7] for the more general SONC and SOSC for equality- and inequality-constrained problems. As in the equality-constrained case, the FONC for inequality- and equality-constrained problems also have a regularity requirement. We again omit the regularity requirement from the statement of the FONC and instead defer discussion of this requirement to Section 4.5.3.2.

First-Order Necessary Condition for Equality- and Inequality-Constrained Nonlinear Optimization Problems: Consider an equality- and inequality-constrained nonlinear optimization problem of the form:

$$\min_{x \in \mathbb{R}^n} \; f(x)$$

$$\text{s.t.} \; h_1(x) = 0$$

$$h_2(x) = 0$$

$$\vdots$$

$$h_m(x) = 0$$

$$g_1(x) \leq 0$$

$$g_2(x) \leq 0$$

$$\vdots$$

$$g_r(x) \leq 0.$$

If x^* is a local minimum of this problem, then there exist $(m + r)$ **Lagrange multipliers**, $\lambda_1^*, \lambda_2^*, \ldots, \lambda_m^*$ and $\mu_1^*, \mu_2^*, \ldots, \mu_r^*$, such that:

$$\nabla f(x^*) + \sum_{i=1}^{m} \lambda_i^* \nabla h_i(x^*) + \sum_{j=1}^{r} \mu_j^* \nabla g_j(x^*) = 0$$

$$\mu_1^* \geq 0$$

$$\mu_2^* \geq 0$$

$$\vdots$$

$$\mu_r^* \geq 0$$

$$\mu_1^* g_1(x^*) = 0$$

$$\mu_2^* g_2(x^*) = 0$$

$$\vdots$$

$$\mu_r^* g_r(x^*) = 0.$$

The FONC for equality-constrained and equality- and inequality-constrained problems are very similar, in that we retain a condition involving the sum of the gradients of the objective and constraint functions and Lagrange multipliers. The FONC for equality- and inequality-constrained problems differ, however, in that the Lagrange multipliers on the *inequality* constraints must be non-negative. The third set of conditions for the equality- and inequality-constrained case is **complementary-slackness**.

To understand the complementary-slackness conditions, we first define what it means for an inequality constraint to be binding as opposed to non-binding at a solution. Note that these definitions follow immediately from analogous definitions given for linear inequality constraints in Section 2.7.6. Consider the inequality constraint:

$$g_j(x) \leq 0.$$

We say that this constraint is **non-binding** at x^* if:

$$g_j(x^*) < 0.$$

Thus, a non-binding constraint has the property that when we substitute x^* into it, there is a difference or slack between the two sides of the constraint. Conversely, this constraint is said to be **binding** at x^* if:

$$g_j(x^*) = 0.$$

Let us now examine the complementary slackness condition, taking the case of the jth inequality constraint in the following discussion. The condition requires that:

$$\mu_j^* g_j(x^*) = 0.$$

In other words, we must have $\mu_j^* = 0$ (*i.e.*, the Lagrange multiplier associated with the jth constraint is equal to zero), $g_j(x^*) = 0$ (*i.e.*, the jth inequality constraint is binding), or both. This complementary-slackness condition is analogous to the complementary-slackness conditions for linear optimization problems, which are introduced in Section 2.7.6.

The complementary-slackness condition is sometimes abbreviated as:

$$\mu_j^* \geq 0 \perp g_j(x_j^*) \leq 0.$$

What this condition says is that $\mu_j^* \geq 0$ and $g_j(x_j^*) \leq 0$ (as before). Moreover, the $\perp$ says that μ_j^* must be perpendicular to $g_j(x_j^*)$, in the sense that their product is zero. Thus, the FONC for equality- and inequality-constrained problems are often written more compactly as:

$$\nabla f(x^*) + \sum_{i=1}^{m} \lambda_i^* \nabla h_i(x^*) + \sum_{j=1}^{r} \mu_j^* \nabla g_j(x^*) = 0$$

$$h_1(x^*) = 0$$

$$h_2(x^*) = 0$$

$$\vdots$$

$$h_m(x^*) = 0$$

$$\mu_1^* \geq 0 \perp g_1(x^*) \leq 0$$

$$\mu_2^* \geq 0 \perp g_2(x^*) \leq 0$$

$$\vdots$$

$$\mu_r^* \geq 0 \perp g_r(x^*) \leq 0.$$

We finally note that the FONC for equality- and inequality-constrained problems are often referred to as the **Karush-Kuhn-Tucker (KKT) conditions**. The KKT conditions are named after the three people who discovered the result. Karush first formulated the KKT conditions in his M.S. thesis. Quite a few years later Kuhn and Tucker rediscovered them independently.

Example 4.21 Consider the equality- and inequality-constrained problem (that only has inequality constraints):

$$\min_{x} \; f(x) = 2x_1^2 + 2x_1 x_2 + 2x_2^2 + x_1 + x_2$$

$$\text{s.t. } g_1(x) = x_1^2 + x_2^2 - 9 \leq 0$$

$$g_2(x) = -x_1 + 2x_2 + 1 \leq 0.$$

To write out the KKT conditions we define two Lagrange multipliers, μ_1 and μ_2, associated with the two inequality constraints. The KKT conditions and the constraints of the original problem are then:

$$4x_1 + 2x_2 + 1 + 2\mu_1 x_1 - \mu_2 = 0$$

$$2x_1 + 4x_2 + 1 + 2\mu_1 x_2 + 2\mu_2 = 0$$

$$\mu_1 \geq 0 \perp x_1^2 + x_2^2 - 9 \leq 0$$

$$\mu_2 \geq 0 \perp -x_1 + 2x_2 + 1 \leq 0.$$

Note that these conditions are considerably more difficult to work with than the FONC in the unconstrained and equality-constrained cases. This is because we now have a system of equations and inequalities, the latter coming from the inequality constraints and the non-negativity restrictions on the Lagrange multipliers associated with them.

As such, we approach equality- and inequality-constrained problems by conjecturing which of the inequality constraints are binding and non-binding, and then solving the resulting the KKT conditions. We must examine all combinations of binding and non-binding constraints until we find all solutions to the KKT conditions.

With the problem at hand, let us first consider the case in which neither of the inequality constraints are binding. The complementary-slackness conditions then imply that $\mu_1 = 0$ and $\mu_2 = 0$. The gradient conditions are then simplified to:

$$4x_1 + 2x_2 + 1 = 0$$

$$2x_1 + 4x_2 + 1 = 0.$$

Solving the two equations gives:

$$(x_1 \ x_2) = (-1/6 \ -1/6),$$

meaning we have found as a possible KKT point:

$$(x_1 \ x_2 \ \mu_1 \ \mu_2) = (-1/6 \ -1/6 \ 0 \ 0).$$

However, we only found these values of x and μ by assuming which of the constraints are binding and non-binding (to determine the value of μ) and then solving for x in the gradient conditions. We must still check to ensure that these values satisfy all of the other conditions. If we do so, we see that the second inequality constraint is violated, meaning that this is not a solution to the KKT conditions.

We next consider the case in which the first inequality constraint is binding and the second is non-binding. The complementary-slackness conditions then imply that $\mu_2 = 0$ whereas we cannot make any determination about μ_1. Thus, the gradient conditions become:

$$4x_1 + 2x_2 + 1 + 2\mu_1 x_1 = 0$$

$$2x_1 + 4x_2 + 1 + 2\mu_1 x_2 = 0.$$

This is a system of two equations with three unknowns. We, however, have one additional equality that x must satisfy, which is the first inequality constraint. Because we are assuming in this case that this constraint is binding, we impose it as a third equation:

$$x_1^2 + x_2^2 - 9 = 0.$$

Solving this system of equations gives:

$$(x_1 \ x_2 \ \mu_1) \approx (-2.12 \ -2.12 \ -2.76),$$

$$(x_1 \ x_2 \ \mu_1) \approx (1.86 \ -2.36 \ -1.00),$$

and:

$$(x_1 \ x_2 \ \mu_1) \approx (-2.36 \ 1.86 \ -1.00),$$

meaning that:

$$(x_1 \ x_2 \ \mu_1 \ \mu_2) \approx (-2.12 \ -2.12 \ -2.76 \ 0),$$

$$(x_1 \ x_2 \ \mu_1 \ \mu_2) \approx (1.86 \ -2.36 \ -1.00 \ 0),$$

and:

$$(x_1 \ x_2 \ \mu_1 \ \mu_2) \approx (-2.36 \ 1.86 \ -1.00 \ 0),$$

are candidate KKT points. However, because μ_1 is negative in all three of these vectors, these are not KKT points.

The third case that we examine is the one in which the first inequality constraint is non-binding and the second inequality is binding. The complementary-slackness conditions imply that $\mu_1 = 0$ whereas we cannot make any determination regarding the value of μ_2. Thus, the simplified gradient conditions and the second inequality constraint (which we impose as an equality) are:

$$4x_1 + 2x_2 + 1 - \mu_2 = 0$$

$$2x_1 + 4x_2 + 1 + 2\mu_2 = 0$$

$$-x_1 + 2x_2 + 1 = 0.$$

Solving these three equations gives:

$$(x_1 \ x_2 \ \mu_2) = (1/14 \ -13/28 \ 5/14),$$

meaning that we have:

$$(x_1 \ x_2 \ \mu_1 \ \mu_2) = (1/14 \ -13/28 \ 0 \ 5/14),$$

as a possible solution KKT point. Moreover, when we check the remaining conditions, we find that they are all satisfied, meaning that this is indeed a KKT point.

The last possible case that we examine is the one in which both of the inequality constraints are binding. In this case, complementary slackness does not allow us to fix any of the μ's equal to zero. Thus, we solve the following system of equations:

$$4x_1 + 2x_2 + 1 + 2\mu_1 x_1 - \mu_2 = 0$$

$$2x_1 + 4x_2 + 1 + 2\mu_1 x_2 + 2\mu_2 = 0$$

$$x_1^2 + x_2^2 - 9 = 0$$

$$-x_1 + 2x_2 + 1 = 0,$$

which has the solutions:

$$(x_1 \ x_2 \ \mu_1 \ \mu_2) \approx (-2.45 \ -1.73 \ -2.66 \ 0.81),$$

and:

$$(x_1 \ x_2 \ \mu_1 \ \mu_2) \approx (2.85 \ 0.93 \ -2.94 \ -2.49).$$

Clearly neither of these are KKT points, because both of them have negative values for μ_1.

Thus, the only solution to the KKT conditions and the only candidate point that could be a local minimum is $(x_1^*, x_2^*) = (1/14, -13/28)$. $\qquad\qquad\square$

Below we give an algorithm for finding KKT points. We first, in Step 2, conjecture which inequalities are binding and non-binding. We next, in Step 3, fix the μ's associated with non-binding constraints equal to zero (due to the complementary-slackness requirement). Next, in Step 4, we solve the system of equations given by the gradient conditions, all of the equality constraints, and any inequality constraints that are assumed to be binding. Inequalities that are assumed to be binding are written as equal-to-zero constraints. We finally check in Step 5 that x satisfies the inequality constraints that we assume to be non-binding in Step 2 (and which are, thus, ignored when solving for x and μ). We also check that μ is non-negative. If both of these conditions are true, then the values found for x and μ in Step 4 of the algorithm give a KKT point. Otherwise, they do not and the point is discarded from further consideration.

Algorithm for Finding KKT Points

1: **procedure** KKT FIND
2:     Fix which inequalities are binding and non-binding
3:     Fix μ's associated with inequalities assumed to be non-binding to zero
4:     Solve system of equations given by gradient conditions, equality constraints, and binding inequalities (written as equalities)
5:     Check that x satisfies constraints assumed to be non-binding and μ is non-negative
6: **end procedure**

Although the Algorithm for Finding KKT Points provides an efficient way to handle the inequalities in the KKT conditions, it is still quite cumbersome. This is because finding all KKT points typically requires the process be repeated for *every* possible combination of binding and non-binding inequality constraints. The problem in Example 4.21 has two inequality constraints, which gives us four cases to check. A problem with r inequality constraints would require checking 2^r cases. Clearly, even a small problem can require many cases to be tested to find all KKT points. For instance, with $r = 10$ we must check 1024 cases whereas a 100-inequality problem would require about 1.27×10^{30} cases to be examined.

There are two ways that we can reduce the search process. First, we use the fact that for a convex equality- and inequality-constrained problem, the KKT conditions are sufficient for a global minimum. This implies that as soon as a KKT point is found, we need not search any further. This is because we know the point that we have is a global minimum. The second is that we can use knowledge of a problem's structure to determine if a set of constraints would be binding or not in an optimum. We begin by first discussing the convex case.

Sufficient Condition for Convex Equality- and Inequality-Constrained Nonlinear Optimization Problems: Consider an equality- and inequality-constrained nonlinear optimization problem of the form:

$$\min_{x \in \mathbb{R}^n} \ f(x)$$

$$\text{s.t. } h_1(x) = 0$$

$$h_2(x) = 0$$

$$\vdots$$

$$h_m(x) = 0$$

$$g_1(x) \leq 0$$

$$g_2(x) \leq 0$$

$$\vdots$$

$$g_r(x) \leq 0.$$

> If the equality constraint functions, $h_1(x), h_2(x), \ldots, h_m(x)$, are all linear in x and the inequality-constraint functions, $g_1(x), g_2(x), \ldots, g_r(x)$, and the objective function, $f(x)$, are all convex on the feasible region then the KKT condition is sufficient for a point to be a global minimum.

Example 4.22 Consider the following equality- and inequality-constrained problem, which is given in Example 4.21:

$$\min_x \ f(x) = 2x_1^2 + 2x_1x_2 + 2x_2^2 + x_1 + x_2$$

$$\text{s.t. } g_1(x) = x_1^2 + x_2^2 - 9 \leq 0$$

$$g_2(x) = -x_1 + 2x_2 + 1 \leq 0.$$

Note that the Hessian of the objective function is:

$$\nabla^2 f(x) = \begin{bmatrix} 4 & 2 \\ 2 & 4 \end{bmatrix},$$

which is positive definite, meaning that the objective function is a convex function. Moreover, the second inequality constraint is linear, which we know defines a convex feasible region. The Hessian of the first inequality-constraint function is:

$$\nabla^2 f(x) = \begin{bmatrix} 2 & 0 \\ 0 & 2 \end{bmatrix},$$

which is also positive definite, meaning that this constraint function is also convex. Thus, this problem is convex and any KKT point that we find is guaranteed to be a global minimum. This means that once we find the KKT point $(x_1, x_2, \mu_1, \mu_2) = (1/14, -13/28, 0, 5/14)$, we can stop and ignore the fourth case, because we have a global minimum.

It is also worth noting that when we have a convex equality- and inequality-constrained problem, our goal is to find a KKT point as quickly as possible. Because we first try the case in which neither constraint is binding and find that the second constraint is violated, it could make sense to next examine the case in which the second constraint is binding and the first constraint is non-binding (the third case that we examine in Example 4.21). This shortcut—assuming that violated constraints are binding in an optimal solution—will often yield a KKT point more quickly than randomly examining different combinations of binding and non-binding inequality constraints.   □

Another approach to reducing the number of cases to examine in finding KKT points is to use knowledge of a problem's structure to determine if some constraints are binding or non-binding in an optimum. We demonstrate this approach with the following example.

Example 4.23 Consider the Packing-Box Problem, which is formulated as:

$$\max_{h,w,d} \; hwd$$

$$\text{s.t. } 2wh + 2dh + 6wd \le 60$$

$$w, h, d \ge 0,$$

in Section 4.1.1.1.

This problem has four inequality constraints, thus exhaustively checking all combinations of binding and non-binding inequalities would result in examining 16 cases. As opposed to doing this, let us argue that some of the constraints must be binding or non-binding in an optimal solution. We begin by arguing that each of the nonnegativity constraints must be non-binding. To see this, note that if any of h, w, or d equals zero, then we have a box with a volume of $0 \, \text{cm}^3$. Setting each of h, w, and d equal to one gives a box with a larger volume and does not violate the restriction on the amount of cardboard that can be used. Thus, a box with a volume of zero cannot be optimal. Knowing that these three constraints must be non-binding in an optimum has reduced the number of cases that we must examine from 16 to two.

We can, further, argue that the first constraint must be binding, which gives us only a single case to examine. To see this, note that if the constraint is non-binding, this means that there is unused cardboard. In such a case, we can increase the value of any one of h, w, or d by a small amount so as not to violate the $60 \, \text{cm}^2$ restriction, and at the same time increase the volume of the box. Thus, a box that does not use the full $60 \, \text{cm}^2$ of cardboard cannot be optimal. Knowing this, the number of cases that we must examine is reduced to one.

To solve for an optimal solution, we convert the problem to standard form, which is:

$$\min_{h,w,d} \; f(h, w, d) = -hwd$$

$$\text{s.t. } g_1(h, w, d) = 2wh + 2dh + 6wd - 60 \le 0$$

$$g_2(h, w, d) = -h \le 0$$

$$g_3(h, w, d) = -w \le 0$$

$$g_3(h, w, d) = -d \le 0.$$

If we assign four Lagrange multipliers, μ_1, μ_2, μ_3, and μ_4, to the inequality constraints, the KKT conditions are:

$$-wd + \mu_1 \cdot (2w + 2d) - \mu_2 = 0$$

$$-hd + \mu_1 \cdot (2h + 6d) - \mu_3 = 0$$

$$-hw + \mu_1 \cdot (2h + 6w) - \mu_4 = 0$$

$$\mu_1 \geq 0$$

$$\mu_2 \geq 0$$

$$\mu_3 \geq 0$$

$$\mu_4 \geq 0$$

$$\mu_1 \cdot (2wh + 2dh + 6wd - 60) = 0$$

$$-\mu_2 h = 0$$

$$-\mu_3 w = 0$$

$$-\mu_4 d = 0$$

$$2wh + 2dh + 6wd - 60 \leq 0$$

$$-h \leq 0$$

$$-w \leq 0$$

$$-d \leq 0.$$

Based on the argument just presented, we must only consider one case in which the first constraint is binding and the others non-binding. The complementary-slackness and gradient conditions and binding constraint give us the following system of equations:

$$-wd + \mu_1 \cdot (2w + 2d) = 0$$

$$-hd + \mu_1 \cdot (2h + 6d) = 0$$

$$-hw + \mu_1 \cdot (2h + 6w) = 0$$

$$2wh + 2dh + 6wd - 60 = 0,$$

which has the solution:

$$(h \ w \ d \ \mu_1) \approx (5.48 \ 1.83 \ 1.83 \ 0.46).$$

Because this problem has a bounded feasible region and the objective does not asymptote, the point:

$$(h \ w \ d) \approx (5.48 \ 1.83 \ 1.83),$$

which is the only candidate local minimum, must be a local and global minimum of this problem.   □

4.5.3.1 Geometric Interpretation of the Karush-Kuhn-Tucker Condition for Equality- and Inequality-Constrained Problems

It is helpful to provide some intuition behind the KKT condition for equality- and inequality-constrained problems. We specifically examine the gradient and complementary-slackness conditions and the sign restriction on Lagrange multipliers for inequality constraints. Thus, we examine the case of an equality- and inequality-constrained problem that only has inequality constraints.

To understand how the KKT condition is derived, consider a two-variable problem with two inequality constraints:

$$\min_{x} \ f(x)$$

$$\text{s.t. } g_1(x) \le 0$$

$$g_2(x) \le 0,$$

and suppose that x^* is a local minimum of this problem. Figure 4.25 shows the constraints and feasible region of this problem and where x^* lies in relation to them. As shown in the figure, the first constraint is binding at x^*. This is because x^* is on the boundary defined by the first constraint, meaning that there is no slack in the two sides of the constraints. Conversely, the second constraint is non-binding at x^*. This is because x^* is not on the boundary of the constraint. Thus, there is slack between the two sides of the constraint. Because x^* is a local minimum, we know that there is a neighborhood of feasible points around x^* with the property that x^* has the smallest objective-function value on this neighborhood. This neighborhood is denoted by the dotted circle centered around x^* in Figure 4.25. All of the points in the shaded region that are within the dotted circle give objective-function values that are greater than or equal to $f(x^*)$.

Now consider the problem

$$\min_{x} \ f(x)$$

$$\text{s.t. } g_1(x) = 0.$$

Figure 4.26 shows the feasible region of this problem and x^*. This problem has the same objective function as the problem shown in Figure 4.25, but the feasible region differs. Specifically, the binding constraint from the original problem is now an equality constraint and the non-binding constraint is removed. Figure 4.26 shows the same neighborhood of points around x^*, denoted by the dotted circle. Note that if x^* gives the best objective-function value in the neighborhood shown in Figure 4.25 then it also gives the best objective-function value in the neighborhood shown in

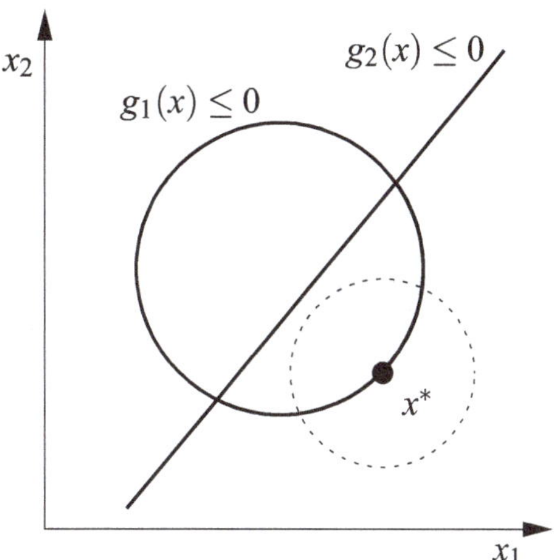

Fig. 4.25 A local minimum of a two-variable problem with two inequality constraints

Figure 4.26. This is because the neighborhood in Figure 4.26 has fewer points (only those on the boundary where $g_1(x) = 0$, which is highlighted in red in Figure 4.26) and the objective function of the two problems are identical.

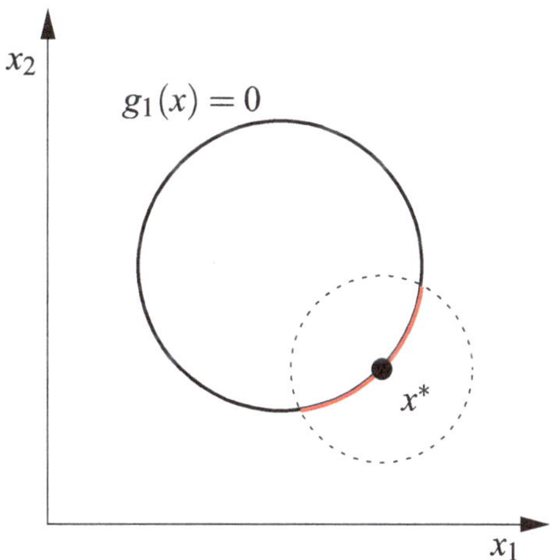

Fig. 4.26 A local minimum of a two-variable problem with one equality constraint that is equivalent to the problem illustrated in Figure 4.25

We, thus, conclude that if x^* is a local minimum of the problem:

$$\min_{x} \ f(x)$$

$$\text{s.t. } g_1(x) \leq 0$$

$$g_2(x) \leq 0,$$

and that only the first constraint is binding at x^*, then it must also be a local minimum of the problem:

$$\min_x \; f(x)$$

$$\text{s.t. } g_1(x) = 0.$$

This second problem is an equality-constrained problem. Thus, we can apply the FONC for equality-constrained problems, which is discussed in Section 4.5.2, to it. Doing so gives:

$$\nabla f(x^*) + \mu_1^* g_1(x^*) = 0,$$

where we are letting μ_1^* denote the Lagrange multiplier on the equality constraint. If we define $\mu_2^* = 0$, we can write this gradient condition as:

$$\nabla f(x^*) + \mu_1^* g_1(x^*) + \mu_2^* g_2(x^*) = \nabla f(x^*) + \sum_{j=1}^{r} \mu_r^* g_r(x^*) = 0, \qquad (4.15)$$

which is the gradient condition we have if we apply the KKT condition to the original equality- and inequality-constrained problem. We further have the complementary-slackness condition that the KKT condition requires. This is because we fix $\mu_2^* = 0$ when deriving equation (4.15). Note, however, that μ_2^* is the Lagrange multiplier on the second inequality constraint. Moreover, the second inequality constraint is the one that is non-binding at the point x^*, as shown in Figure 4.25.

Thus, the gradient and complementary-slackness requirements of the KKT condition can be derived by applying FONC to the equivalent equality-constrained problem.

We can also provide some intuition around the sign restriction on Lagrange multipliers by conducting this type of analysis. Again, if we take the two-constraint problem:

$$\min_x \; f(x)$$

$$\text{s.t. } g_1(x) \le 0$$

$$g_2(x) \le 0,$$

then the gradient condition is:

$$\nabla f(x^*) + \mu_1^* g_1(x^*) + \mu_2^* g_2(x^*) = \nabla f(x^*) + \mu_1^* g_1(x^*) = 0,$$

because we are assuming that the second inequality constraint is non-binding and by complementary slackness we have that $\mu_2^* = 0$. This condition can be rewritten as:

$$\nabla f(x^*) = -\mu_1^* g_1(x^*).$$

This condition has the same interpretation that is discussed for equality-constrained problems in Section 4.5.2.1. Namely, it says that the gradient of the objective function must be linearly dependent with the gradient of the binding inequality constraint at a local minimum. However, if we further restrict $\mu_1^* \geq 0$, then the gradient condition further says that the gradient of the objective function must be a *non-positive* multiple of the gradient of the binding inequality constraint at a local minimum. Figures 4.27 and 4.28 show why this sign restriction is important.

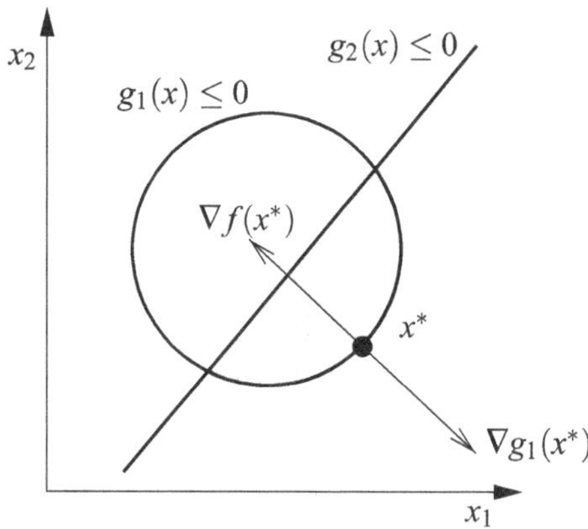

Fig. 4.27 $\nabla f(x^*)$ and $\nabla g_1(x^*)$ of a two-variable problem with two inequality constraints if $\mu_1^* \geq 0$

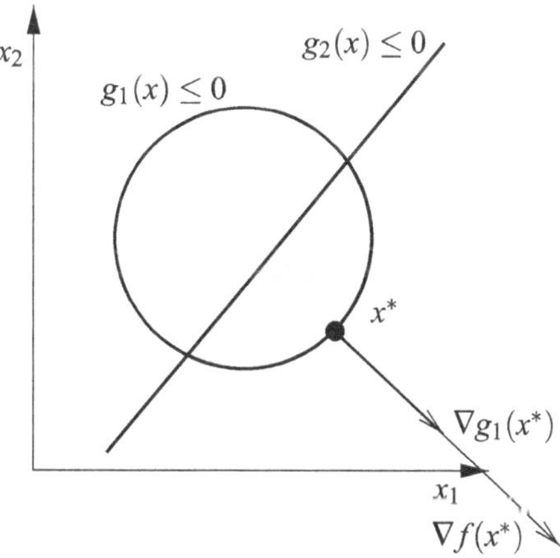

Fig. 4.28 $\nabla f(x^*)$ and $\nabla g_1(x^*)$ of a two-variable problem with two inequality constraints if $\mu_1^* \leq 0$

Figure 4.27 shows the gradient of the binding inequality constraint at the local minimum. We know that this gradient points outward from the feasible region, because that is the direction in which $g_1(x)$ increases. Recall that when we provide a geometric interpretation of the Lagrange multipliers for equality-constrained problems in Section 4.5.2.1, we find that the only directions in which we can move away from a local minimum are perpendicular to the gradient of the constraint function. This is no longer true when we have inequality constraints. Indeed, we can move in any direction away from x^* *into* the shaded region that is shown in Figure 4.27. We know, however, that we *cannot* move away from x^* in the direction of $\nabla g_1(x^*)$. This is because $g_1(x^*) = 0$ (because the first constraint is binding at x^*) and because $\nabla g_1(x^*)$ is a direction in which $g_1(x)$ increases. Thus, moving in the direction of $\nabla g_1(x^*)$ would violate the constraint.

Figure 4.27 also shows that if $\mu_1^* \geq 0$, then the gradient of $f(x^*)$ is pointing inward to the feasible region. This is desirable because we know that $\nabla f(x^*)$ is a direction in which the objective function increases, meaning that $-\nabla f(x^*)$ is a direction in which the objective function decreases. To see this, note that if we move in a direction, $d = -\nabla f(x^*)$, away from x^*, the first-order Taylor approximation of the objective function at the new point is:

$$f(x^* + d) = f(x^* - \nabla f(x^*)) \approx f(x^*) - \nabla f(x^*)^\mathsf{T} \nabla f(x^*) < f(x^*).$$

However, because $-\nabla f(x^*)$ points in the same direction as $\nabla g_1(x^*)$, this direction that decreases the objective function is an infeasible direction to move in.

Figure 4.28 also shows the gradient of the binding inequality constraint at the local minimum. It further shows that if $\mu_1^* \leq 0$, then $\nabla f(x^*)$ and $\nabla g_1(x^*)$ point in the same direction. However, x^* cannot be a local minimum in this case. The reason is that if we move in the direction of $-\nabla f(x^*)$ away from x^* (which is a feasible direction to move in, because $\nabla f(x^*)$ and $\nabla g_1(x^*)$ now point in the same direction), the objective function decreases.

Finally, we can use this same kind of analysis to show the complementary-slackness condition in another way. Figure 4.29 shows a problem with the same feasible region as that shown in Figures 4.25–4.28, however the objective function is now different and the local minimum, x^*, is interior to both inequality constraints. The gradient condition for this problem would be:

$$\nabla f(x^*) + \mu_1^* g_1(x^*) + \mu_2^* g_2(x^*) = \nabla f(x^*) = 0,$$

because we are assuming that both inequality constraints are non-binding and by complementary slackness we have that $\mu_1^* = 0$ and $\mu_2^* = 0$. In some sense, the complementary-slackness condition says that the gradient condition should ignore the two non-binding inequality constraints and find a point at which the gradient of the objective function is equal to zero. Figure 4.29 shows the logic of this condition by supposing that $\nabla f(x^*) \neq 0$. If the gradient of the objective function is as shown in the figure, then x^* cannot be a local minimum, because moving a small distance in the direction $d = -\nabla f(x^*)$ away from x^* (which is a feasible direction to move in) reduces the objective function compared to x^*. This, however, contradicts the definition of a local minimum.

All of these derivations can be generalized to problems with any number of variables and equality and inequality constraints. However, we focus on problems with two variables and only two inequality constraints to simplify this discussion.

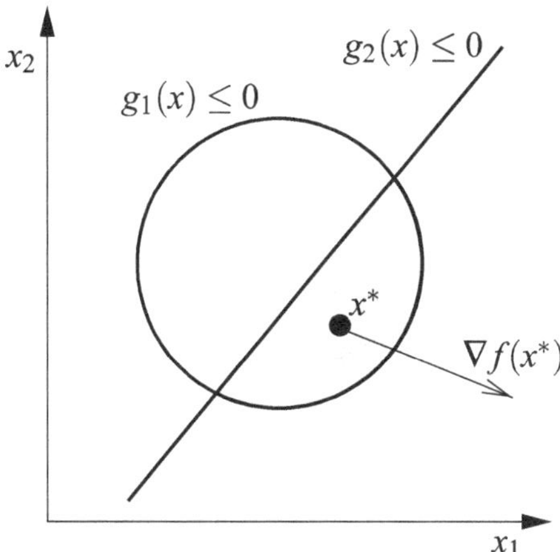

Fig. 4.29 $\nabla f(x^*)$ when inequality constraints are non-binding

4.5.3.2 Regularity and the Karush-Kuhn-Tucker Condition

When applied to equality- and inequality-constrained problems, the KKT condition has the same regularity requirement that we have with equality-constrained problems. However, the definition of a regular point is slightly different when we have inequality constraints. We say that a point, x^*, is **regular** in an equality- and inequality-constrained problem if the gradients of the equality constraints and all of the inequality constraints that are *binding* at x^* are linearly independent.

As with equality-constrained problems, we can have equality- and inequality-constrained problems, such as in Example 4.20, that have local minima that do not satisfy this regularity condition. In such a case, the local minimum may not satisfy the KKT condition. Interested readers are referred to more advanced texts [2] that further discuss this regularity issue.

4.6 Sensitivity Analysis

The subject of **sensitivity analysis** is concerned with estimating how changes to a nonlinear optimization problem affect the optimal objective-function value. Thus, this analysis is akin to that carried out in Section 2.6 for linear optimization problems. The following Sensitivity Property explains how this sensitivity analysis is conducted with nonlinear problems.

Sensitivity Property: Consider an equality- and inequality-constrained nonlinear optimization problem of the form:

$$\min_{x \in \mathbb{R}^n} \; f(x)$$

$$\text{s.t. } h_1(x) = 0$$
$$h_2(x) = 0$$
$$\vdots$$
$$h_m(x) = 0$$
$$g_1(x) \le 0$$
$$g_2(x) \le 0$$
$$\vdots$$
$$g_r(x) \le 0.$$

Suppose x^* is a local minimum and $\lambda_1^*, \lambda_2^*, \ldots, \lambda_m^*, \mu_1^*, \mu_2^*, \ldots, \mu_r^*$ are Lagrange multipliers associated with the equality and inequality constraints.

Consider the alternate equality- and inequality-constrained nonlinear optimization problem:

$$\min_{x \in \mathbb{R}^n} \; f(x)$$

$$\text{s.t. } h_1(x) = u_1$$
$$h_2(x) = u_2$$
$$\vdots$$
$$h_m(x) = u_m$$
$$g_1(x) \le v_1$$
$$g_2(x) \le v_2$$
$$\vdots$$
$$g_r(x) \le v_r,$$

and let $\hat{x}$ be a local minimum of this problem. So long as $|u_1|, |u_2|, \ldots, |u_m|, |v_1|, |v_2|, \ldots, |v_r|$ are sufficiently small, we can estimate the objective-function value of the new problem as:

$$f(\hat{x}) \approx f(x^*) - \sum_{i=1}^{m} \lambda_i^* u_i - \sum_{j=1}^{r} \mu_j^* v_j.$$

The Sensitivity Property says that the Lagrange multipliers found in the FONC of constrained nonlinear optimization problems provide the same sensitivity information that the sensitivity vector (which are equal to dual variables) give for linear optimization problems. It is also important to stress that although the Sensitivity Property is stated for problems with both equality and inequality constraints, it can clearly be applied to problems with only one type of constraint. A problem with only equality constraints would not have any μ's, because those Lagrange multipliers are associated with inequality constraints. It can similarly be applied to problems with only inequality constraints.

The Sensitivity Property does require the changes to the right-hand sides of the constraints to be small in magnitude, however it does not specify how large a value of u and v can be used. This is an unfortunate limitation of the theorem and is a difference compared to sensitivity analysis for linear optimization problems. For linear optimization problems, we can explicitly determine how much the right-hand side of constraints can change before the optimal basis changes using condition (2.49). We have no such result for nonlinear problems.

We now demonstrate the use of the Sensitivity Property with an example.

Example 4.24 Consider the Packing-Box Problem, which is examined in Example 4.23:

$$\max_{h,w,d} \ hwd$$

$$\text{s.t. } 2wh + 2dh + 6wd \le 60$$

$$h \ge 0$$

$$w \ge 0$$

$$d \ge 0.$$

In standard form this problem is:

$$\min_{h,w,d} \ f(h, w, d) = -hwd$$

$$
\begin{aligned}
\text{s.t. } & g_1(h, w, d) = 2wh + 2dh + 6wd - 60 \le 0 && (\mu_1)\\
& g_2(h, w, d) = -h \le 0 && (\mu_2)\\
& g_3(h, w, d) = -w \le 0 && (\mu_3)\\
& g_3(h, w, d) = -d \le 0, && (\mu_4)
\end{aligned}
$$

where the Lagrange multiplier associated with each constraint is indicated in the parentheses to right of it. We know from the analysis in Example 4.23 that:

$$(h \ w \ d \ \mu_1 \ \mu_2 \ \mu_3 \ \mu_4) \approx (5.48 \ 1.83 \ 1.83 \ 0.46 \ 0 \ 0 \ 0),$$

is the unique solution to the KKT condition and is a local and global minimum of the problem.

We wish to know the effect of increasing the amount of cardboard available to $62\,\mathrm{cm}^2$ and requiring the box to be at least $0.4\,\mathrm{cm}$ wide. In other words, we want to estimate the optimal objective-function value of the following problem:

$$\max_{h,w,d} \ hwd$$

$$\text{s.t. } 2wh + 2dh + 6wd \le 62$$
$$h \ge 0$$
$$w \ge 0.4$$
$$d \ge 0.$$

To apply the Sensitivity Property to answer this question, we must convert the constraints of this problem to have the same left-hand sides as the standard-form problem that is solved in Example 4.23. This is because the Sensitivity Property only tells us how to estimate the effect of changes to the right-hand side of constraints. The objective function must also be changed to a minimization. We can write the problem with additional cardboard and the minimum-width requirement as:

$$\min_{h,w,d} \ f(h, w, d) = -hwd$$

$$\text{s.t. } g_1(h, w, d) = 2wh + 2dh + 6wd - 60 \le 2$$
$$g_2(h, w, d) = -h \le 0$$
$$g_3(h, w, d) = -w \le -0.4$$
$$g_3(h, w, d) = -d \le 0.$$

Applying the Sensitivity Property, we can estimate the new optimal objective-function value as:

$$f(\hat{x}) \approx f(x^*) - 2\mu_1^* - 0\mu_2^* + 0.4\mu_3^* - 0\mu_4^* = -18.26 - 0.92 = -19.18.$$

The Sensitivity Property shows the objective function decreasing when we increase the amount of available cardboard. Recall, however, that the original problem is a maximization. We change the objective function to a minimization by multiplying the objective through by -1 to apply the KKT condition. Thus, when we take this into account, we conclude that the volume of the box *increases* by approximately $0.92\,\mathrm{cm}^3$ when we add the cardboard and impose the minimum-width requirement.
□

4.6.1 *Further Interpretation of the Karush-Kuhn-Tucker Condition*

As a final note, we can use the Sensitivity Property to gain some more insights into the complementary-slackness and sign restrictions in the KKT condition for equality-

and inequality-constrained problems. For this discussion, consider a simple problem with one constraint:

$$\min_{x \in \mathbb{R}^n} \ f(x)$$

$$\text{s.t. } g_1(x) \leq 0.$$

Suppose that we have a local minimum, x^*, and a Lagrange multiplier, μ_1^*, that satisfy the KKT condition. Suppose that we change the right-hand side of the constraint so the problem becomes:

$$\min_{x \in \mathbb{R}^n} \ f(x)$$

$$\text{s.t. } g_1(x) \leq v_1,$$

where $v_1 < 0$ but $|v_1|$ is small (*i.e.*, we change the right-hand side to a negative number that is small in magnitude). We can intuitively determine what happens to the objective-function value when we make this change.

First consider the case in which the constraint is non-binding in the original problem. If we change the constraint to $g_1(x) \leq v_1$ where v_1 is sufficiently small in magnitude, then the same x^* is still feasible and optimal in the new problem. Thus, the objective-function value does not change at all. The Sensitivity Property tells us that we can estimate the change in the objective-function value as:

$$f(\hat{x}) \approx f(x^*) - \mu_1^* v_1.$$

Because we reasoned that the objective-function value is the same, we must have $\mu_1^* v_1 = 0$ or $\mu_1^* = 0$. This, however, is precisely what complementary slackness requires. Because the constraint is non-binding, the Lagrange multiplier associated with it must be zero. The Sensitivity Property further tells us that changing the right-hand side of a constraint that is not binding by a small amount will not change the optimal objective-function value.

Now, consider the case in which the constraint is binding in the original problem. If we change the constraint to $g_1(x) \leq v_1$ the objective-function value must get worse (*i.e.*, larger). This is because the feasible region is reduced in size when the right-hand side of the constraint is changed. Before the constraint is changed, solutions for which $g_1(x) = 0$ are feasible. These solutions are no longer feasible when the right-hand side of the constraint is changed. Thus, the objective function cannot be better when we make this change. Again, the Sensitivity Property tells us that we can estimate the change in the objective-function value as:

$$f(\hat{x}) \approx f(x^*) - \mu_1^* v_1.$$

Because we reasoned that the objective-function value gets worse when we make the change, this means $-\mu_1^* v_1 \geq 0$ or $\mu_1^* \geq 0$, because we have that $v_1 < 0$. This, however, is the sign restriction that the KKT condition places on Lagrange multipliers associated with inequality constraints. Thus, the Sensitivity Property tells us that when we change the right-hand sides of inequality constraints that are binding, the objective function changes in a specific direction.

This interpretation of the complementary-slackness property required by the KKT condition is analogous to the derivation of complementary slackness between a primal linear optimization problem and its dual in Section 2.7.6.

4.7 Final Remarks

This chapter introduces analytic methods of solving nonlinear optimization problems. These rely on analyzing optimality conditions, which are a powerful tool for certain types of problems. However, in some cases optimality conditions may yields systems of equations or inequalities that are too difficult to solve. For this reason, iterative solution algorithms, which is the topic of Chapter 5, are often used. These algorithms are implemented in software packages and can be likened to using the Simplex method to solve linear optimization problems.

Our discussion of optimality conditions does not include the more general second-order conditions for constrained problems. Such conditions are beyond the level of this book. Interested readers are referred to more advanced texts for a treatment of such conditions [2, 7]. More advanced texts [1] also provide alternate optimality conditions to the KKT condition that can handle problems that do not satisfy the regularity requirement and specialized treatment of optimality conditions for convex optimization problems [3].

4.8 GAMS Codes

This section provides GAMS [4] codes for the main problems considered in this chapter. GAMS can use a variety of different software packages, among them MINOS [8], CONOPT [6], and KNITRO [5], to actually solve an NLPP.

4.8.1 Packing-Box Problem

The Packing-Box Problem, which is introduced in Section 4.1.1.1, has the following GAMS formulation:

```
1  variable z;
2  positive variables h, w, d;
3  equations of, cardBoard;
4  of .. z =e= h*w*d;
```

```
5   cardBoard .. 2*w*h+2*h*d+6*w*d =l= 60;
6   model box /all/;
7   solve box using nlp maximizing z;
```

Lines 1 and 2 declare variables, Line 3 gives names to the model equations, Line 4 defines the objective function, Line 5 specifies the constraint, Line 6 defines the model, and Line 7 directs GAMS to solve it.

The GAMS output that provides information about the optimal solution is:

```
1                          LOWER      LEVEL      UPPER      MARGINAL

3   ---- VAR z            -INF       18.257      +INF          .
4   ---- VAR h              .         5.477      +INF          .
5   ---- VAR w              .         1.826      +INF          .
6   ---- VAR d              .         1.826      +INF          .
```

4.8.2 Awning Problem

An instance of the Awning Problem, which is introduced in Section 4.1.1.2, which has $h = 2$ and $w = 3$ has the following GAMS formulation:

```
1   scalars h /2/, w /3/;
2   variable z;
3   positive variables x, y;
4   equations of, box;
5   of .. z =e= sqrt(x**2+y**2);
6   box .. y-w*y/x =g= h;
7   model awning /all/;
8   solve awning using nlp minimizing z;
```

Line 1 declares and sets the values of the scalar parameters, Lines 2 and 3 declare variables, Line 4 gives names to the model equations, Line 5 defines the objective function, Line 6 specifies the constraint, Line 7 defines the model, and Line 8 directs GAMS to solve it.

The GAMS output that provides information about the optimal solution is:

```
1                          LOWER      LEVEL      UPPER      MARGINAL

3   ---- VAR z            -INF        7.023      +INF          .
4   ---- VAR x              .         5.289      +INF          .
5   ---- VAR y              .         4.621      +INF          .
```

4.8.3   Facility-Location Problem

An instance of the Facility-Location Problem, which is introduced in Section 4.1.1.3, that has three retail locations at coordinates $(1, 1)$, $(-1, 2)$, and $(3, 0)$ that each receive 10 trucks, has the following GAMS formulation:

```
set ret /1*3/;
parameters
    V(ret)
        /1  10
         2  10
         3  10/
    x(ret)
        /1  1
         2 -1
         3  3/
    y(ret)
        /1  1
         2  2
         3  0/;
variables z, a, b;
equations of;
of .. z =e= sum(ret,2*V(ret)*sqrt(power((x(ret)-a),2)+
    power((y(ret)-b),2)));
model facLoc /all/;
solve facLoc using nlp minimizing z;
```

Line 1 declares and defines the set of retail locations. Sets are a construct in GAMS that allow us to create data, variables, or constraints that are assigned to different entities being modeled. For instance, in the Facility-Location Problem each retail location has a pair of coordinates and a fixed number of trucks that must make deliveries to it as model data. The set allows these to be modeled without having to individually write out each piece of data individually in the model. Lines 2–14 declare and set the values of the problem parameters, Line 15 declares variables, Line 16 gives a name to the model equation, Line 17 defines the objective function, Line 18 defines the model, and Line 19 directs GAMS to solve it.

The GAMS output that provides information about the optimal solution is:

```
                        LOWER       LEVEL       UPPER      MARGINAL

---- VAR z              -INF       89.443       +INF          .
---- VAR a              -INF        1.000       +INF          .
---- VAR b              -INF        1.000       +INF          .
```

4.8.4 *Cylinder Problem*

An instance of the Cylinder Problem, which is introduced in Section 4.1.1.4, in which $N = 10$, $c_1 = 2$ and $c_2 = 0.5$, has the following GAMS formulation:

```
1  scalars N /10/, c1 /2/, c2 /0.5/;
2  variable z;
3  positive variables r, h;
4  equations of;
5  of .. z =e= N*pi*h*r**2-c1*pi*r**2-c2*pi*h*(pi*r**2+2*
      pi*r*h)*r**2;
6  model cyl /all/;
7  solve cyl using nlp maximizing z;
```

Line 1 declares and sets the values of scalar parameters, Lines 2 and 3 declare variables, Line 4 gives a name to the model equation, Line 5 defines the objective function, Line 6 defines the model, and Line 7 directs GAMS to solve it.

The GAMS output that provides information about the optimal solution is:

```
1                          LOWER      LEVEL      UPPER      MARGINAL

3  ---- VAR z             -INF       12.174     +INF          .
4  ---- VAR r               .         0.826     +INF          .
5  ---- VAR h               .         1.721     +INF          .
```

4.8.5 *Machining-Speed Problem*

An instance of the Machining-Speed Problem, that is introduced in Section 4.1.2.1, in which $p = 10$, $m = 1$, $t_p = 1$, $\lambda = 0.1$, $t_c = 1.1$, $C = 1$, $n = 2$, and $h = 0.4$, has the following GAMS formulation:

```
1  scalars p /10/, m /1/, tp /1/, lambda /0.1/, tc /1.1/, C /1/, n
      /2/, h /0.4/;
2  variable z;
3  positive variable v;
4  equations of;
5  of .. z =e= (p-m)/(tp+lambda/v+tc*((C/v)**(1/n))/(lambda/v))-h;
6  model machine /all/;
7  solve machine using nlp maximizing z;
```

Line 1 declares and sets the values of scalar parameters, Lines 2 and 3 declare variables, Line 4 gives a name to the model equation, Line 5 defines the objective function, Line 6 defines the model, and Line 7 directs GAMS to solve it.

The GAMS output that provides information about the optimal solution is:

		LOWER	LEVEL	UPPER	MARGINAL
---- VAR z		-INF	1.286	+INF	.
---- VAR v		.	0.069	+INF	.

4.8.6 Hanging-Chain Problem

An instance of the Hanging-Chain Problem, which is introduced in Section 4.1.2.2, in which the chain has 10 links and $L = 4$, has the following GAMS formulation:

```
set links /1*10/;
scalars g /9.80665/, L /4/;
variables z, y(links);
equations of, height, width;
of .. z =e= 50*g*sum(links,(card(links)-ord(links)
    +0.5)*y(links));
height .. sum(links,y(links)) =e= 0;
width .. sum(links,sqrt(10-power(y(links),2))) =e= L;
model chain /all/;
solve chain using nlp minimizing z;
```

Line 1 declares and defines the set of chain links, Line 2 declares and sets the values of scalar parameters, Line 3 declares variables, Line 4 gives names to the model equations, Line 5 defines the objective function, Lines 6 and 7 declare the constraints, Line 8 defines the model, and Line 9 directs GAMS to solve it.

The GAMS output that provides information about the optimal solution is:

		LOWER	LEVEL	UPPER	MARGINAL
---- VAR z		-INF	-3.859E+4	+INF	.
---- VAR y					

	LOWER	LEVEL	UPPER	MARGINAL
1	-INF	-3.160	+INF	.
2	-INF	-3.158	+INF	.
3	-INF	-3.154	+INF	.
4	-INF	-3.139	+INF	.
5	-INF	-2.969	+INF	.
6	-INF	2.969	+INF	.
7	-INF	3.139	+INF	.
8	-INF	3.154	+INF	.
9	-INF	3.158	+INF	.
10	-INF	3.160	+INF	.

4.8.7 Return-Maximization Problem

The Return-Maximization Problem, which is introduced in Section 4.1.3.1, has the following GAMS formulation:

```
1   set asset;
2   alias(asset,assetA);
3   parameters ret(asset), cov(asset,asset);
4   scalar s;
5   variable z;
6   positive variable w(asset);
7   equations of, risk, alloc;
8   of .. z =e= sum(asset,ret(asset)*w(asset));
9   risk .. sum((asset,assetA),cov(asset,assetA)*w(asset)*
         w(assetA)) =l= s;
10  alloc .. sum(asset,w(asset)) =e= 1;
11  model invest /all/;
12  solve invest using nlp maximizing z;
```

Line 1 declares the set of assets and Line 2 declares an alias of this set. Lines 3 and 4 declare the problem parameters, Lines 5 and 6 declare variables, Line 7 gives names to the model equations, Line 8 defines the objective function, Lines 9 and 10 declare the constraints, Line 11 defines the model, and Line 12 directs GAMS to solve it.

Note that this GAMS code will not compile without values being assigned to the set asset and to the parameters ret(asset), cov(asset,asset), and s.

4.8.8 Variance-Minimization Problem

The Variance-Minimization Problem, which is introduced in Section 4.1.3.2, has the following GAMS formulation:

```
1   set asset;
2   alias(asset,assetA);
3   parameters ret(asset), cov(asset,asset);
4   scalar R;
5   variable z;
6   positive variable w(asset);
7   equations of, earn, alloc;
8   of .. z =e= sum((asset,assetA),cov(asset,assetA)*w(asset)*
         w(assetA));
9   earn .. sum(asset,ret(asset)*w(asset)) =g= R;
10  alloc .. sum(asset,w(asset)) =e= 1;
11  model invest /all/;
12  solve invest using nlp minimizing z;
```

Line 1 declares the set of assets and Line 2 declares an alias of this set. Lines 3 and 4 declare the problem parameters, Lines 5 and 6 declare variables, Line 7 gives names to the model equations, Line 8 defines the objective function, Lines 9 and 10 declare the constraints, Line 11 defines the model, and Line 12 directs GAMS to solve it.

Note that this GAMS code will not compile without values being assigned to the set `asset` and to the parameters `ret(asset)`, `cov(asset,asset)`, and R.

4.8.9 Inventory-Planning Problem

The Inventory-Planning Problem, which is introduced in Section 4.1.3.3, has the following GAMS formulation:

```
1  variable z;
2  positive variables xs, xm, xl;
3  equations of;
4  of .. z =e= 10*(xs-(xs**2)/6000)+12*(xm-(xm**2)/6000)
       +13*(xl-(xl**2)/6000)-xs-2*xm-4*xl;
5  model inventory /all/;
6  solve inventory using nlp maximizing z;
```

Lines 1 and 2 declare variables, Line 3 gives a name to the model equation, Line 4 defines the objective function, Line 5 defines the model, and Line 6 directs GAMS to solve it.

The GAMS output that provides information about the optimal solution is:

```
1                        LOWER      LEVEL      UPPER     MARGINAL

3  ---- VAR z           -INF    33996.154     +INF          .
4  ---- VAR xs            .       2700.000     +INF          .
5  ---- VAR xm            .       2500.000     +INF          .
6  ---- VAR xl            .       2076.923     +INF          .
```

4.8.10 Economic-Dispatch Problem

The Economic-Dispatch Problem, which is introduced in Section 4.1.4.1, has the following GAMS formulation:

```
1  set node;
2  alias(node,nodeA);
3  parameters a0(node), a1(node), a2(node), D(node), Y(node,node),
       link(node,node), L(node,node), minQ(node), maxQ(node);
4  variables z, theta(node), f(node,node), q(node);
5  positive variable q(node);
6  equations of, demand, balance, flow;
```

```
7   of ..  z =e=  sum(node,a0(node)+a1(node)*q(node)+a2(node)*q(node)
        **2);
8   demand(node) ..  D(node) =e= q(node)  + sum(nodeA$(link(node,
        nodeA)),f(nodeA,node));
9   balance ..  sum(node,D(node)) =e= sum(node,q(node));
10  flow(node,nodeA) ..  f(node,nodeA)$(link(node,nodeA)) =e= Y(node
        ,nodeA)*sin(theta(node)-theta(nodeA));
11  f.up(node,nodeA) = L(node,nodeA);
12  q.lo(node) = minQ(node);
13  q.up(node) = maxQ(node);
14  model dispatch /all/;
15  solve dispatch using nlp minimizing z;
```

Line 1 declares the set of nodes and Line 2 declares an alias of this set. Line 3 declares the problem parameters, Lines 4 and 5 declare variables, Line 6 gives names to the model equations, Line 7 defines the objective function, and Lines 8–10 declare the constraints. Note that the constraint in Line 8 only adds flow from nodeA to node in determining the supply/demand balance constraint for node if the two nodes are directly linked by a transmission line (which is what the parameter link(node,nodeA) indicates. Similarly, the constraint in Line 10 only computes the flow on lines that are directly linked. Line 11 imposes the upper bounds on the flow variables and Lines 12 and 13 impose the lower and upper bounds on production at each node. Line 14 defines the model and Line 15 directs GAMS to solve it.

Note that this GAMS code will not compile without values being assigned to the set node and to the parameters a0(node), a1(node), a2(node), D(node), Y(node,node), link(node,node), L(node,node), minQ(node), and maxQ(node).

4.9 Exercises

4.1 Jose builds electrical cable using two types of metallic alloys. Alloy 1 is 55% aluminum and 45% copper, while alloy 2 is 75% aluminum and 25% copper. The prices at which Jose can buy the two alloys depends on the amount he purchases. The total cost of buying x_1 tons of alloy 1 is given by:

$$5x_1 + 0.01x_1^2,$$

and the total cost of buying x_2 tons of alloy 2 is given by:

$$4x_2 + 0.02x_2^2.$$

Formulate a nonlinear optimization problem to determine the cost-minimizing quantities of the two alloys that Jose should use to produce 10 tons of cable that is at least 30% copper.

4.2 Emma is participating in an L km bicycle race. She is planning on carrying a hydration bladder on her back to keep herself hydrated during the race. If we let v denote her average speed during the race in km/h and w the volume of the hydration bladder in liters, then she consumes water at an average rate of $cv^3 \cdot (w + 1)^2$ liters per hour. Formulate a nonlinear optimization problem to determine how much water Emma should carry and the average speed at which she should bike to minimize her race time.

4.3 Vishnu has \$35 to spend on any combination of three different goods—apples, oranges, and bananas. Apples cost \$2 each, oranges \$1.50 each, and bananas \$5 each. Vishnu measures his happiness from consuming apples, oranges, and bananas using a utility function. If Vishnu consumes x_a apples, x_o oranges, and x_b bananas, then his utility is given by:

$$3 \log(x_a) + 0.4 \log(x_o + 2) + 2 \log(x_b + 3).$$

Formulate a nonlinear optimization problem to determine how Vishnu should spend his \$35.

4.4 Convert the models formulated for Exercises 4.1–4.3 into standard form for the type of nonlinear optimization problems that they are.

4.5 Is the model formulated for the Facility-Location Problem that is introduced in Section 4.1.1.3 a convex optimization problem?

4.6 Is a local minimum of Exercise 4.1 guaranteed to be a global minimum? Explicitly explain why or why not.

4.7 What difficulties could arise in applying FONC, SONC, and SOSC to the model formulated for the Facility-Location Problem that is in Section 4.1.1.3?

4.8 Find all of the KKT points for the model formulated in Exercise 4.3. Are any of these KKT points guaranteed to be global optima? Explicitly explain why or why not.

4.9 Using the solution to Exercise 4.8, approximate how much Vishnu's utility increases if he has an additional \$1.25 to spend and must purchase at least one orange. Compare your approximation to the actual change in Vishnu's utility.

4.10 Write a GAMS code for the model formulated in Exercise 4.1.

References

1. Bazaraa MS, Sherali HD, Shetty CM (2006) Nonlinear programming: theory and algorithms, 3rd edn. Wiley-Interscience, Hoboken
2. Bertsekas D (2016) Nonlinear programming, 3rd edn. Athena Scientific, Belmont

3. Boyd S, Vandenberghe L (2004) Convex optimization. Cambridge University Press, Cambridge
4. Brook A, Kendrick D, Meeraus A (1988) GAMS – a user's guide. ACM, New York. www.gams.com
5. Byrd RH, Nocedal J, Waltz RA (2006) KNITRO: an integrated package for nonlinear optimization. In: di Pillo G, Roma M (eds) Large-scale nonlinear optimization, pp 35–59. www.artelys.com/en/optimization-tools/knitro
6. Drud AS (1994) CONOPT – a large-scale GRG code. ORSA J Comput 6(2):207–216. www.conopt.com
7. Luenberger DG, Ye Y (2016) Linear and nonlinear programming, 4th edn. Springer, New York
8. Murtagh BA, Saunders MA (1995) MINOS 5.4 user's guide. Technical report SOL 83-20R, Systems Optimization Laboratory, Department of Operations Research, Stanford University, Stanford, California. www.sbsi-sol-optimize.com/asp/sol_product_minos.htm

Chapter 5
Iterative Solution Algorithms
for Nonlinear Optimization

Chapter 4 introduces optimality conditions to solve nonlinear optimization problems. Optimality conditions have the benefit that they allow us to find all points that are candidate local minima. Working with optimality conditions can be quite cumbersome, however. This is because they can require solving large systems of nonlinear equations. Moreover, in the case of equality- and inequality-constrained problems, the Karush–Kuhn–Tucker (KKT) condition can require a massive combinatorial search, in which all possible cases of which constraints are binding and non-binding are examined. For these reasons, in many practical cases nonlinear programming problems (NLPPs) are solved using iterative algorithms that are implemented on a computer. Many of these iterative solution algorithms rely on the properties that the optimality conditions introduced in Chapter 4 tell us that an optimal solution should exhibit.

In this chapter we begin by first introducing a generic iterative algorithm for solving an unconstrained NLPP. We then provide details on the steps of this generic algorithm. We next explain how the generic iterative algorithm can be applied to solve a constrained NLPP. We introduce two approaches to solving a constrained NLPP. The first implicitly accounts for the constraints. This is done by removing the constraints from the problem and instead adding terms to the objective function that 'penalize' constraint violations. These penalties are intended to drive the algorithm to a feasible solution. The second approach explicitly accounts for the constraints. While this approach is guaranteed to find a feasible solution, it can only be applied to problems with relatively simple constraints.

5.1 Iterative Solution Algorithm for Unconstrained Nonlinear Optimization Problems

In this section we introduce a basic iterative solution algorithm for unconstrained nonlinear optimization problems. Throughout this section we assume that our unconstrained problem has the form:

© Springer International Publishing AG 2017
R. Sioshansi and A.J. Conejo, *Optimization in Engineering*,
Springer Optimization and Its Applications 120, DOI 10.1007/978-3-319-56769-3_5

$$\min_{x \in \mathbb{R}^n} f(x),$$

where $f(x) : \mathbb{R}^n \to \mathbb{R}$ is the objective function being minimized.

Below we give a high-level overview of a Generic Algorithm for Unconstrained Nonlinear Optimization Problems. This is an example of an **iterative solution algorithm**. In Lines 2 and 3 we initialize the algorithm by setting k, the counter for the number of iterations completed, to 0 and picking a starting point, x^0. We use the notational convention here that the superscript on x denotes the value of a variable after that a number of iterations are completed. Thus, x^0 is a vector of decision variables after zero iterations have been completed (*i.e.*, it is an initial guess). Lines 4 through 9 are the iterative procedure. In Line 4 we check to see whether certain termination criteria are met. If so, we stop and output the current point that we have, x^k. Otherwise, we proceed with an additional iteration. Each iteration consists of two main procedures. In Line 5 we find a **search direction** in which to move away from the current point, x^k. In Line 6 we conduct what is known as a **line search** to determine how far to move in the search direction that is found in Line 5. The value, α^k, that we find is called the **step size**. Finally, in Line 7 we update our point based on the search direction and step size found in Lines 5 and 6, respectively, and we update our iteration counter in Line 8.

Generic Algorithm for Unconstrained Nonlinear Optimization Problems

```
1: procedure GENERIC UNCONSTRAINED ALGORITHM
2:     k ← 0                                          ▷ Set iteration counter to 0
3:     Fix x^0                                        ▷ Fix a starting point
4:     while Termination criteria are not met do
5:         Find direction, d^k, to move in
6:         Determine step size, α^k
7:         x^{k+1} ← x^k + α^k d^k                    ▷ Update point
8:         k ← k + 1                                  ▷ Update iteration counter
9:     end while
10: end procedure
```

The major steps of this iterative algorithm are very similar to those in the Simplex method, which is used to solve linear optimization problems. The Simplex method consists of two major steps. First, we check whether any of the coefficients in the objective-function row of the tableau are negative. If so, we increase the value of the associated non-basic variable from zero. This is akin to finding a search direction in Line 5 of the Generic Algorithm for Unconstrained Nonlinear Optimization Problems. This is because in both algorithms we are moving away from the point that we are currently at to improve the objective-function value. The other major step of the Simplex method is to conduct a ratio test, which determines how much to increase the value of the non-basic variable that enters the basis. This is akin to the line search that is conducted in Line 6 of the Generic Algorithm for Unconstrained Nonlinear

Optimization Problems. This is because in both algorithms we are determining how far to move away from the point that we are currently at.

To use the Generic Algorithm for Unconstrained Nonlinear Optimization Problems, there are three important details that must be addressed. The first is how to determine the search direction. The second is what termination criteria should be used. The third is how to conduct the line search. We now address each of these in turn.

5.2 Search Direction

Although there are a multitude of search-direction methods available, we limit our discussion to two classic methods: steepest descent and Newton's method. These are two of the easiest search-direction methods to grasp and implement with real-world problems. Interested readers are referred to other texts [2, 7] that discuss more advanced search-direction methods.

Because we focus on finding the search direction here, we ignore the step size parameter (*i.e.*, α^k) in the following discussion. Thus, we assume in the following discussion that the points we generate after each iteration are defined as:

$$x^{k+1} = x^k + d^k.$$

After concluding the discussion of the search-direction methods here, we reintroduce the step size parameter in Section 5.4, where we focus exclusively on line search procedures.

5.2.1 Steepest Descent

The method of steepest descent determines the search direction by attempting to make the objective function after each iteration as small as possible. Put another way, the idea behind steepest descent is that we choose the search direction, d^k, to solve the following minimization problem:

$$\min_{d^k \in \mathbb{R}^n} f(x^k + d^k).$$

Note that solving this minimization problem is as difficult as solving the original unconstrained minimization. If we could solve it easily we would not need an iterative solution algorithm in the first place! Thus, instead of minimizing $f(x^k + d^k)$ we instead minimize its first-order Taylor approximation, which is:

$$f(x^k + d^k) \approx f(x^k) + d^{k^\top} \nabla f(x^k).$$

Note, however, that this objective function is linear in d^k. If we attempt to solve an unconstrained minimization problem with a linear objective function, the problem will necessarily be unbounded. To address this unboundedness, we add a constraint that the direction vector, d^k, must have squared length equal to one. It may seem arbitrary to restrict the direction vector to have length one. We 'get away' with adding this restriction because all we are concerned with at this point is finding a search direction in which to move. After we find a search direction, we then conduct a line search to determine the step size, which tells us how far to move in the direction found. Thus, the length of the d^k vector doesn't matter—the value of α^k will be adjusted to ensure that we move the 'correct' distance in whatever direction is identified.

With this constraint, our optimization problem becomes:

$$\min_{d^k} f(x^k) + {d^k}^\top \nabla f(x^k)$$

$$\text{s.t. } {d^k}^\top d^k = 1,$$

which can be expanded out to:

$$\min_{d^k} f(x^k) + d_1^k \frac{\partial}{\partial x_1} f(x^k) + d_2^k \frac{\partial}{\partial x_2} f(x^k) + \cdots + d_n^k \frac{\partial}{\partial x_n} f(x^k)$$

$$\text{s.t. } {d_1^k}^2 + {d_2^k}^2 + \cdots + {d_n^k}^2 - 1 = 0.$$

We can treat this as an equality-constrained nonlinear optimization problem, and if we introduce a Lagrange multiplier, λ, the First-Order Necessary Condition (FONC) for Equality-Constrained Nonlinear Optimization Problems is:

$$\frac{\partial}{\partial x_1} f(x^k) + 2\lambda d_1^k = 0$$

$$\frac{\partial}{\partial x_2} f(x^k) + 2\lambda d_2^k = 0$$

$$\vdots$$

$$\frac{\partial}{\partial x_n} f(x^k) + 2\lambda d_n^k = 0.$$

We can write the FONC more compactly as:

$$\nabla f(x^k) + 2\lambda d^k = 0.$$

Solving for d^k gives:

$$d^k = -\frac{1}{2\lambda} \nabla f(x^k). \tag{5.1}$$

Equation (5.1) has two possible solutions. One, in which $\lambda > 0$, has us move in a direction that is opposite the gradient. The other, in which $\lambda < 0$, has us move in a direction that is the same as the gradient. Note, however, that if $\lambda > 0$ then we have:

$$f(x^k) + d^{k^\top} \nabla f(x^k) = f(x^k) - \frac{1}{2\lambda} \nabla f(x^k)^\top \nabla f(x^k) \leq f(x^k),$$

because:

$$\nabla f(x^k)^\top \nabla f(x^k) \geq 0.$$

Conversely, if $\lambda < 0$ then we have:

$$f(x^k) + d^{k^\top} \nabla f(x^k) = f(x^k) - \frac{1}{2\lambda} \nabla f(x^k)^\top \nabla f(x^k) \geq f(x^k).$$

From this, we conclude that the solution in which $\lambda > 0$ is the minimum and from this observation we obtain the Steepest Descent Rule, which says to always move in a direction that is opposite the gradient. This is stated in the following Steepest Descent Rule.

Steepest Descent Rule: Given an unconstrained nonlinear optimization problem and a current point, x^k, the **steepest descent** search direction is:

$$d^k = -\nabla f(x^k).$$

The purpose of the $1/(2\lambda)$ term in Equation (5.1) is to scale the gradient so that the direction vector has a squared length of one. We can ignore this scaling, however, because the restriction that d^k have a squared length of one is arbitrarily added when we minimize the first-order Taylor approximation of $f(x^k + d^k)$. As discussed above, the length of the direction vector is ultimately unimportant because we always conduct a line search to determine how far to move in the search direction found.

Example 5.1 Consider the unconstrained nonlinear optimization problem:

$$\min_x \; f(x) = (x_1 - 3)^2 + (x_2 - 2)^2.$$

Starting from the point, $x^0 = (1, 1)^\top$, we wish to find the steepest descent direction. To do so, we first compute the gradient of the objective function as:

$$\nabla f(x) = \begin{pmatrix} 2(x_1 - 3) \\ 2(x_2 - 2) \end{pmatrix}.$$

Thus, the steepest descent direction at $x^0 = (1, 1)^\top$ is:

$$d^0 = -\nabla f(x^0) = \begin{pmatrix} 4 \\ 2 \end{pmatrix}.$$

$\square$

5.2.2 Newton's Method

Newton's method takes a fundamentally different approach compared to steepest descent in finding a search direction. Instead of looking for a direction that minimizes the objective function, Newton's method uses the knowledge that only stationary points can be minima of unconstrained nonlinear optimization problems. Thus, the underlying premise of Newton's method is that we choose a search direction to make $\nabla f(x^k + d^k) = 0$. We note at the outset of this chapter that one of difficulties with using optimality conditions, such as the FONC, is that they may involve solving large systems of nonlinear equations. Thus, finding d^k to make $\nabla f(x^k + d^k)$ equal zero may be difficult. Instead of trying to make the gradient equal zero, Newton's method aims to make the first-order Taylor approximation of the gradient equal to zero.

To derive the Newton's method direction, we first use Taylor's theorem to approximate the first-order partial derivative of $f(x)$ with respect to the ith variable at $x^k + d^k$ as:

$$\frac{\partial}{\partial x_i} f(x^k + d^k) \approx \frac{\partial}{\partial x_i} f(x^k) + d^{k^\top} \nabla \frac{\partial}{\partial x_i} f(x^k)$$

$$= \frac{\partial}{\partial x_i} f(x^k) + (d_1^k, d_2^k, \dots, d_n^k) \begin{pmatrix} \frac{\partial}{\partial x_1} \frac{\partial}{\partial x_i} f(x^k) \\ \frac{\partial}{\partial x_2} \frac{\partial}{\partial x_i} f(x^k) \\ \vdots \\ \frac{\partial}{\partial x_n} \frac{\partial}{\partial x_i} f(x^k) \end{pmatrix}$$

$$= \frac{\partial}{\partial x_i} f(x^k) + (d_1^k, d_2^k, \dots, d_n^k) \begin{pmatrix} \frac{\partial^2}{\partial x_1 x_i} f(x^k) \\ \frac{\partial^2}{\partial x_2 x_i} f(x^k) \\ \vdots \\ \frac{\partial^2}{\partial x_n x_i} f(x^k) \end{pmatrix}.$$

Thus, we can approximate the gradient of $f(x)$ at $x^k + d^k$ as:

$$\nabla f(x^k + d^k) = \begin{pmatrix} \frac{\partial}{\partial x_1} f(x^k + d^k) \\ \frac{\partial}{\partial x_2} f(x^k + d^k) \\ \vdots \\ \frac{\partial}{\partial x_n} f(x^k + d^k) \end{pmatrix}$$

$$\approx \begin{pmatrix} \frac{\partial}{\partial x_1} f(x^k) + (d_1^k, d_2^k, \ldots, d_n^k) \begin{pmatrix} \frac{\partial^2}{\partial x_1 x_1} f(x^k) \\ \frac{\partial^2}{\partial x_2 x_1} f(x^k) \\ \vdots \\ \frac{\partial^2}{\partial x_n x_1} f(x^k) \end{pmatrix} \\ \frac{\partial}{\partial x_2} f(x^k) + (d_1^k, d_2^k, \ldots, d_n^k) \begin{pmatrix} \frac{\partial^2}{\partial x_1 x_2} f(x^k) \\ \frac{\partial^2}{\partial x_2 x_2} f(x^k) \\ \vdots \\ \frac{\partial^2}{\partial x_n x_2} f(x^k) \end{pmatrix} \\ \vdots \\ \frac{\partial}{\partial x_n} f(x^k) + (d_1^k, d_2^k, \ldots, d_n^k) \begin{pmatrix} \frac{\partial^2}{\partial x_1 x_n} f(x^k) \\ \frac{\partial^2}{\partial x_2 x_n} f(x^k) \\ \vdots \\ \frac{\partial^2}{\partial x_n x_n} f(x^k) \end{pmatrix} \end{pmatrix} .$$

We can break this Taylor approximation into two terms, which gives:

$$\begin{pmatrix} \frac{\partial}{\partial x_1} f(x^k) \\ \frac{\partial}{\partial x_2} f(x^k) \\ \vdots \\ \frac{\partial}{\partial x_n} f(x^k) \end{pmatrix} + \begin{pmatrix} (d_1^k, d_2^k, \ldots, d_n^k) \begin{pmatrix} \frac{\partial^2}{\partial x_1 x_1} f(x^k) \\ \frac{\partial^2}{\partial x_2 x_1} f(x^k) \\ \vdots \\ \frac{\partial^2}{\partial x_n x_1} f(x^k) \end{pmatrix} \\ (d_1^k, d_2^k, \ldots, d_n^k) \begin{pmatrix} \frac{\partial^2}{\partial x_1 x_2} f(x^k) \\ \frac{\partial^2}{\partial x_2 x_2} f(x^k) \\ \vdots \\ \frac{\partial^2}{\partial x_n x_2} f(x^k) \end{pmatrix} \\ \vdots \\ (d_1^k, d_2^k, \ldots, d_n^k) \begin{pmatrix} \frac{\partial^2}{\partial x_1 x_n} f(x^k) \\ \frac{\partial^2}{\partial x_2 x_n} f(x^k) \\ \vdots \\ \frac{\partial^2}{\partial x_n x_n} f(x^k) \end{pmatrix} \end{pmatrix} . \qquad (5.2)$$

The first term in (5.2) is the gradient, $\nabla f(x^k)$. Making this substitution and factoring d^k out of the second term gives:

$$\nabla f(x^k) + \begin{bmatrix} \frac{\partial^2}{\partial x_1 x_1} f(x^k) & \frac{\partial^2}{\partial x_2 x_1} f(x^k) & \cdots & \frac{\partial^2}{\partial x_n x_1} f(x^k) \\ \frac{\partial^2}{\partial x_1 x_2} f(x^k) & \frac{\partial^2}{\partial x_2 x_2} f(x^k) & \cdots & \frac{\partial^2}{\partial x_n x_2} f(x^k) \\ \vdots & \vdots & \ddots & \vdots \\ \frac{\partial^2}{\partial x_1 x_n} f(x^k) & \frac{\partial^2}{\partial x_2 x_n} f(x^k) & \cdots & \frac{\partial^2}{\partial x_n x_n} f(x^k) \end{bmatrix} \begin{pmatrix} d_1^k \\ d_2^k \\ \vdots \\ d_n^k \end{pmatrix}. \tag{5.3}$$

The second term in (5.3) is the Hessian multiplied by d^k. Thus, the Taylor approximation of the gradient is:

$$\nabla f(x^k + d^k) \approx \nabla f(x^k) + \nabla^2 f(x^k)d^k.$$

If we set this approximation equal to zero, we have:

$$\nabla f(x^k) + \nabla^2 f(x^k)d = 0,$$

which gives:

$$d^k = -\left[\nabla^2 f(x^k)\right]^{-1} \nabla f(x^k),$$

so long as $\nabla^2 f(x^k)$ is invertible. This is stated in the following Newton's Method Rule.

Newton's Method Rule: Given an unconstrained nonlinear optimization problem and a current point, x^k, the **Newton's method** search direction is:

$$d^k = -\left[\nabla^2 f(x^k)\right]^{-1} \nabla f(x^k),$$

so long as $\nabla^2 f(x^k)$ is invertible. Otherwise, the Newton's method search direction is not defined.

Unlike with steepest descent, Newton's method may not yield a well-defined search direction. This happens if we are at a point where the Hessian of the objective function is not invertible. It is important to stress that depending on the objective function that we are minimizing, the invertibility problem may only arise at certain iterations in the overall algorithm. Thus, algorithms that implement Newton's method typically include a check to ensure that the Hessian is invertible in each iteration. If so, the Newton search direction is used. Otherwise, another search direction (such as the steepest descent direction) is used in that iteration. In subsequent iterations, we may move to points at which the Hessian is invertible and Newton's method directions are again used.

Example 5.2 Consider the unconstrained nonlinear optimization problem:

$$\min_{x} \; f(x) = (x_1 - 3)^2 + (x_2 - 2)^2,$$

which is introduced in Example 5.1. Starting from the point, $x^0 = (1, 1)^\top$, we wish to find the Newton's method direction. We already know that the gradient of the objective function is:

$$\nabla f(x) = \begin{pmatrix} 2(x_1 - 3) \\ 2(x_2 - 2) \end{pmatrix}.$$

We can further compute the Hessian as:

$$\nabla^2 f(x) = \begin{bmatrix} 2 & 0 \\ 0 & 2 \end{bmatrix}.$$

Because the Hessian is invertible, we can compute the Newton's method direction at $x^0 = (1, 1)^\top$ as:

$$d^0 = -\left[\nabla^2 f(x^0)\right]^{-1} \nabla f(x^0) = \begin{pmatrix} 2 \\ 1 \end{pmatrix}.$$

$\square$

5.3 Termination Criteria

Both steepest descent and Newton's method use the same termination criteria, and that is to stop once reaching a stationary point. If the current point is stationary, both the steepest descent and Newton's method search directions will be equal to zero. That is, both search methods tell us not to move away from the current point that we are at. This termination criterion makes intuitive sense. We know from the FONC for Unconstrained Nonlinear Optimization Problems, which is given in Section 4.5.1, that a minimum must be a stationary point. From this perspective, steepest descent and Newton's method terminating at stationary points is desirable. However, as discussed throughout Section 4.5, with the exception of convex optimization problems, points that satisfy the FONC are not guaranteed to be local or global minima. If we have a convex optimization problem, then both steepest descent and Newton's method terminate at global minima. Otherwise, for more general problems, all we know is that they provide us with stationary points that could be local minima or may not be.

In practice, software packages include other termination criteria. For instance, most packages have an upper limit on the number of iterations to conduct. This is because it may take more effort and time to try finding a stationary point than the user would like the computer to expend. Indeed, the Generic Algorithm for Unconstrained Nonlinear Optimization Problems may never reach a stationary point if it is given an unbounded optimization problem. Typically, after the maximum number of iterations,

$\bar{k}$, is conducted, the software package will report the final solution, $x^{\bar{k}}$, and indicate to the user that the algorithm terminated because it reached the maximum number of iterations (as opposed to finding a stationary point). The use could then choose to use the final point that is reported, conduct more iterations starting from the final point that is reported, or try solving the problem in another way (*e.g.*, using a different algorithm or starting point).

Another issue that comes up in practice is that software packages typically numerically approximate partial derivatives when computing the gradient and Hessian. As such, the solver may be at a stationary point, but the computed gradient is not equal to zero due to approximation errors. Thus, most software packages include a parameter that allows some tolerance in looking for a stationary point. For instance, the algorithm may terminate if $||\nabla f(x^k)|| < \varepsilon$, where ε is a user-specified tolerance.

5.4 Line Search

Once we determine a direction in which to move, using steepest descent, Newton's method, or another technique, the next question is how far to move in the direction that is identified. This is the purpose of a line search. We call this process a line search because what we are doing is 'looking' along a line in the search direction. Looking along this line, we determine how far to move in the direction, with the aim of reducing the objective-function value.

As with search directions, there are many line-search methods available. We introduce three of them and refer interested readers to other texts [2, 7] that discuss other more advanced methods.

5.4.1 *Exact Line Search/Line Minimization*

An **exact line search** or **line minimization** solves the minimization problem:

$$\min_{\alpha^k \geq 0} \; f(x^k + \alpha^k d^k),$$

to determine the optimal choice of step size. The advantage of an exact line search is that it provides the best choice of step size. This comes at a cost, however, which is that we must solve an optimization problem. This is always a single-variable problem, however, and in some instances the exact-line-search problem can be solved relatively easily. Indeed, the exact-line-search problem typically simplifies to solving:

$$\frac{\partial}{\partial \alpha^k} f(x^k + \alpha^k d^k) = 0,$$

because we can often treat the exact line search as being an unconstrained optimization problem. Nevertheless, an exact line search may be impractical for some problems. We demonstrate its use with an example.

Example 5.3 Consider the unconstrained problem:

$$\min_{x} \ f(x) = (x_1 - 3)^2 + (x_2 - 2)^2,$$

which is introduced in Example 5.1. We know that if we start from the point $x^0 = (1, 1)^\top$ the initial steepest descent direction is:

$$d^0 = -\nabla f(x^0) = \begin{pmatrix} 4 \\ 2 \end{pmatrix}.$$

To conduct an exact line search we solve the following minimization problem:

$$\begin{aligned}
\min_{\alpha^0 \geq 0} \ f\left(x^0 + \alpha^0 d^0\right) &= f\left(\begin{pmatrix} 1 \\ 1 \end{pmatrix} + \alpha^0 \begin{pmatrix} 4 \\ 2 \end{pmatrix}\right) \\
&= f\left(\begin{pmatrix} 1 + 4\alpha^0 \\ 1 + 2\alpha^0 \end{pmatrix}\right) \\
&= (4\alpha^0 - 2)^2 + (2\alpha^0 - 1)^2.
\end{aligned}$$

To solve this unconstrained minimization, we use the FONC for Unconstrained Nonlinear Optimization Problems, which is:

$$\frac{d}{d\alpha^0}\left[(4\alpha^0 - 2)^2 + (2\alpha^0 - 1)^2\right] = 8(4\alpha^0 - 2) + 4(2\alpha^0 - 1) = 0,$$

and which gives $\alpha^0 = 1/2$. We further have that:

$$\frac{d^2}{d\alpha^{0^2}}\left[(4\alpha^0 - 2)^2 + (2\alpha^0 - 1)^2\right] = 40 > 0,$$

meaning that this value of α^0 is a global minimum. Thus, our new point is:

$$x^1 = x^0 + \alpha^0 d^0 = (3, 2)^\top.$$

From this new point we now conduct another iteration. The steepest descent direction is:

$$d^1 = -\nabla f(x^1) = \begin{pmatrix} 0 \\ 0 \end{pmatrix},$$

meaning that we are at a stationary point. Thus, we terminate the algorithm at this point.   □

5.4.2 Armijo Rule

The **Armijo rule** is a heuristic line search that is intended to give objective-function improvement without the overhead of having to solve an optimization problem, which is required in an exact line search. The Armijo rule works by specifying fixed scalars, s, β, and σ, with $s > 0$, $0 < \beta < 1$, and $0 < \sigma < 1$. We then have:

$$\alpha^k = \beta^{m^k} s,$$

where m^k is the smallest non-negative integer for which:

$$f(x^k) - f(x^k + \beta^{m^k} s d^k) \geq -\sigma \beta^{m^k} s d^{k^\top} \nabla f(x^k).$$

The idea behind the Armijo rule is that we start with an initial guess, s, of what a good step size is. If this step size gives us sufficient improvement in the objective function, which is measured by $f(x^k) - f(x^k + s d^k)$, then we use this as our step size (*i.e.*, we have $\alpha^k = s$). Otherwise, we reduce the step size by a factor of β and try this new reduced step size. That is, we check $f(x^k) - f(x^k + \beta s d^k)$. We continue doing this (*i.e.*, reducing the step size by a factor of β) until we get sufficient improvement in the objective function. The term on the right-hand side of the inequality in the Armijo rule, $-\sigma \beta^{m^k} s d^{k^\top} \nabla f(x^k)$, says that we want the objective to improve by some fraction of the direction times the gradient. The direction times the gradient is the first-order Taylor approximation of the change in the objective-function value.

Usually, σ is chosen close to zero. Values of $\sigma \in [10^{-5}, 10^{-1}]$ are common. The reduction factor, β, is usually chosen to be between $1/2$ and $1/10$. The initial step size guess usually depends on what we know about the objective function. An initial guess of $s = 1$ can be used if we have no knowledge of the behavior of the objective function.

Example 5.4 Consider the unconstrained problem:

$$\min_{x} \ f(x) = (x_1 - 3)^2 + (x_2 - 2)^2,$$

which is introduced in Example 5.1. We know that if we start from the point, $x^0 = (1, 1)^\top$, the initial steepest descent direction is:

$$d^0 = -\nabla f(x^0) = \begin{pmatrix} 4 \\ 2 \end{pmatrix}.$$

We next conduct a line search using the Armijo rule with $\sigma = 10^{-1}$, $\beta = 1/4$ and $s = 1$. To conduct the line search, we compute the objective-function value at the current point as:

$$f(x^0) = 5,$$

and the objective-function value if $\alpha^0 = s = 1$ as:

$$f(x^0 + sd^0) = 5.$$

We also compute:

$$-\sigma sd^{0\top} \nabla f(x^0) = 2.$$

We then check to see if:

$$f(x^0) - f(x^0 + sd^0) \geq -\sigma sd^{0\top} \nabla f(x^0).$$

Finding that it is not, we reduce the step size by a factor of $\beta = 1/4$. We now compute the objective function if $\alpha^0 = \beta s = 1/4$ as:

$$f(x^0 + \beta sd^0) = \frac{5}{4}.$$

We also compute:

$$-\sigma \beta sd^{0\top} \nabla f(x^0) = \frac{1}{2}.$$

We then check to see if:

$$f(x^0) - f(x^0 + \beta sd^0) \geq -\sigma \beta sd^{0\top} \nabla f(x^0).$$

Finding that it is we set $\alpha^0 = \beta s = 1/4$, meaning that our new point is $x^1 = x^0 + \alpha^0 d^0 = (2, 3/2)^\top$.

At this new point the steepest descent direction is:

$$d^1 = -\nabla f(x^1) = \begin{pmatrix} 2 \\ 1 \end{pmatrix},$$

meaning that we are not at a stationary point. Thus, we must continue conducting iterations of the Generic Algorithm for Unconstrained Nonlinear Optimization Problems to solve this problem.

$\square$

5.4.3 Pure Step Size

If we are using the Newton search direction, there is a third step size available to us, which is a **pure step size**. The pure step size rule is to simply fix $\alpha^k = 1$. It is important to stress that the pure step size should only be used with Newton's method. If it is used with steepest descent (or another search direction) there is no guarantee how the algorithm will perform. The pure step size seems completely arbitrary on the surface, but it can perform well in many practical applications. In fact, it can be shown [2] that if we are currently at a point, x^k, that is sufficiently close to a stationary point, then $\alpha^k = 1$ is the step size that we would find from using the Armijo rule, so long as $s = 1$ and $\sigma < 1/2$. As a result, some Newton's method-based software packages skip the line search and use the pure step size instead.

5.5 Performance of Search Directions

A common question that arises is which of steepest descent or Newton's method is a 'better' search direction to use. The answer to this question is somewhat ambiguous, because they have advantages and disadvantages relative to one another. First of all, it is clear that Newton's method involves more work in each iteration of the Generic Algorithm for Unconstrained Nonlinear Optimization Problems. Steepest descent requires the gradient vector to be computed or approximated in each iteration, whereas Newton's method requires the gradient and the Hessian. Thus, we must compute n partial derivatives per iteration with steepest descent as opposed to $n \cdot (n + 1)$ with Newton's method. Moreover, we must invert the $n \times n$ Hessian matrix[1] in each iteration of Newton's method, which can be very computationally expensive.

As the following examples demonstrate, however, in many instances Newton's method provides greater improvement in the objective function in each iteration.

Example 5.5 Consider the unconstrained nonlinear optimization problem:

$$\min_x \ f(x) = (x_1 - 3)^2 + (x_2 - 2)^2,$$

which is introduced in Example 5.1. If we start from the point $x^0 = (1, 1)^\top$ and conduct one iteration of steepest descent with an exact line search, our new point is $x^1 = (3, 2)^\top$, which we show in Example 5.3 to be a stationary point at which the algorithm terminates. We can further show that this point is a local and global minimum of the problem.

If we instead conduct one iteration of Newton's method with an exact line search stating from the point $x^0 = (1, 1)^\top$, our initial search direction is:

$$d^0 = -\left[\nabla^2 f(x^0)\right]^{-1} \nabla f(x^0) = \begin{pmatrix} 2 \\ 1 \end{pmatrix}.$$

Using this direction, we next conduct an exact line search by solving the following minimization problem:

---

[1]In practice, software packages often factorize or decompose the Hessian matrix (as opposed to inverting it) to find the Newton's method direction.

$$\min_{\alpha^0} \, f\left(x^0 + \alpha^0 d^0\right) = f\left(\begin{pmatrix} 1 + 2\alpha^0 \\ 1 + 1\alpha^0 \end{pmatrix}\right),$$

which gives $\alpha^0 = 1$ as a solution. We can also easily show that $f(x^0 + \alpha^0 d^0)$ is convex in α^0, thus we know that $\alpha^0 = 1$ is a global minimum. This means that our new point, when we use Newton's method, is also $x^1 = (3, 2)^\top$.

$\square$

In this example, steepest descent and Newton's method provide the same amount of objective-function improvement in one iteration. This means that the added work involved in computing the Newton's method direction is wasted. However, the following example shows that this does not hold true for all problems.

Example 5.6 Consider the unconstrained nonlinear optimization problem:

$$\min_{x} \, f(x) = 10(x_1 - 3)^2 + 2(x_2 - 2)^2.$$

Starting from the point, $x^0 = (1, 1)^\top$ we first conduct one iteration of steepest descent with an exact line search. To do so, we compute the gradient of the objective function as:

$$\nabla f(x) = \begin{pmatrix} 20(x_1 - 3) \\ 4(x_2 - 2) \end{pmatrix}.$$

Thus, the steepest descent direction is:

$$d^0 = -\nabla f(x^0) = \begin{pmatrix} 40 \\ 4 \end{pmatrix}.$$

Solving the line minimization problem gives $\alpha^0 = 101/2004$, meaning that $x^1 \approx (3.02, 1.20)^\top$. Note that this is not a stationary point, because:

$$\nabla f(x^1) \approx \begin{pmatrix} 0.4 \\ -3.2 \end{pmatrix}.$$

Next, we conduct one iteration of Newton's method with an exact line search, starting from $x^0 = (1, 1)^\top$. To do so, we compute the Hessian of the objective function as:

$$\nabla^2 f(x) = \begin{bmatrix} 20 & 0 \\ 0 & 4 \end{bmatrix}.$$

Thus, the Newton's method direction is:

$$d^0 = -\left[\nabla^2 f(x^0)\right]^{-1} \nabla f(x^0) = \begin{pmatrix} 2 \\ 1 \end{pmatrix}.$$

Solving the line minimization problem gives $\alpha^0 = 1$. Thus, $x^1 = (3, 2)^\top$. It is simple to show that this point is stationary and a local and global minimum. $\qquad\square$

In this example, Newton's method performs much better than steepest descent, finding a stationary point and terminating after a single iteration. It is worth noting that in both Examples 5.5 and 5.6 the starting point is the exact same distance from the stationary point of the objective function (we start at the same point in both examples and the objective functions in both examples have the same stationary points). Nevertheless, steepest descent is unable to find the stationary point in a single iteration with the latter objective function whereas Newton's method is.

To understand why this happens, Figures 5.1 and 5.2 show the contour plots of the objective functions in Examples 5.5 and 5.6, respectively. The contour plot of the objective function in Example 5.5 is a set of concentric circles centered around the stationary point at $x^* = (3, 2)^\top$. This means that starting from any point (not just the choice of $x^0 = (1, 1)^\top$ that is given in the example), the gradient points directly away from the stationary point and the steepest descent direction points directly at the stationary point. Thus, conducting a single iteration of steepest descent with an exact line search on the problem in Example 5.5, starting from any point, gives the stationary point.

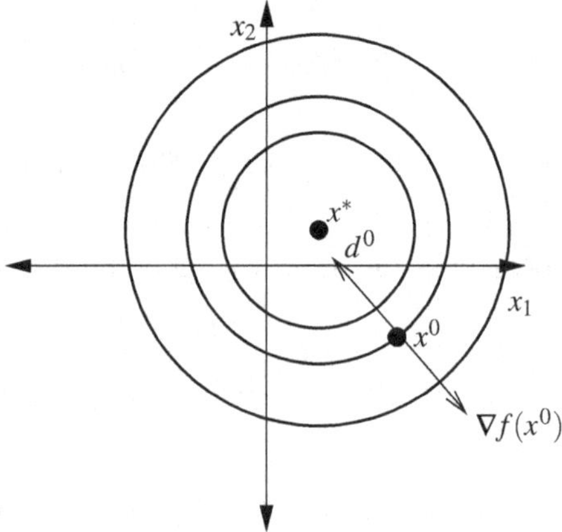

Fig. 5.1 Contour plot of the objective function in Example 5.5

Conversely, the contour plot of the objective function in Example 5.6 is a set of concentric ellipses centered around the stationary point at $x^* = (3, 2)^\top$. This means that starting from almost any point (including the choice of $x^0 = (1, 1)^\top$ that is given in Example 5.6), the gradient does not point directly away from the stationary point. Thus, the steepest descent direction does not point directly at the stationary point. This means that an exact line search is not able to find the stationary point in one iteration. Of course, if we are fortunate and choose a starting point which is on the major or minor axis of the concentric ellipses, then the steepest descent direction will point directly toward the stationary point. However, such luck is typically rare.

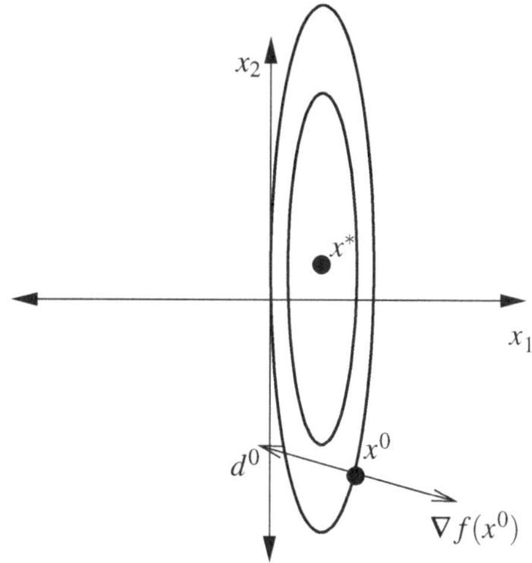

Fig. 5.2 Contour plot of the objective function in Example 5.6

The differences between the objective functions in the two examples, which result in steepest descent performing poorly, are the scaling factors on the x_1 and x_2 terms. Generally speaking, how well steepest descent performs in one iteration depends on the ratio between the largest and smallest eigenvalues of the Hessian matrix. The closer this ratio is to one, the better steepest descent performs. Indeed this ratio is equal to one for the objective function in Example 5.5, which is why steepest descent converges in a single iteration. As this ratio gets higher, steepest descent tends to perform worse. For example, the ratio for the objective in Example 5.6 is five. Newton's method uses second derivative information, which is encoded in the Hessian matrix, to mitigate these scaling issues.

Although Newton's method handles poorly scaled problems better than steepest descent, it has drawbacks (in addition to it requiring more work per iteration). This is illustrated in the following example.

Example 5.7 Consider the unconstrained nonlinear optimization problem:

$$\min_{x} \; f(x) = -(x_1 - 3)^2 - 2(x_2 - 2)^2.$$

Starting from the point, $x^0 = (1, 1)^\top$ we use Newton's method with a pure step size to solve the problem. To find the Newton's method direction we first compute the gradient and Hessian of the objective function, which are:

$$\nabla f(x) = \begin{pmatrix} -2(x_1 - 3) \\ -4(x_2 - 2) \end{pmatrix},$$

and:

$$\nabla^2 f(x) = \begin{bmatrix} -2 & 0 \\ 0 & -4 \end{bmatrix}.$$

Thus, the Newton's method direction is:

$$d^0 = -\left[\nabla^2 f(x^0)\right]^{-1} \nabla f(x^0) = \begin{pmatrix} 2 \\ 1 \end{pmatrix}.$$

A pure step size means that $\alpha^0 = 1$, meaning that our new point is $x^1 = (3, 2)^\top$. If we attempt to conduct another iteration from this incumbent point we have $\nabla f(x^1) = (0, 0)^\top$, meaning that we are at a stationary point and the algorithm terminates.

Note, however, that after conducting this iteration the objective function has gotten worse. We went from an objective-function value of $f(x^0) = -6$ to $f(x^1) = 0$. Indeed, it is easy to show that the stationary point that we find after a single iteration using the Newton's method direction is a local and global *maximum* of $f(x)$. $\square$

Example 5.7 demonstrates a major *caveat* in blindly using Newton's method. Recall that the underlying premise of Newton's method is that we are trying to find a direction, d^k, to move away from our current point, x^k, to make the first-order Taylor approximation of the gradient equal to zero. The rationale behind this is our knowledge from the FONC for Unconstrained Nonlinear Optimization Problems that any local minimum must be a stationary point. However, we also see in a number of examples and the discussion in Section 4.5.1 that local maxima and saddle points are also stationary points. Newton's method attempts to find a stationary point, without distinguishing between what type of stationary point it is.

A question that this raises is whether we can determine *a priori* whether Newton's method will move in a direction that gives an objective-function improvement or not. To answer this question, we examine a first-order Taylor approximation of the objective-function value after conducting a Newton iteration, which is:

$$f(x^k + \alpha^k d^k) \approx f(x^k) + \alpha^k d^{k\top} \nabla f(x^k) = f(x^k) - \alpha^k \nabla f(x^k)^\top \left[\nabla^2 f(x^k)\right]^{-1} \nabla f(x^k).$$

We can rewrite this as:

$$f(x^k) - f(x^k + \alpha^k d^k) \approx \alpha^k \nabla f(x^k)^\top \left[\nabla^2 f(x^k)\right]^{-1} \nabla f(x^k).$$

To get an objective-function improvement we want $f(x^k) - f(x^k + \alpha^k d^k) \geq 0$ or, using the Taylor approximation, we want:

$$\nabla f(x^k)^\top \left[\nabla^2 f(x^k)\right]^{-1} \nabla f(x^k) \geq 0,$$

after dividing through by α^k and noting that $\alpha^k > 0$. This inequality is guaranteed to hold if $\left[\nabla^2 f(x^k)\right]^{-1}$ is positive definite. Moreover, because the eigenvalues of

$\left[\nabla^2 f(x^k)\right]^{-1}$ are the reciprocals of the eigenvalues of $\nabla^2 f(x^k)$, a sufficient condition to guarantee that the objective function improves after conducting an iteration of Newton's method is to have $\nabla^2 f(x^k)$ be positive definite.

The Hessian of the objective function of the problem in Example 5.7 is negative definite, which explains why the objective-function value does not improve after conducting the iteration of Newton's method. As a result of this property, most software packages that use Newton's method test whether the Hessian of the objective function is positive definite at each iteration. If the matrix is positive definite, the algorithm uses the Newton's method direction. Otherwise, another search direction (such as steepest descent) is used. There are also other methods that build off of the concept of Newton's method, which is to use second derivative information to deal with scaling issues and get better objective-function improvements in each iteration. These so-called quasi-Newton methods reduce the amount of work involved in computing the search direction and avoid cases of the objective function getting worse when its Hessian is not positive definite. More advanced textbooks [2, 7] provide details of these types of methods.

Unlike Newton's method, steepest descent is guaranteed to give an objective-function improvement in each iteration, so long as the step size is chosen appropriately. We can see this from the first-order Taylor approximation of the objective function after conducting an iteration, which is:

$$f(x^k + \alpha^k d^k) \approx f(x^k) + \alpha^k d^{k\top} \nabla f(x^k) = f(x^k) - \alpha^k \nabla f(x^k)^\top \nabla f(x^k).$$

This can be rewritten as:

$$f(x^k) - f(x^k + \alpha^k d^k) \approx \alpha^k \nabla f(x^k)^\top \nabla f(x^k),$$

meaning that we get an objective-function improvement so long as:

$$\nabla f(x^k)^\top \nabla f(x^k) \geq 0,$$

which is guaranteed to be true. Indeed, the steepest descent direction starting from x^0 in Example 5.7 is $d^0 = (-4, -2)^\top$. It is easy to show that the objective function improves in this direction, as opposed to the Newton's method direction.

Thus, steepest descent is an example of what is known as a **descent algorithm**—the objective function is guaranteed to improve at each iteration of the Generic Algorithm for Unconstrained Nonlinear Optimization Problems. We can further draw some conclusions about what kinds of stationary points steepest descent converges to. So long as we do not start at a point, x^0, that is a local maximum, if steepest descent converges to a stationary point that point will either be a saddle point or a local minimum. On its own, Newton's method has no such guarantee. If, however, we use Newton's method only at points where the Hessian is positive definite and use steepest descent at other points, we will also have a descent algorithm. This hybrid algorithm will also only stop at stationary points that are local minima or saddle points (so long as it does not start at a local maximum).

5.6 Iterative Solution Algorithms for Constrained Nonlinear Optimization Problems

Applying iterative solution algorithms to constrained nonlinear optimization problems is more difficult than doing so for unconstrained problems. This is because we must be careful of the constraints to ensure that the final solution is feasible. We take two approaches to addressing this difficulty.

The first is to relax the constraints from the problem and instead add terms to the objective function that penalize us for violating them. When we remove or relax the constraints, we can treat the problem as being unconstrained and apply any of the methods discussed in Sections 5.1 through 5.4. The important detail that we must pay attention to with these methods is to ensure that we add terms to the objective function that properly penalize constraint violations. This method can be applied to nonlinear optimization problems with any types of constraints.

The second approach explicitly accounts for the constraints. In this approach we modify the search direction and line search to ensure that after we conduct each iteration the point that we move to is feasible. This method can only easily be applied to problems with linear constraints.

5.6.1 Relaxation-Based Methods

The strength of relaxation-based methods is that they can be applied to problems with any types of constraints. There are two relaxation-based methods that we discuss. We begin with the penalty-based and then discuss the multiplier method, which builds off of the penalty-based method. We use a generic equality- and inequality-constrained nonlinear optimization problem, of the form:

$$\min_{x \in \mathbb{R}^n} f(x)$$

$$\text{s.t. } h_1(x) = 0$$

$$h_2(x) = 0$$

$$\vdots$$

$$h_m(x) = 0$$

$$g_1(x) \le 0$$

$$g_2(x) \le 0$$

$$\vdots$$

$$g_r(x) \le 0,$$

in explaining these methods.

5.6.1.1 Penalty-Based Method

We begin with a discussion of the penalty-based method. The first step to applying the method is to convert the inequality constraints to equalities. We know from Section 2.2.2.1 how to convert inequalities into equalities when applying the Simplex method to linear optimization problems. This is done by adding or subtracting non-negative slack or surplus variables from the left-hand side of the constraints. We follow the same approach here. If we introduce slack variables, $z_1, z_2, \ldots, z_r$ (*i.e.*, one for each inequality constraint), our problem becomes:

$$\min_{x \in \mathbb{R}^n, z \in \mathbb{R}^r, z \ge 0} f(x)$$

$$\text{s.t.} \quad h_1(x) = 0$$
$$h_2(x) = 0$$
$$\vdots$$
$$h_m(x) = 0$$
$$g_1(x) + z_1 = 0$$
$$g_2(x) + z_2 = 0$$
$$\vdots$$
$$g_r(x) + z_r = 0.$$

It is important to remember that the slack variables are now variables that must be listed under the min operator in the problem formulation.

After converting our inequality constraints to equalities, we next add terms to the objective function to penalize constraint violations. Because we have equal-to-zero constraints, we want to penalize having the left-hand sides of any of the constraints be negative or positive. One way to do this is to add terms to the objective function with the absolute value of the left-hand side of each constraint. Absolute value is not desirable, however, because it is not a differentiable function. The methods that are discussed in Section 5.2 rely on derivatives to determine search directions. The alternative, which we use, is to add terms to the objective with the left-hand side of each constraint squared. When we do this, our penalized objective function is:

$$f(x) + \sum_{i=1}^{m} \rho_i \cdot (h_i(x))^2 + \sum_{j=1}^{r} \phi_j \cdot (g_j(x) + z_j)^2,$$

where $\rho_1, \rho_2, \ldots, \rho_m, \phi_1, \phi_2, \ldots, \phi_r > 0$ are fixed coefficients that determine how much weight is placed on violating a particular constraint. Instead of solving the original constrained problem, we solve the problem:

$$\min_{x \in \mathbb{R}^n, z \in \mathbb{R}^r, z \geq 0} f(x) + \sum_{i=1}^{m} \rho_i \cdot (h_i(x))^2 + \sum_{j=1}^{r} \phi_j \cdot (g_j(x) + z_j)^2, \qquad (5.4)$$

and relax the original equality and inequality constraints.

We solve this problem in two steps, by first minimizing with respect to z. This is because we can write Problem (5.4) as:

$$\min_{x \in \mathbb{R}^n} \left\{ \min_{z \in \mathbb{R}^r, z \geq 0} \left\{ f(x) + \sum_{i=1}^{m} \rho_i \cdot (h_i(x))^2 + \sum_{j=1}^{r} \phi_j \cdot (g_j(x) + z_j)^2 \right\} \right\}$$

$$= \min_{x \in \mathbb{R}^n} \left\{ f(x) + \sum_{i=1}^{m} \rho_i \cdot (h_i(x))^2 + \min_{z \in \mathbb{R}^r, z \geq 0} \left\{ \sum_{j=1}^{r} \phi_j \cdot (g_j(x) + z_j)^2 \right\} \right\}.$$

$$(5.5)$$

We can write the problem in this way because the first two terms in (5.4) do not depend on z. If we examine the third term in (5.5), which is:

$$\min_{z \in \mathbb{R}^r, z \geq 0} \sum_{j=1}^{r} \phi_j \cdot (g_j(x) + z_j)^2, \qquad (5.6)$$

we note that this is a quadratic function of z. We can minimize this function with respect to z using the KKT condition, which is:

$$2\phi_1 \cdot (g_1(x) + z_1) - \mu_1 = 0$$

$$\vdots$$

$$2\phi_r \cdot (g_r(x) + z_r) - \mu_r = 0$$

$$0 \leq z_1 \perp \mu_1 \geq 0$$

$$\vdots$$

$$0 \leq z_r \perp \mu_r \geq 0,$$

where μ_j is the Lagrange multiplier on the jth non-negativity constraint (*i.e.*, the constraint $z_j \geq 0$). Solving the KKT condition gives:

$$z_j^* = \begin{cases} -g_j(x), & \text{if } g_j(x) < 0; \\ 0, & \text{otherwise.} \end{cases}$$

We can write this more compactly as $z_j^* = \max\{0, -g_j(x)\}$. We further know that $z \geq 0$ defines a convex feasible region for Problem (5.6) and that the Hessian of its objective function is:

$$\nabla_z^2 \left[\sum_{j=1}^{r} \phi_j \cdot (g_j(x) + z_j)^2 \right] = \begin{bmatrix} 2\phi_1 & 0 & \cdots & 0 \\ 0 & 2\phi_2 & \cdots & 0 \\ \vdots & \vdots & \ddots & \vdots \\ 0 & 0 & \cdots & 2\phi_r \end{bmatrix},$$

where ∇_z^2 denotes the Hessian with respect to z. This Hessian matrix is positive definite. Thus, Problem (5.6) is convex and $z_j^* = \max\{0, -g_j(x)\}$ is its global minimum.

Because we have an explicit closed-form solution for z, we can substitute this into (5.4) and instead solve the following unconstrained minimization problem:

$$\min_{x \in \mathbb{R}^n} \mathscr{F}_{\rho,\phi}(x) = f(x) + \sum_{i=1}^{m} \rho_i \cdot (h_i(x))^2 + \sum_{j=1}^{r} \phi_j \cdot (\max\{0, g_j(x)\})^2, \qquad (5.7)$$

using any of the methods that are discussed in Sections 5.1 through 5.4. The penalized objective function, including the max term in it, is continuously differentiable. However, the Hessian is discontinuous at any x where $g_j(x) = 0$. Thus, the steepest descent direction can be used to solve the penalized objective function without any problem. Moreover, Newton's method can be used, except at these points where the Hessian is discontinuous.

A question at this point is what values to set the ρ's and ϕ's equal to. A naïve answer is to set them to arbitrarily large values. By doing so, we ensure that we have enough weight on constraint violations to find a feasible solution. The problem with setting the ρ's and ϕ's to very high values is that doing so can introduce scaling problems. We know from Example 5.6 that poorly scaled problems can be difficult to solve using iterative algorithms.

Thus, in practice, we use the following Penalty-Based Algorithm for Constrained Nonlinear Optimization Problems. Lines 2 through 4 initialize the algorithm by setting the iteration counter to zero, picking starting penalty weights, and a starting guess for a solution. If we have information about the problem, that can help in picking starting values for the penalty weights. Otherwise, values of one are reasonable starting values.

Penalty-Based Algorithm for Constrained Nonlinear Optimization Problems

```
1: procedure PENALTY- BASED ALGORITHM
2:    k ← 0                                                  ▷ Set iteration counter to 0
3:    Fix ρ₁, ρ₂, . . . , ρₘ, φ₁, φ₂, . . . , φᵣ to positive values   ▷ Initialize penalty weights
4:    Fix x⁰                                                 ▷ Fix a starting point
5:    while Termination criteria are not met do
6:        Find direction, dᵏ, to move in
7:        Determine step size, αᵏ
8:        xᵏ⁺¹ ← xᵏ + αᵏdᵏ                                   ▷ Update point
9:        for i ← 1, . . . , m do
10:           if hᵢ(xᵏ⁺¹) ≠ 0 then
11:               Increase ρᵢ      ▷ Increase penalty weight if constraint is not satisfied
12:           end if
13:       end for
14:       for j ← 1, . . . , r do
15:           if gⱼ(xᵏ⁺¹) > 0 then
16:               Increase φⱼ      ▷ Increase penalty weight if constraint is not satisfied
17:           end if
18:       end for
19:       k ← k + 1                                          ▷ Update iteration counter
20:   end while
21: end procedure
```

Lines 5 through 20 are the main iterative loop, much of which is the same as in the Generic Algorithm for Unconstrained Nonlinear Optimization Problems. Lines 6 and 7 determine the search direction and conduct a line search, using any of the methods that are outlined in Sections 5.2 and 5.4. It is important to stress that these steps should be conducted using:

$$\min_{x\in\mathbb{R}^n} \mathscr{F}_{\rho,\phi}(x) = f(x) + \sum_{i=1}^{m} \rho_i \cdot (h_i(x))^2 + \sum_{j=1}^{r} \phi_j \cdot (\max\{0, g_j(x)\})^2,$$

as the objective function, as opposed to simply using $f(x)$. If we use $f(x)$ alone as the objective function, then we are *completely* ignoring the constraints of the original problem. After we update our point in Line 8 we next go through each constraint in Lines 9 through 18. If we find that any constraint is still violated at our new point, then we increase the penalty weight associated with that constraint. Otherwise, if the constraint is satisfied the weight is kept the same. This process continues until we satisfy our termination criteria in Line 5.

The typical termination criterion that we use is that we have a solution, x^k, that satisfies all of the constraints of the original problem and that is a stationary point of $\mathscr{F}_{\rho,\phi}(x)$. However, as discussed in Section 5.3, we may include other termination criteria, such as a maximum number of iterations, or tolerances in the feasibility and gradient conditions.

Example 5.8 Consider the problem:

$$\min_{x} f(x) = (x_1 + 2)^2 + (x_2 - 3)^2$$

$$\text{s.t.} \, h_1(x) = x_1 + 2x_2 - 4 = 0$$

$$g_1(x) = -x_1 \leq 0.$$

We derive the penalized objective function:

$$\mathscr{F}_{\rho_1,\phi_1}(x) = (x_1 + 2)^2 + (x_2 - 3)^2 + \rho_1 \cdot (x_1 + 2x_2 - 4)^2 + \phi_1 \cdot (\max\{0, -x_1\})^2,$$

which becomes:

$$\mathscr{F}_{1,1}(x) = (x_1 + 2)^2 + (x_2 - 3)^2 + (x_1 + 2x_2 - 4)^2 + (\max\{0, -x_1\})^2,$$

if we assume starting weights of one on both constraints. Starting from the point $x^0 = (-1, -1)^\top$, we conduct one iteration of steepest descent using the Armijo rule with $\sigma = 10^{-1}$, $\beta = 1/10$ and $s = 1$.

To conduct the iteration, we must compute the gradient of the objective function. To do so, we note that at $x^0 = (-1, -1)^\top$ the objective function is equal to:

$$\begin{aligned}
\mathscr{F}_{1,1}(x) &= (x_1 + 2)^2 + (x_2 - 3)^2 + (x_1 + 2x_2 - 4)^2 + (\max\{0, -x_1\})^2 \\
&= (x_1 + 2)^2 + (x_2 - 3)^2 + (x_1 + 2x_2 - 4)^2 + (-x_1)^2.
\end{aligned}$$

This step of determining whether:

$$\max\{0, -x_1\},$$

is equal to 0 or $-x_1$ is vitally important, as the gradient that we compute depends on what we substitute for this term. The gradient is equal to:

$$\nabla \mathscr{F}_{1,1}(x) = \begin{pmatrix} 2(x_1 + 2) + 2(x_1 + 2x_2 - 4) + 2x_1 \\ 2(x_2 - 3) + 4(x_1 + 2x_2 - 4) \end{pmatrix},$$

and the search direction is:

$$d^0 = -\nabla \mathscr{F}_{1,1}(x^0) = \begin{pmatrix} 14 \\ 36 \end{pmatrix}.$$

To conduct the line search, we compute the objective-function value at the current point as:

$$\mathscr{F}_{1,1}(x^0) = 67,$$

and the objective-function value if $\alpha^0 = s = 1$ as:

$$\mathscr{F}_{1,1}(x^0 + sd^0) = 7490.$$

We also compute:

$$-\sigma s d^{0\top} \nabla \mathscr{F}_{1,1}(x^0) = 149.2.$$

We then check to see if:

$$\mathscr{F}_{1,1}(x^0) - \mathscr{F}_{1,1}(x^0 + sd^0) \geq -\sigma s d^{0\top} \nabla \mathscr{F}_{1,1}(x^0).$$

Finding that it is not, we reduce the step size by a factor of $\beta = 1/10$. We now compute the objective function if $\alpha^0 = \beta s = 1/10$ as:

$$\mathscr{F}_{1,1}(x^0 + \beta s d^0) = 8.48.$$

We also compute:

$$-\sigma \beta s d^{0\top} \nabla \mathscr{F}_{1,1}(x^0) = 14.92.$$

We then check to see if:

$$\mathscr{F}_{1,1}(x^0) - \mathscr{F}_{1,1}(x^0 + s\beta d^0) \geq -\sigma \beta s d^{0\top} \nabla \mathscr{F}_{1,1}(x^0).$$

Finding that it is we set $\alpha^0 = \beta s = 1/10$, meaning that our new point is $x^1 = (0.4, 2.6)^\top$.

We already know that the value of the penalized objective function improves from this one iteration. We further have that:

$$f(x^0) = 17,$$

and:

$$f(x^1) = 5.92,$$

meaning that the objective function of the original problem improves in this one iteration. We also have that:

$$h_1(x^0) = -7,$$

$$h_1(x^1) = 1.6,$$

$$g_1(x^0) = 1,$$

and:

$$g_1(x^1) = -0.4,$$

meaning that the extent to which the two constraints are violated is reduced in conducting this one iteration. The $h_1(x) = 0$ constraint is still violated, but by less than before. The $g_1(x) \leq 0$ constraint is satisfied by the new point. Moreover, the point $x^1 = (0.4, 2.6)^\top$ is not a stationary point of $\mathscr{F}_{1,1}(x)$, because we have that:

$$\nabla \mathscr{F}_{1,1}(x^1) = \begin{pmatrix} 8 \\ 5.6 \end{pmatrix}.$$

Thus, we must continue the Penalty-Based Algorithm for Constrained Nonlinear Optimization Problems. We increase the value of ρ_1 *only* (*i.e.*, ϕ_1 is kept at the same value), because the equality constraint is violated while the inequality constraint is satisfied at x^1

$\qquad\qquad\qquad\qquad\qquad\qquad\qquad\qquad\qquad\qquad\qquad\qquad\qquad\square$

Assuming that the original constrained problem is feasible, the Penalty-Based Algorithm for Constrained Nonlinear Optimization Problems will typically find a feasible solution, so long as the penalty weights are sufficiently large. We can come to this conclusion by relying on the following Penalty-Function Property.

Penalty-Function Property: Suppose that the equality- and inequality-constrained problem:

$$\min_{x \in \mathbb{R}^n} f(x)$$

$$\text{s.t. } h_1(x) = 0$$
$$h_2(x) = 0$$
$$\vdots$$
$$h_m(x) = 0$$
$$g_1(x) \leq 0$$
$$g_2(x) \leq 0$$
$$\vdots$$
$$g_r(x) \leq 0,$$

is feasible. Then so long as $\rho_1, \rho_2, \ldots, \rho_m, \phi_1, \phi_2, \ldots, \phi_r$ are sufficiently large, a global minimum of:

$$\min_{x \in \mathbb{R}^n} \mathscr{F}_{\rho,\phi}(x) = f(x) + \sum_{i=1}^{m} \rho_i \cdot (h_i(x))^2 + \sum_{j=1}^{r} \phi_j \cdot (\max\{0, g_j(x)\})^2,$$

is optimal and feasible in the original equality- and inequality-constrained problem.

Because we can choose $\rho_1, \rho_2, \ldots, \rho_m, \phi_1, \phi_2, \ldots, \phi_r$ to be arbitrarily large, we can examine the case in which the penalty weights go to $+\infty$. Let x^* be a global minimum of:

$$\lim_{(\rho_1,\rho_2,\ldots,\rho_m,\phi_1,\phi_2,\ldots,\phi_r)\to+\infty} \mathscr{F}_{\rho,\phi}(x).$$

If x^* is infeasible in the original constrained problem, then that means at least one of $h_i(x^*) \neq 0$ or $g_j(x^*) > 0$. However, in this case we have that:

$$\lim_{(\rho_1,\rho_2,\ldots,\rho_m,\phi_1,\phi_2,\ldots,\phi_r)\to+\infty} \mathscr{F}_{\rho,\phi}(x^*) = +\infty.$$

On the other hand, if we choose a feasible value for x, we would have:

$$\lim_{(\rho_1,\rho_2,\ldots,\rho_m,\phi_1,\phi_2,\ldots,\phi_r)\to+\infty} \mathscr{F}_{\rho,\phi}(x) = f(x) < +\infty.$$

This shows that x^* must be feasible in the original problem.

Moreover, because we know that x^* must be feasible in the original constrained problem, we know that $\mathscr{F}_{\rho,\phi}(x^*) = f(x^*)$. This is because all of the penalty terms in $\mathscr{F}_{\rho,\phi}(x^*)$ equal zero. Now, suppose that x^* is not optimal in the original problem. That means there is a different solution, $\hat{x}$, that is optimal in the original problem with the property that $f(\hat{x}) < f(x^*)$. We also know that $\hat{x}$ is feasible in the original problem. Thus, we have that $\mathscr{F}_{\rho,\phi}(\hat{x}) = f(\hat{x}) < f(x^*) = \mathscr{F}_{\rho,\phi}(x^*)$, which means that x^* cannot be a global minimum of:

$$\lim_{(\rho_1,\rho_2,\ldots,\rho_m,\phi_1,\phi_2,\ldots,\phi_r)\to+\infty} \mathscr{F}_{\rho,\phi}(x).$$

What the Penalty-Function Property says is that so long as the penalty weights are large enough, at some point minimizing $\mathscr{F}_{\rho,\phi}(x)$ will start giving solutions that are feasible in the original problem. The Penalty-Based Algorithm for Constrained Nonlinear Optimization Problems keeps increasing the penalty weights until x^k is feasible. Moreover, in each iteration of the algorithm we choose a search direction and conduct a line search to reduce $\mathscr{F}_{\rho,\phi}(x)$. These two properties of the algorithm work together to ensure that we eventually arrive at variable values that satisfy the constraints of the original problem. Moreover, if the algorithm finds a global minimum of $\mathscr{F}_{\rho,\phi}(x)$, then that solution is also a global minimum of the original constrained problem.

5.6.1.2 Multiplier Method

The multiplier method builds off of the penalty-based method for solving a constrained nonlinear optimization problem. This is because it takes the same approach of relaxing the constraints and adding the same types of terms to the objective function to penalize constraint violations. However, the multiplier method also makes use of Lagrange multiplier theory to further penalize constraint violations beyond what is done in the penalty-based method.

Although we can derive the multiplier method for a generic problem with a mixture of equality and inequality constraints, the steps are more straightforward for a problem with only equality constraints. As such, we begin by first showing another way (which is different from that used in the penalty-based method) to convert any generic equality- and inequality-constrained optimization problem of the form:

$$\min_{x \in \mathbb{R}^n} f(x)$$

$$\text{s.t. } h_1(x) = 0$$
$$h_2(x) = 0$$
$$\vdots$$
$$h_m(x) = 0$$
$$g_1(x) \le 0$$
$$g_2(x) \le 0$$
$$\vdots$$
$$g_r(x) \le 0,$$

into a problem with equality constraints only. We again do this using slack variables, which we denote $z_1, \ldots, z_r$. However, instead of constraining the slack variables to be non-negative, we instead square them when adding them to the left-hand sides of the inequality constraints. Doing so ensures that the added slacks are non-negative. Adding these variables gives us the equality-constrained problem:

$$\min_{x \in \mathbb{R}^n, z \in \mathbb{R}^r} f(x)$$

$$\text{s.t. } h_1(x) = 0$$
$$h_2(x) = 0$$
$$\vdots$$
$$h_m(x) = 0$$
$$g_1(x) + z_1^2 = 0$$
$$g_2(x) + z_2^2 = 0$$

$$\vdots$$

$$g_r(x) + z_r^2 = 0,$$

which is equivalent to our starting problem. Because we can convert any equality- and inequality-constrained problem to an equivalent problem with equality constraints only, we hereafter in this discussion restrict our attention to the following generic equality-constrained problem:

$$\min_{x \in \mathbb{R}^n} f(x) \tag{5.8}$$

$$\text{s.t. } h_1(x) = 0 \tag{5.9}$$

$$h_2(x) = 0 \tag{5.10}$$

$$\vdots$$

$$h_m(x) = 0. \tag{5.11}$$

We next show an alternate way to derive the FONC for equality-constrained problems (*cf.* Section 4.5.2). This alternative derivation of the FONC is used in explaining the concept behind the multiplier method. The alternate derivation of the FONC begins by defining what is known as the **Lagrangian function** for the equality-constrained problem, which is:

$$\mathscr{L}(x, \lambda) = f(x) + \sum_{i=1}^{m} \lambda_i h_i(x).$$

We now show the very useful Lagrangian-Function Property relating this Lagrangian function to the original equality-constrained problem.

Lagrangian-Function Property: Any solution to the FONC of the unconstrained optimization problem:

$$\min_{x, \lambda} \mathscr{L}(x, \lambda) = f(x) + \sum_{i=1}^{m} \lambda_i h_i(x), \tag{5.12}$$

is a feasible solution to the FONC for equality-constrained problem (5.8)–(5.11) and vice versa. Moreover, a global minimum of (5.12) is a feasible solution to the FONC for equality-constrained problem (5.8)–(5.11).

We show this by writing the FONC for (5.12). Because both x and λ are variables in this minimization problem, the FONC is:

$$\nabla_x \mathscr{L}(x, \lambda) = \nabla f(x) + \sum_{i=1}^{m} \lambda_i \nabla h_i(x) = 0, \tag{5.13}$$

$$\nabla_{\lambda_1} \mathscr{L}(x, \lambda) = h_1(x) = 0, \tag{5.14}$$

$$\nabla_{\lambda_2} \mathscr{L}(x, \lambda) = h_2(x) = 0, \tag{5.15}$$

$$\vdots$$

$$\nabla_{\lambda_m} \mathscr{L}(x, \lambda) = h_m(x) = 0, \tag{5.16}$$

where ∇_x and ∇_{λ_j} indicate the gradient with respect to x and λ_j, respectively.

Note that condition (5.13) is exactly the FONC (*cf.* Section 4.5.2) for equality-constrained problem (5.8)–(5.11) and that conditions (5.14)–(5.16) are equality constraints (5.9)–(5.11). Thus, the FONC of the Lagrangian function is equivalent to the FONC and constraints for the equality-constrained problem. As a result, a solution to either set of conditions is a solution to the other.

The second part of the property, regarding a global minimum of (5.12), immediately follows. If $x^*, \lambda_1^*, \lambda_2^*, \ldots, \lambda_m^*$ are a global minimum of (5.12), then they must satisfy the FONC of (5.12). Thus, they must also satisfy the FONC of problem (5.8)–(5.11) and x^* must be feasible in problem (5.8)–(5.11).

The Lagrangian-Function Property is useful because it provides a different way of thinking about how we solve an equality-constrained problem. Namely, we can solve an equality-constrained problem by minimizing its Lagrangian function, which we do by solving the FONC of (5.12). Using this observation, the multiplier method solves the equality-constrained problem by relaxing the problem constraints and solving the unconstrained problem:

$$\min_{x \in \mathbb{R}^n} \mathscr{A}_{\lambda, \rho}(x) = f(x) + \sum_{i=1}^{m} \lambda_i h_i(x) + \sum_{i=1}^{m} \rho_i \cdot (h_i(x))^2.$$

$\mathscr{A}_{\lambda, \rho}(x)$ is called the **augmented Lagrangian function**, because it adds the penalty terms, which we use in the penalty-based method, to the Lagrangian function. The idea behind the multiplier method is that we solve the original equality-constrained problem by increasing the penalty weights, $\rho_1, \rho_2, \ldots, \rho_m$. At the same time, we also try and get the Lagrange multipliers, $\lambda_1, \lambda_2, \ldots, \lambda_m$, to go to the values,

$\lambda_1^*, \lambda_2^*, \ldots, \lambda_m^*$, that would be obtained from solving problem (5.8)–(5.11) using the FONC. By increasing the ρ's and getting the λ's to the values that solve the FONC, the multiplier method typically solves a constrained problem much faster than the penalty-based method does.

The only challenge in using the multiplier method is that we typically do not know the values of $\lambda_1^*, \lambda_2^*, \ldots, \lambda_m^*$ *a priori*. We would normally only know their values from solving the FONC directly. However, if the FONC is already solved, there is no need to use an iterative solution algorithm! The multiplier method gets around this problem by iteratively updating the Lagrange multipliers as new values of x are found.

To understand how this updating is done, we must show another property relating the Lagrangian function to the original constrained problem. We also stress that we provide only an outline of the derivation of the multiplier-updating rule. Interested readers are referred to more advanced textbooks [2] that provide all of the details of the derivation.

Lagrangian-Optimality Property: Let x^* be a global minimum of equality-constrained problem (5.8)–(5.11). Then x^* is also a global minimum of:

$$\min_{x \in \mathbb{R}^n} \mathscr{L}(x, \hat{\lambda}) = f(x) + \sum_{i=1}^{m} \hat{\lambda}_i h_i(x) \tag{5.17}$$

$$\text{s.t. } h_1(x) = 0 \tag{5.18}$$

$$h_2(x) = 0 \tag{5.19}$$

$$\vdots \tag{5.20}$$

$$h_m(x) = 0, \tag{5.21}$$

where $\hat{\lambda}_1, \hat{\lambda}_2, \ldots, \hat{\lambda}_m$ are constants.

Before showing the Lagrangian-Optimality Property, let us briefly comment on problem (5.17)–(5.21). This problem has the same constraints as the original equality-constrained problem, but has the Lagrangian function as its objective function. However, the values of the λ's are fixed in this problem, and we are minimizing with respect to x only. What this property tells us is that if we fix the values of the λ's but impose the same constraints as the original problem, then minimizing the Lagrangian function gives us the same global minimum that minimizing the original objective function does. This result should not be too surprising in light of the Lagrangian-Function Property.

We show this result by contradiction. Suppose that the Lagrangian-Optimality Property is not true. This means that there is a vector, which we call $\hat{x}$, with the property that $\mathscr{L}(\hat{x}, \hat{\lambda}) < \mathscr{L}(x^*, \hat{\lambda})$. Because problem (5.17)–(5.21) includes the constraints of the original equality-constrained problem, we know that both $\hat{x}$ and x^* are feasible in the original equality-constrained problem. Thus, we have that:

$$\mathscr{L}(\hat{x}, \hat{\lambda}) = f(\hat{x}) + \sum_{i=1}^{m} \hat{\lambda}_i h_i(\hat{x}) = f(\hat{x}),$$

and:

$$\mathscr{L}(x^*, \hat{\lambda}) = f(x^*) + \sum_{i=1}^{m} \hat{\lambda}_i h_i(x^*) = f(x^*).$$

Combining these relationships tells us that:

$$f(\hat{x}) = \mathscr{L}(\hat{x}, \hat{\lambda}) < \mathscr{L}(x^*, \hat{\lambda}) = f(x^*),$$

which means that x^* cannot be a global minimum of problem (5.8)–(5.11), because $\hat{x}$ is feasible in this problem and gives a smaller objective-function value than x^* does.

We can now derive the rule to update the value of λ in the multiplier method. In doing so, we assume that we currently have a value, which we denote as λ^k, for the multipliers in the augmented Lagrangian after conducting k iterations. Thus, our goal is to determine how to update λ^k to a new value, which we call λ^{k+1}, for the next iteration.

We begin by examining the original equality-constrained problem:

$$\min_{x \in \mathbb{R}^n} f(x)$$

$$\text{s.t. } h_1(x) = 0$$

$$h_2(x) = 0$$

$$\vdots$$

$$h_m(x) = 0.$$

We know that x^* is a global minimum of this problem. Furthermore, the FONC tells us that there are Lagrange multipliers, $\lambda_1^*, \lambda_2^*, \ldots, \lambda_m^*$, such that:

$$\nabla f(x^*) + \sum_{i=1}^{m} \lambda_i^* \nabla h_i(x^*) = 0. \tag{5.22}$$

We next examine the following problem that keeps the same equality constraints, but minimizes the Lagrangian function:

$$\min_{x \in \mathbb{R}^n} \mathscr{L}(x, \lambda^k) = f(x) + \sum_{i=1}^{m} \lambda_i^k h_i(x) \tag{5.23}$$

$$\text{s.t. } h_1(x) = 0 \tag{5.24}$$

$$h_2(x) = 0 \tag{5.25}$$

$$\vdots \tag{5.26}$$

$$h_m(x) = 0, \tag{5.27}$$

where λ is fixed equal to λ^k. From the Lagrangian-Optimality Property, we know that x^* is a global minimum of this problem. Thus, the FONC tells us that there exist Lagrange multipliers, $\hat{\lambda}_1, \hat{\lambda}_2, \ldots, \hat{\lambda}_m$, such that:

$$\nabla f(x^*) + \sum_{i=1}^{m} \lambda_i^k \nabla h_i(x^*) + \sum_{i=1}^{m} \hat{\lambda}_i \nabla h_i(x^*) = 0. \tag{5.28}$$

Subtracting (5.28) from (5.22) gives us:

$$\sum_{i=1}^{m} (\lambda_i^* - \lambda_i^k - \hat{\lambda}_i) \nabla h_i(x^*) = 0.$$

The regularity requirement of the FONC states that the gradients of the equality constraints must be linearly independent at x^*. Thus, the only way to make a linear combination of the $h_i(x^*)$'s equal to zero is to set all of the scalar coefficients equal to zero, meaning that:

$$\lambda_i^* - \lambda_i^k - \hat{\lambda}_i = 0, \forall i = 1, \ldots, m,$$

or:

$$\hat{\lambda}_i = \lambda_i^* - \lambda_i^k, \forall i = 1, \ldots, m. \tag{5.29}$$

We now examine the following unconstrained optimization problem:

$$\min_{x \in \mathbb{R}^n} \mathscr{A}_{\lambda^k, \rho}(x) = f(x) + \sum_{i=1}^{m} \lambda_i^k h_i(x) + \sum_{i=1}^{m} \rho_i \cdot (h_i(x))^2, \tag{5.30}$$

in which the augmented Lagrangian function is minimized and where λ is fixed equal to λ^k. Let x^{k+1} denote the global minimum of $\mathscr{A}_{\lambda^k, \rho}(x)$, which is the value for x that is obtained after conducting $(k + 1)$ iterations of the solution algorithm. Applying the FONC to unconstrained problem (5.30) gives:

$$\nabla f(x^{k+1}) + \sum_{i=1}^{m} \lambda_i^k \nabla h_i(x^{k+1}) + \sum_{i=1}^{m} 2\rho_i h_i(x^{k+1}) \nabla h_i(x^{k+1}) = 0. \qquad (5.31)$$

We finally substitute x^{k+1} in place of x^* in (5.28), which gives:

$$\nabla f(x^{k+1}) + \sum_{i=1}^{m} \lambda_i^k \nabla h_i(x^{k+1}) + \sum_{i=1}^{m} \hat{\lambda}_i \nabla h_i(x^{k+1}) = 0. \qquad (5.32)$$

Subtracting equation (5.31) from (5.32) gives:

$$\sum_{i=1}^{m} (\hat{\lambda}_i - 2\rho_i h_i(x^{k+1})) \nabla h_i(x^{k+1}) = 0.$$

We can, again, rely on the regularity requirement to simplify this to:

$$\hat{\lambda}_i - 2\rho_i h_i(x^{k+1}) = 0, \forall i = 1, \ldots, m,$$

or:
$$\hat{\lambda}_i = 2\rho_i h_i(x^{k+1}), \forall i = 1, \ldots, m. \qquad (5.33)$$

Combining equations (5.29) and (5.33) gives:

$$\lambda_i^* - \lambda_i^k = 2\rho_i h_i(x^{k+1}), \forall i = 1, \ldots, m,$$

which we can rewrite as:

$$\lambda_i^* = \lambda_i^k + 2\rho_i h_i(x^{k+1}), \forall i = 1, \ldots, m.$$

This equation gives our multiplier-updating rule, which is:

$$\lambda_i^{k+1} \leftarrow \lambda_i^k + 2\rho_i h_i(x^{k+1}), \forall i = 1, \ldots, m,$$

which says that the Lagrange multiplier is updated based on the size of the constraint violation in each iteration.

We now summarize the steps of the Multiplier-Based Algorithm for Constrained Nonlinear Optimization Problems. Lines 2 through 5 initialize the algorithm by first setting the iteration counter to zero, fixing starting values for the penalty weights and Lagrange multipliers, and fixing an initial guess for x. Lines 6 through 17 are the main iterative loop. Lines 7 and 8 find a search direction and conduct a line search and Line 9 updates x based on the direction and step size. Note that the search direction and line search should be conducted using the augmented Lagrangian function:

$$\mathscr{A}_{\lambda^k,\rho}(x) = f(x) + \sum_{i=1}^{m} \lambda_i^k h_i(x) + \sum_{i=1}^{m} \rho_i \cdot (h_i(x))^2,$$

as the objective function, as opposed to $f(x)$ alone. If the search direction and line search are conducted using $f(x)$ alone, this would result in completely ignoring the original constraints.

Multiplier-Based Algorithm for Constrained Nonlinear Optimization Problems

```
1: procedure MULTIPLIER- BASED ALGORITHM
2:     k ← 0                                              ▷ Set iteration counter to 0
3:     Fix ρ₁, ρ₂, ..., ρₘ to positive values            ▷ Initialize penalty weights
4:     Fix λ⁰                                             ▷ Initialize Lagrange multipliers
5:     Fix x⁰                                             ▷ Fix a starting point
6:     while Termination criteria are not met do
7:         Find direction, dᵏ, to move in
8:         Determine step size, αᵏ
9:         xᵏ⁺¹ ← xᵏ + αᵏdᵏ                               ▷ Update point
10:        for i ← 1, ..., m do
11:            λᵢᵏ⁺¹ ← λᵢᵏ + 2ρᵢhᵢ(xᵏ⁺¹)                  ▷ Update multiplier
12:            if hᵢ(xᵏ⁺¹) ≠ 0 then
13:                Increase ρᵢ        ▷ Increase penalty weight if constraint is not satisfied
14:            end if
15:        end for
16:        k ← k + 1                                       ▷ Update iteration counter
17:    end while
18: end procedure
```

Lines 10 through 15 go through each problem constraint and first update the associated Lagrange multiplier in Line 11. Next, in Lines 12 through 14 the penalty weight of any constraint that is not satisfied by x^{k+1} is increased. This iterative process repeats until we satisfy a termination criterion in Line 6. We typically use the same termination criterion as in the Penalty-Based Algorithm for Constrained Nonlinear Optimization Problems. Namely, we stop once we find a solution, x^k, that satisfies all of the original problem constraints and is a stationary point of $\mathscr{A}_{\lambda^k,\rho}(x)$.

Before demonstrating the multiplier method, we note that we can argue that this method will also give us a feasible solution to the original problem, so long as the penalty weights are sufficiently large. This comes from the following Augmented-Lagrangian Property.

Augmented-Lagrangian Property: Suppose that equality-constrained problem (5.23)–(5.27) is feasible. Then so long as $\rho_1, \rho_2, \ldots, \rho_m$ are sufficiently large, a global minimum of:

$$\min_{x \in \mathbb{R}^n} \mathscr{A}_{\hat{\lambda}, \rho}(x) = f(x) + \sum_{i=1}^{m} \hat{\lambda}_i h_i(x) + \sum_{i=1}^{m} \rho_i \cdot (h_i(x))^2,$$

is optimal and feasible in the original equality-constrained problem, where $\hat{\lambda}_1, \hat{\lambda}_2, \ldots, \hat{\lambda}_m$ are constants.

We essentially follow the same logic that is used to show the Penalty-Function Property. Because we can choose $\rho_1, \rho_2, \ldots, \rho_m$ to be arbitrarily large, we can examine the case in which the penalty weights go to $+\infty$. Let x^* be a global minimum of:

$$\lim_{(\rho_1, \rho_2, \ldots, \rho_m) \to +\infty} \mathscr{A}_{\hat{\lambda}, \rho}(x).$$

If x^* is infeasible in the original constrained problem, then that means at least one of $h_i(x^*) \neq 0$. However, in this case we have that:

$$\lim_{(\rho_1, \rho_2, \ldots, \rho_m) \to +\infty} \mathscr{A}_{\hat{\lambda}, \rho}(x^*) = +\infty.$$

On the other hand, if we choose any feasible value for x, we would have:

$$\lim_{(\rho_1, \rho_2, \ldots, \rho_m) \to +\infty} \mathscr{A}_{\hat{\lambda}, \rho}(x) = f(x) < +\infty.$$

This shows that x^* must be feasible in the original problem.

Moreover, because we know that x^* must be feasible in the original constrained problem, we know that $\mathscr{A}_{\hat{\lambda}, \rho}(x^*) = f(x^*)$. Now, suppose that x^* is not optimal in the original problem. That means there is a different solution, which we call $\hat{x}$, such that $f(\hat{x}) < f(x^*)$. We also know that $\hat{x}$ is feasible in the original problem. Thus, we have that $\mathscr{A}_{\hat{\lambda}, \rho}(\hat{x}) = f(\hat{x}) < f(x^*) = \mathscr{A}_{\hat{\lambda}, \rho}(x^*)$, which means that x^* cannot be a global minimum of:

$$\lim_{(\rho_1, \rho_2, \ldots, \rho_m) \to +\infty} \mathscr{A}_{\hat{\lambda}, \rho}(x).$$

Example 5.9 Consider the problem:

$$\min_{x} f(x) = (x_1 + 2)^2 + (x_2 - 3)^2$$
$$\text{s.t. } h_1(x) = x_1 + 2x_2 - 4 = 0$$
$$g_1(x) = -x_1 \leq 0,$$

which is introduced in Example 5.8. To solve this problem using the multiplier method, we first convert it to the equality-constrained problem:

$$\min_{x} f(x) = (x_1 + 2)^2 + (x_2 - 3)^2$$
$$\text{s.t. } h_1(x) = x_1 + 2x_2 - 4 = 0$$
$$h_2(x) = -x_1 + x_3^2 = 0.$$

We next derive the augmented Lagrangian function:

$$\mathscr{A}_{\lambda,\rho}(x) = (x_1 + 2)^2 + (x_2 - 3)^2 + \lambda_1 \cdot (x_1 + 2x_2 - 4)$$
$$+ \lambda_2 \cdot (x_3^2 - x_1) + \rho_1 \cdot (x_1 + 2x_2 - 4)^2 + \rho_2 \cdot (x_3^2 - x_1)^2,$$

which becomes:

$$\mathscr{A}_{(1,1)^\top,(1,1)^\top}(x) = (x_1 + 2)^2 + (x_2 - 3)^2 + (x_1 + 2x_2 - 4)$$
$$+ (x_3^2 - x_1) + (x_1 + 2x_2 - 4)^2 + (x_3^2 - x_1)^2,$$

if we assume starting Lagrange multiplier and penalty-weight values of one.

Starting from the point $x^0 = (-1, -1, -1)^\top$, we conduct one iteration of steepest descent using the Armijo rule with $\sigma = 10^{-1}$, $\beta = 1/10$ and $s = 1$. The gradient of the augmented Lagrangian function is:

$$\nabla \mathscr{A}_{(1,1)^\top,(1,1)^\top}(x) = \begin{pmatrix} 2(x_1 + 2) + 2(x_1 + 2x_2 - 4) - 2(x_3^2 - x_1) \\ 2(x_2 - 3) + 2 + 4(x_1 + 2x_2 - 4) \\ 2x_3 + 4x_3 \cdot (x_3^2 - x_1) \end{pmatrix},$$

which gives us a search direction of:

$$d^0 = -\nabla \mathscr{A}_{(1,1)^\top,(1,1)^\top}(x^0) = \begin{pmatrix} 16 \\ 34 \\ 10 \end{pmatrix}.$$

Applying the Armijo rule, which we exclude for sake of brevity, results in our setting $\alpha^0 = \beta s = 1/10$, meaning that our new point is $x^1 = (0.6, 2.4, 0)^\top$.

We can now examine the changes in the objective- and constraint-function values of our original problem. We have that:

$$f(x^0) = 17,$$

$$h_1(x^0) = -7,$$

$$g_1(x^0) = 1,$$

and:

$$f(x^1) = 7.12,$$

$$h_1(x^1) = 1.4,$$

$$g_1(x^1) = -2.4.$$

Thus we see that in this one iteration the objective-function value of the original problem improves and that the second constraint is no longer violated. Although the first constraint is violated, the amount by which it is violated has come down. That being said, we cannot terminate the Multiplier-Based Algorithm for Constrained Nonlinear Optimization Problems, because both termination requirements are violated (x^1 is not feasible in the original problem and it is not a stationary point of the augmented Lagrangian function). We increase the penalty weight on the first constraint only to conduct the next iteration (*i.e.*, ρ_2 remains the same, because the second constraint is satisfied by x^1). We also update the Lagrange multipliers to:

$$\lambda_1^1 \leftarrow \lambda_1^0 + 2\rho_1 h_1(x^1) = 3.8,$$

and:

$$\lambda_2^1 \leftarrow \lambda_2^0 + 2\rho_2 h_2(x^1) = -0.2.$$

$\square$

5.6.2 *Feasible-Directions Method*

The feasible-directions method differs from relaxation-based methods in that the problem constraints are *explicitly* accounted for throughout the algorithm. In the relaxation-based methods we relax the constraints and account for them implicitly by adding penalty terms to the objective function. The terms that we add to the objective function penalize constraint violations. The benefit of the feasible-directions method is that it ensures that each point found is feasible in the original problem constraints. Thus, there is no need to adjust penalty weights to ensure feasibility. The feasible-directions method is limited, however, as it can only easily be applied to problems with relatively simple constraints.

In this discussion, we take the case of problems with linear constraints. Thus, without loss of generality, we assume that we have a linearly constrained nonlinear optimization problem of the form:

$$\min_{x \in \mathbb{R}^n} f(x)$$

$$\text{s.t.} \ Ax - b \leq 0,$$

where A is an $m \times n$ matrix of constraint coefficients and b is an $m \times 1$ vector of constants. This problem structure only has inequality constraints. Note, however, that an equality constraint of the form:

$$h(x) = 0,$$

can be written as two inequalities:

$$h(x) \leq 0,$$

and:

$$h(x) \geq 0.$$

Note, however, that converting equality constraints into inequalities in this way can create numerical difficulties in practice. However, the problem structure that we assume is generic in that the feasible-directions method can be applied to a problem with any combination of linear equality and inequality constraints.

The feasible-directions method follows the same approach as the Generic Algorithm for Unconstrained Nonlinear Optimization Problems that is outlined in Section 5.1, with four major differences. First, the starting point x^0 must be feasible in the constraints. Second, the updating formula used in each iteration has the form:

$$x^{k+1} \leftarrow x^k + \alpha^k \cdot (d^k - x^k).$$

As seen in the following discussion, this updating formula is used because it allows us to easily ensure that the point that we find after each iteration is feasible. Third, we must also change our direction and line searches conducted in each iteration to ensure that the new point found after each iteration is feasible. Fourth, we must also modify our termination criteria. This is because when constraints are included in a problem, a point being stationary is no longer necessary for it to be a local minimum.

We now provide details on how to find a feasible starting point and how the search direction and line search are modified in the feasible-directions method.

5.6.2.1 Finding a Feasible Starting Point

Finding a feasible starting point means finding a point, x^0, that satisfies the constraints of the problem. In other words, we want to find a vector, x^0, such that $Ax^0 - b \leq 0$. We can do this by solving the linear optimization problem:

$$\min_{x \in \mathbb{R}^n} 0^\top x \tag{5.34}$$

$$\text{s.t.}\ \ Ax - b \le 0, \tag{5.35}$$

using the Simplex method. We arbitrarily set objective function (5.34) of this problem to $0^\top x = 0$. This is because the sole purpose of employing the Simplex method is to find a feasible starting point for the feasible-directions method. We are not concerned with what feasible point we begin the feasible-directions method from. Indeed, we do not even need to solve Problem (5.34)–(5.35) to optimality. As soon as we have a tableau with a non-negative $\tilde{b}$, we can terminate the Simplex method because we have a feasible solution at that point (*cf.* Section 2.5 for further details on the Simplex method). Note that the Simplex method requires a problem to be written in standard form (*i.e.*, with structural equality constraints and non-negative variables). The steps that are outlined in Section 2.2.2.1 can be employed to convert Problem (5.34)–(5.35) to this standard form.

5.6.2.2 Finding a Feasible Search Direction

We find a feasible search direction by following the same underlying logic as in the steepest descent method. Namely, we want to find a direction that gives the greatest improvement in the objective function. As in the derivation of the search directions for unconstrained nonlinear problems, we ignore the step size parameter when conducting the direction search. Thus, in the following discussion we assume that $\alpha^k = 1$ and that our updated point has the form:

$$x^{k+1} = x^k + (d^k - x^k).$$

Later, when we conduct the line search, we reintroduce the step size parameter.

Thus, we can write a first-order Taylor approximation of the objective-function value at the new point as:

$$f(x^{k+1}) = f(x^k + (d^k - x^k)) \approx f(x^k) + (d^k - x^k)^\top \nabla f(x^k).$$

Our goal is to choose d^k to minimize this Taylor approximation. Because we have the linear constraints in our problem, however, we want to also choose d^k in such a way that guarantees that the new point is feasible. We do this by imposing the constraints:

$$Ax^{k+1} - b = A \cdot (x^k + (d^k - x^k)) - b = Ad^k - b \le 0.$$

Thus, in the feasible-directions method we always find our search direction by solving the following linear optimization problem, which is outlined in the Feasible-Directions-Search-Direction Rule.

Feasible-Directions-Search-Direction Rule: Given a linearly constrained nonlinear optimization problem and a current point, x^k, the **feasible-directions** search direction is found by solving the linear optimization problem:

$$\min_{d^k \in \mathbb{R}^n} \; f(x^k) + (d^k - x^k)^\top \nabla f(x^k) \tag{5.36}$$

$$\text{s.t.} \quad Ad^k - b \leq 0. \tag{5.37}$$

It is important to stress that the search direction in the feasible-directions method is found in a very different way than the steepest descent and Newton's method directions for unconstrained problems. The steepest descent and Newton's method directions are found by applying simple rules. That is, those directions are found by simply using derivative information. The feasible-directions method requires a linear optimization problem to be solved at each iteration to find the search direction. This is typically easy to do, however, as the Simplex method can efficiently solve very large linear optimization problems.

5.6.2.3   Termination Criteria

The termination criteria with the feasible-directions are typically not the same as for an unconstrained problem. This is because with an unconstrained problem local minima are stationary points. Thus, iterative algorithms applied to unconstrained problems terminate at stationary points. Local minima of constrained problems are not typically stationary points. Rather, they are points that satisfy the KKT condition, which involve the gradients of the objective *and* constraint functions.

The termination criteria in the feasible-directions method depend on the value of the objective function from the search-direction problem. Specifically, we examine the sign of the $(d^k - x^k)^\top \nabla f(x^k)$ term in objective function (5.36). Note that $d^k = x^k$ is feasible in constraints (5.37). Substituting $d^k = x^k$ into (5.36) gives:

$$f(x^k) + (d^k - x^k)^\top \nabla f(x^k) = f(x^k) + (x^k - x^k)^\top \nabla f(x^k) = f(x^k).$$

We can conclude from this that when the search-direction problem is solved, the optimal value of $f(x^k) + (d^k - x^k)^\top \nabla f(x^k)$ must be no greater than $f(x^k)$.
If:

$$f(x^k) + (d^k - x^k)^\top \nabla f(x^k) = f(x^k),$$

which means that:

$$(d^k - x^k)^\top \nabla f(x^k) = 0,$$

then the search-direction problem is not able to find a feasible direction to move in that improves the objective function. Thus, we terminate the feasible-directions method. Otherwise, if:

$$f(x^k) + (d^k - x^k)^\top \nabla f(x^k) < f(x^k),$$

which means that:

$$(d^k - x^k)^\top \nabla f(x^k) < 0,$$

then the search-direction problem has found a feasible direction that improves the objective function and we continue the algorithm.

5.6.2.4 Conducting a Feasible Line Search

With the feasible-directions method, we can use either of the exact line search or Armijo rule methods that are discussed in Section 5.4, so long as we restrict the step size to be less than or equal to one. If α^k is between zero and one, we can show that the new point is guaranteed to be feasible. To see this, note that for the new point to be feasible we must have:

$$A \cdot (x^k + \alpha^k \cdot (d^k - x^k)) - b \le 0.$$

By distributing the product and collecting terms, we can rewrite the left-hand side of this inequality as:

$$
\begin{aligned}
A \cdot (x^k + \alpha^k \cdot (d^k - x^k)) - b &= Ax^k + \alpha^k Ad^k - \alpha^k Ax^k - b \\
&= (1 - \alpha^k)Ax^k - (1 - \alpha^k)b + \alpha^k Ad^k - \alpha^k b \\
&= (1 - \alpha^k)(Ax^k - b) + \alpha^k \cdot (Ad^k - b).
\end{aligned}
$$

We next note that because our current point, x^k, is feasible, $Ax^k - b \le 0$. Moreover, constraint (5.37), which we use in finding the search direction, ensures that $Ad^k - b \le 0$. Combining these observations with the fact that $\alpha^k \ge 0$ and $1 - \alpha^k \ge 0$, tells us that:

$$A \cdot (x^k + \alpha^k(d^k - x^k)) - b = (1 - \alpha^k)(Ax^k - b) + \alpha^k \cdot (Ad^k - b) \le 0,$$

meaning that x^{k+1} is feasible, so long as we have $0 \le \alpha^k \le 1$.

We now illustrate the use of the feasible-directions with the following example.

Example 5.10 Consider the linearly constrained nonlinear optimization problem:

$$\min_{x} f(x) = (x_1 + 1/2)^2 + (x_2 - 2)^2$$

$$\text{s.t. } x_1 \le 0$$
$$x_2 \le 0$$
$$- x_1 - 10 \le 0$$
$$- x_2 - 10 \le 0.$$

Starting from the point $x^0 = (-1, -1)^\top$, we use the feasible-directions method with an exact line search to solve the problem.

To find a feasible direction to move away from x^0 in, we compute the gradient of the objective function, which is:

$$\nabla f(x) = \begin{pmatrix} 2(x_1 + 1/2) \\ 2(x_2 - 2) \end{pmatrix}.$$

Thus, we find the search direction at $x^0 = (-1, -1)^\top$ by solving the linear optimization problem:

$$\min_{d^0} f(x^0) + (d^0 - x^0)^\top \nabla f(x^0) = \frac{37}{4} - (d_1^0 + 1) - 6(d_2^0 + 1)$$

$$\text{s.t. } d_1^0 \le 0$$
$$d_2^0 \le 0$$
$$- d_1^0 - 10 \le 0$$
$$- d_2^0 - 10 \le 0.$$

We have $d^0 = (0, 0)^\top$ as an optimal solution to this problem. We further have that $(d^0 - x^0)^\top \nabla f(x^0) = -7$. Thus, we indeed have a feasible direction that improves the objective function. This means that we should proceed with an iteration and not yet terminate the algorithm.

We conduct the exact line search by solving the problem:

$$\min_{\alpha^0} f(x^0 + \alpha^0 \cdot (d^0 - x^0)) = f\left(\begin{pmatrix} \alpha^0 - 1 \\ \alpha^0 - 1 \end{pmatrix} \right) = \left(\alpha^0 - \frac{1}{2} \right)^2 + (\alpha^0 - 3)^2$$

$$\text{s.t. } 0 \le \alpha^0 \le 1.$$

If we solve:

$$\frac{\partial}{\partial \alpha^0} f(x^0 + \alpha^0 \cdot (d^0 - x^0)) = 0,$$

this gives $\alpha^0 = 7/4$. However, we know that α^0 must be no greater than 1. If we substitute $\alpha^0 = 1$ into $f(x^0 + \alpha^0 \cdot (d^0 - x^0))$, we find that this gives the optimal step size. Thus, we have $\alpha^0 = 1$ and our new point after one iteration is $x^1 = (0, 0)^\top$.

To conduct another iteration of the feasible-directions method from our new point, x^1, we find the search direction by solving the following linear optimization problem:

$$\min_{d^1} f(x^1) + (d^1 - x^1)^\top \nabla f(x^1) = \frac{17}{4} + d_1^1 - 4d_2^1$$
$$\text{s.t. } d_1^1 \le 0$$
$$d_2^1 \le 0$$
$$- d_1^1 - 10 \le 0$$
$$- d_2^1 - 10 \le 0.$$

We have $d^1 = (-10, 0)^\top$ as an optimal solution and $(d^1 - x^1)^\top \nabla f(x^1) = -10$. Thus, we must continue conducting another iteration of the feasible-directions method, by solving the exact-line-search problem:

$$\min_{\alpha^1} f(x^1 + \alpha^1 \cdot (d^1 - x^1)) = f\left(\begin{pmatrix} -10\alpha^1 \\ 0 \end{pmatrix}\right) = \left(-10\alpha^1 + \frac{1}{2}\right)^2 + (-2)^2$$
$$\text{s.t. } 0 \le \alpha^1 \le 1.$$

Solving:

$$\frac{\partial}{\partial \alpha^1} f(x^1 + \alpha^1 \cdot (d^1 - x^1)) = 0,$$

gives $\alpha^1 = 1/20$. We can further verify that:

$$\frac{\partial^2}{\partial \alpha^{1^2}} f(x^1 + \alpha^1 \cdot (d^1 - x^1)) = 200,$$

which is positive definite, meaning that $\alpha^1 = 1/20$ is a global minimum. Thus, our new point is $x^2 = (-1/2, 0)^\top$.

To conduct a third iteration of the feasible-directions method starting from x^2, we find the search direction by solving the following linear optimization problem:

$$\min_{d^2} f(x^2) + (d^2 - x^2)^\top \nabla f(x^2) = 4 - 4d_2^2$$
$$\text{s.t. } d_1^2 \le 0$$
$$d_2^2 \le 0$$
$$- d_1^2 - 10 \le 0$$
$$- d_2^2 - 10 \le 0.$$

We have $d^2 = (d_1^2, 0)^\top$, where d_1^2 can be any value between -10 and 0 as an optimal solution. Moreover, we find that $(d^2 - x^2)^\top \nabla f(x^2) = 0$ (regardless of the value of d_1^2 chosen), meaning that we are not able to find a feasible direction to move in that improves the objective function. Thus, we should terminate the algorithm with $x^2 = (-1/2, 0)^\top$ as our final solution.

$\square$

5.6.2.5   Unbounded Search-Direction Problem

One issue that can come up when applying the feasible-directions method is that the linear optimization problem used to find a search direction can be unbounded. This can obviously happen if the nonlinear optimization problem that we begin with is unbounded. However, this can also happen on occasion even if the original problem is bounded.

If the linear optimization problem used to find a search direction is unbounded, then we proceed by picking any direction in which the linear optimization problem is unbounded. This is because the ultimate purpose of the search-direction problem is to find *any* feasible direction in which the objective function of the original nonlinear optimization problem improves. If the search-direction problem is unbounded, that tells us that there are directions in which the Taylor approximation of the original objective function improves without any limit. Any such feasible direction of unboundedness will suffice for the feasible-directions method. We then proceed to conduct a line search, but remove the restriction that $\alpha^k \leq 1$ (for that iteration only). We illustrate this with the following example.

Example 5.11 Consider the linearly constrained nonlinear optimization problem:

$$\min_x \; f(x) = (x_1 - 3)^2 + (x_2 - 2)^2$$

$$\text{s.t.} \; -x_1 \leq 0$$

$$-x_2 \leq 0.$$

We can easily confirm, using the KKT condition, that $x^* = (3, 2)^\top$ is a global optimum of this problem. Thus, this problem is not unbounded.

Let us now start from the point $x^0 = (0, 0)^\top$ and attempt to conduct one iteration of the feasible-directions method. To do so, we first compute the gradient of the objective function, which is:

$$\nabla f(x) = \begin{pmatrix} 2(x_1 - 3) \\ 2(x_2 - 2) \end{pmatrix}.$$

Thus, we find the search direction by solving the linear optimization problem:

$$\min_{d^0} f(x^0) + (d^0 - x^0)^\top \nabla f(x^0) = 13 - 6d_1^0 - 4d_2^0$$

$$\text{s.t.} \ -d_1^0 \le 0$$

$$-d_2^0 \le 0.$$

This problem is unbounded, because we can make either of $d_1^0 \to +\infty$ or $d_2^0 \to +\infty$, which is feasible and makes the objective function, $13 - 6d_1^0 - 4d_2^0$, arbitrarily small. In this case, we pick any direction in which the objective function is unbounded, and for simplicity, we take $d^0 = (1, 1)^\top$. We next proceed to conducting an exact line search, which we do by solving the problem:

$$\min_{\alpha^0} f(x^0 + \alpha^0 \cdot (d^0 - x^0)) = (\alpha^0 - 3)^2 + (\alpha^0 - 2)^2$$

$$\text{s.t.} \ 0 \le \alpha^0.$$

Note that, as discussed before, we remove the restriction that $\alpha \le 1$. Solving this problem gives $\alpha^0 = 5/2$, meaning that $x^1 = (5/2, 5/2)^\top$. We can confirm that this point is feasible in the constraints. Moreover, we have:

$$f(x^0) = 13,$$

while:

$$f(x^1) = 1/2,$$

confirming that the objective-function value improves in going from x^0 to x^1. Note that because $x^1 \ne (3, 2)^\top$, the feasible-directions algorithm does not terminate at this point. Indeed, it is easy to confirm that there is a new search direction that can be found by solving the new search-direction problem starting from the point x^1. $\square$

We conclude this discussion by outlining the steps of the Feasible-Directions Algorithm for Linearly Constrained Nonlinear Optimization Problems below. Lines 2 through 4 initialize the algorithm by setting the iteration counter and the logical indicator, τ, of whether to terminate the algorithm to zero and fixing the initial point. In Lines 5 through 7 we determine if the starting point is feasible. If not, we solve a linear optimization problem to find a feasible starting point.

Feasible-Directions Algorithm for Linearly Constrained Nonlinear Optimization Problems

---

1: **procedure** FEASIBLE- DIRECTIONS ALGORITHM
2:     $k \leftarrow 0$ ▷ Set iteration counter to 0
3:     $\tau \leftarrow 0$
4:     Fix x^0 ▷ Fix a starting point
5:     **if** $Ax^0 - b \not\leq 0$ **then**   ▷ If starting point is infeasible
6:         $x^0 \leftarrow \arg\min_x \{0^\top x \,|\, Ax - b \leq 0\}$ ▷ Find feasible starting point
7:     **end if**
8:     **repeat**
9:         $d^k \leftarrow \arg\min_x \{f(x^k) + (d^k - x^k)^\top \nabla f(x^k) \,|\, Ad^k - b \leq 0\}$ ▷ Find search
            direction
10:        **if** $(d^k - x^k)^\top \nabla f(x^k) = 0$ **then**
11:            $\tau \leftarrow 1$
12:        **else**
13:            **if** $\min_x \{f(x^k) + (d^k - x^k)^\top \nabla f(x^k) \,|\, Ad^k - b \leq 0\}$ is bounded **then**
14:                Determine step size, $\alpha^k \leq 1$
15:            **else**
16:                Determine step size, α^k
17:            **end if**
18:            $x^{k+1} \leftarrow x^k + \alpha^k \cdot (d^k - x^k)$ ▷ Update point
19:            $k \leftarrow k + 1$ ▷ Update iteration counter
20:        **end if**
21:    **until** $\tau = 1$ **or** other termination criteria met
22: **end procedure**

Lines 8 through 21 are the main iterative loop. Line 9 solves the linear optimization problem that is used to determine the new search direction. Lines 10 and 11 test the standard termination criterion. If $(d^k - x^k)^\top \nabla f(x^k) = 0$ then we set $\tau \leftarrow 1$, which terminates the algorithm in Line 21. Otherwise, we continue with the line search in Lines 13 through 17. If the most recently solved search-direction problem is bounded, the step size is limited to be no greater than 1 (*cf.* Line 14). Otherwise, the step size has no upper bound (*cf.* Line 16). Note that any line search method from Section 5.4 can be employed here, so long as the appropriate bounds on α^k are imposed. Lines 18 and 19 update the point and iteration counter. As with the other iterative algorithms that are discussed in this chapter, other termination criteria can be employed in Line 21.

5.7   Final Remarks

This chapter introduces several generic iterative algorithm to solve unconstrained and constrained nonlinear optimization problems. The true power of these algorithms lies in the methods used to generate search directions and step sizes. We only introduce a small subset of search direction and line-search methods. More advanced texts [2, 7] discuss other techniques that are more complicated but can offer better performance than those discussed here. At their heart, these more advanced search direction and line-search methods are built around the same underlying premises of the methods introduced here. The performance of the generic algorithms introduced in this chapter depends on the performance of both the search direction and line search. This is important to stress, as it may be tempting to overlook the line search when developing a solution algorithm.

We introduce only two methods to solve constrained optimization problems. Other more advanced techniques, which are beyond the scope of this textbook, are covered elsewhere [1–4, 6–8]. We also focus exclusively on nonlinear optimization problems with continuous variables only. It is a straightforward extension of mixed-integer linear and nonlinear optimization to formulate mixed-integer nonlinear optimization problems. However, such problems can be quite difficult to solve. We refer interested readers to the work of Floudas [5], who provides an excellent introduction to the formulation and solution of mixed-integer nonlinear optimization problems.

5.8 Exercises

5.1 Consider the Facility-Location Problem, which is introduced in Section 4.1.1.3 and suppose that you are given the data summarized in Table 5.1 for a three-location instance of this problem. Starting from the point $x^0 = (0, 0)^\top$ conduct iterations of the Generic Algorithm for Unconstrained Nonlinear Optimization Problems using the steepest descent search direction and an exact line search to solve this problem. Is the point at which the algorithm terminates guaranteed to be a local or global optimum? Explicitly justify your answer.

Table 5.1 Data for Exercise 5.1

n	x_n	y_n	V_n
1	7	2	7
2	5	−3	10
3	−6	4	15

5.2 Starting from the point $x^0 = (0, 0)^\top$ solve the instance of the Facility-Location Problem, which is given in Exercise 5.1, using the Generic Algorithm for Unconstrained Nonlinear Optimization Problems using the Newton's method search direction and the Armijo rule with $\sigma = 10^{-1}$, $\beta = 1/2$ and $s = 1$. Is the point at which the algorithm terminates guaranteed to be a local or global optimum? Explicitly justify your answer.

5.3 What difficulties could arise in applying the Generic Algorithm for Unconstrained Nonlinear Optimization Problems with the steepest descent search direction to the model formulated for the Facility-Location Problem in Section 4.1.1.3?

5.4 Write the penalized objective function and augmented Lagrangian function for the Packing-Box Problem that is introduced in Section 4.1.1.1 with generic penalty weights, ρ and ϕ, and generic Lagrange multipliers, λ.

5.5 Solve the Packing-Box Problem, which is introduced in Section 4.1.1.1, using the Penalty-Based Algorithm for Constrained Nonlinear Optimization Problems using the following algorithm rules.

- Use the penalized objective function that is found in Exercise 5.4.
- Start at the point $(h, w, d) = (0, 0, 0)$ and with penalty weights of 1 on all of the constraints.
- Use the steepest descent search direction with an exact line search.
- If after conducting an iteration a constraint is violated, increase the penalty weight on that constraint by 1. Otherwise, keep the penalty weight on that constraint the same for the next iteration.

Can you explain the behavior of the algorithm? Solve the problem using the algorithm starting at a different point.

5.6 Solve the Packing-Box Problem, which is introduced in Section 4.1.1.1, using the Multiplier-Based Algorithm for Constrained Nonlinear Optimization Problems using the following algorithm rules.

- Use the augmented Lagrangian function that is found in Exercise 5.4.
- Start at the two points used in Exercise 5.5 and with penalty weights and Lagrange multipliers of 1 on all of the constraints.
- Use the steepest descent search direction with an exact line search.
- If after conducting an iteration a constraint is violated, increase the penalty weight on that constraint by 1. Otherwise, keep the penalty weight on that constraint the same for the next iteration.

5.7 Solve the model formulated in Exercise 4.1 using the feasible-directions method with an exact line search.

References

1. Bazaraa MS, Sherali HD, Shetty CM (2006) Nonlinear programming: theory and algorithms, 3rd edn. Wiley-Interscience, Hoboken
2. Bertsekas D (2016) Nonlinear programming, 3rd edn. Athena Scientific, Belmont
3. Bonnans JF, Gilbert JC, Lemarechal C, Sagastizabal CA (2009) Numerical optimization: theoretical and practical aspects, 2nd edn. Springer, New York
4. Boyd S, Vandenberghe L (2004) Convex optimization. Cambridge University Press, Cambridge
5. Floudas CA (1995) Nonlinear and mixed-integer optimization: fundamentals and applications. Oxford University Press, Oxford
6. Gill PE, Murray W, Wright MH (1995) Practical optimization. Emerald Group Publishing Limited, Bingley
7. Luenberger DG, Ye Y (2016) Linear and nonlinear programming, 4th edn. Springer, New York
8. Nocedal J, Wright S (2006) Numerical optimization, 2nd edn. Springer, New York

Chapter 6
Dynamic Optimization

In this chapter we introduce a markedly different way to formulate and solve optimization problems, compared to the techniques that are introduced in Chapters 2 through 5. The dynamic optimization methods that we introduce here work by decomposing an optimization problem into a successive set of stages, at which the state of the system being optimized is observed and decisions are made. Dynamic optimization techniques require that an optimization problem have certain important characteristics that allow for this decomposition. However, dynamic optimization is incredibly flexible in that it allows problems with linear and nonlinear terms and logical conditions (*e.g.*, 'if then'-type statements) in the objective function and constraints and discrete decisions (*e.g.*, integer variables).

In this chapter we begin by first introducing a simple problem that can be formulated and solved as a dynamic optimization problem, also known as a dynamic programming problem (DPP). We then discuss the common elements that must be identified when formulating a DPP and the important characteristics that such a problem must have. We finally discuss the standard algorithm used to solve dynamic optimization problems.

6.1 Motivating Example: Production-Planning Problem

In this section, we introduce and then solve a motivating example that illustrates the use of dynamic optimization.

6.1.1 Problem Description

A company must plan its production process for a product. Figure 6.1 illustrates the possible production paths that can be used to convert raw material into the finished product. The nodes in the figure represent different states of the product. Node 1

© Springer International Publishing AG 2017

R. Sioshansi and A.J. Conejo, *Optimization in Engineering*,

Springer Optimization and Its Applications 120, DOI 10.1007/978-3-319-56769-3_6

represents raw material and node 12 represents the finished product. The other numbered nodes represent intermediate states of the product between the raw-material and finished-product states.

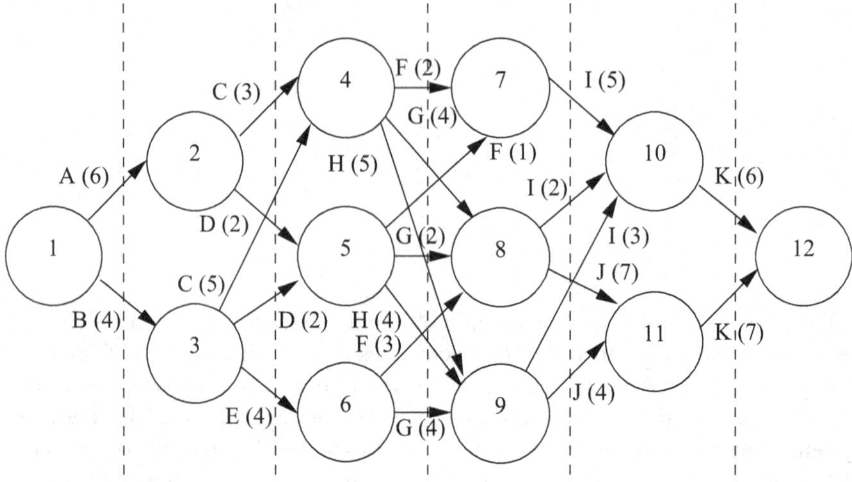

Fig. 6.1 Possible production paths in the Production-Planning Problem

The lettered arcs in Figure 6.1 represent different processes that can be employed. The number in parentheses next to each arc label indicate how many hours of time is required to finish each of the processes. The arcs in the figure are directed because the processes are unidirectional. For instance, process 'F' can be employed to convert the material from state 4 to 7. However, the material cannot be converted 'back' from state 7 to state 4.

The company must only employ a single set of production processes to transform the raw material into the finished product. It wishes to find a set of production processes that do so in the least amount of time possible.

6.1.2 Problem Solution

Before employing dynamic optimization techniques to solve this problem, notice that someone without any knowledge of optimization could solve the problem by conducting an exhaustive search. That is, the processing time of every set of feasible production processes could be computed and the one that takes the least amount of time could be chosen. For this particular problem, such an exhaustive search could be conducted as there are only 22 possible combinations of production processes that could be employed. Let us see, however, how we can reduce the number of computations that must be conducted to solve this problem.

Let us begin by noting that if, by some combination of production decisions, the material is in state 10, it would take 6 more hours to finish production and this would be achieved by choosing process 'K.' Similarly, if the material is in state 11, we can determine that it would take 7 more hours to finish production, which is also achieved by choosing process 'K.' Table 6.1 summarizes the results of these findings for states 10 and 11.

Table 6.1 First set of computations from solving the Production-Planning Problem

State	Optimal Action	Total Hours Remaining
10	K	6
11	K	7

Let us next consider how long it would take to complete production if the material is in each of states 7 through 9. If it is in state 7 the only feasible option is to use process 'I,' which transforms the material into state 10 after 5 hours. We further know, from the calculations that are summarized in Table 6.1, that once the material is in state 10 it will take an additional 6 hours to transform the material into the finished product. This means that from state 7 it will take a total of 11 hours to transform the material into the finished part.

The computations for states 8 and 9 are slightly more involved than the three that are conducted thus far. This is because we have multiple options available to us at these states. Let us first consider state 8. The two options available are processes 'I' and 'J,' which take 2 and 7 hours of time, respectively. After these processes are complete, the material transitions to states 10 and 11, respectively. We know, from the computations that are summarized in Table 6.1, that once the material is in states 10 and 11, there is an additional 6 and 7 hours of processing time, respectively, to transform the material into the finished product. Thus, the decision that we make at state 8 can be summarized as follows:

$$[\text{total time from state 8}] = \min\{[\text{time if 'I' chosen}], [\text{time if 'J' chosen}]\} \qquad (6.1)$$
$$= \min\{[\text{time for 'I'}] + [\text{time from state 10}], [\text{time for 'J'}] + [\text{time from state 11}]\}$$
$$= \min\{2 + 6, 7 + 7\}$$
$$= 8,$$

meaning that from state 8 we would optimally choose process 'I,' because it results in 8 hours of total processing time as opposed to 14 hours with process 'J.'

Before proceeding with the remaining computations, it is important to stress one feature of how we obtain these total processing times. There are two terms that contribute to the total processing time associated with each of the two options that are available at state 8. The first is the amount of time that the immediate process chosen, 'I' or 'J,' entails. This time comes from the numbers on the corresponding arcs in Figure 6.1. The other is the amount of time that remains after the material is transformed into its new state (*i.e.*, into one of states 10 or 11). Although we can easily read these times off of the associated arcs in Figure 6.1, we actually obtain these values from Table 6.1, which summarizes the computations already conducted for states 10 and 11. We see why this is important as we continue solving this problem.

We can now examine the alternatives that are available if the material is in state 9, which results in the following calculations:

$$[\text{total time from state 9}] = \min\{[\text{time if 'I' chosen}], [\text{time if 'J' chosen}]\} \qquad (6.2)$$
$$= \min\{[\text{time for 'I'}] + [\text{time from state 10}], [\text{time for 'J'}] + [\text{time from state 11}]\}$$
$$= \min\{3 + 6, 4 + 7\}$$
$$= 9.$$

These calculations tell us that if we are in state 9, we would optimally choose process 'I,' and the material would require a further 9 hours of processing to yield the final product. Table 6.2 summarizes the results of the calculations that are conducted for states 7 through 9.

Table 6.2 Second set of computations from solving the Production-Planning Problem

State	Optimal Action	Total Hours Remaining
7	I	11
8	I	8
9	I	9

We can now repeat this process for states 4 through 6. Beginning with state 4, we see that there are three options available—processes 'F' through 'H.' These each entail 2, 4, and 5 hours of time, respectively, after which the material transitions to states 7, 8, and 9, respectively. We also know, from Table 6.2, that once the material is in these three states, that there would be, respectively, an additional 11, 8, and 9 hours remaining until we have the finished product. Thus, the decision we make at state 4 is:

$$[\text{total time from state 4}] = \min\{[\text{time from choosing 'F'}], [\text{time from choosing 'G'}],$$
$$[\text{time from choosing 'H'}]\}$$
$$= \min\{2 + 11, 4 + 8, 5 + 9\}$$
$$= 12,$$

meaning that we choose process 'G,' which requires a total of 12 hours to deliver the final product.

Before proceeding, it is important to stress that we again break the time associated with each of the three options into two terms. The first is the time associated with the immediate process chosen (*i.e.*, the time to complete each of processes 'F' through 'H'). These times come from the numbers next to the corresponding arcs in Figure 6.1. The second term is the amount of time that is required after the material transitions into its new state (*i.e.*, into one of states 7, 8, or 9). As before, we could compute these times directly using Figure 6.1. We do not, however, because they are already computed when we examine what to do in states 7, 8, and 9. Thus, we can determine how much time remains from each of these three states using the values that are recorded in Table 6.2. Indeed, as is discussed later, one of the strengths and efficiencies in decomposing and solving a problem as a DPP is that previous computations can be reused. This reduces the amount of computation that must be done to solve the problem. It should also be stressed that the additional times that it takes to complete the product once the material is one of states 7, 8, or 9 come entirely from Table 6.2. We do not need to refer to Table 6.1 at this point. This is because when we compute the total hours remaining in Table 6.2, we account for what happens after process 'I' is conducted (which is the information that is recorded in Table 6.1).

We do not explicitly run through the calculations for states 5 and 6. Instead, we leave it to the reader to repeat the steps that are taken to analyze state 4 to confirm that the values in Table 6.3 are correct.

Table 6.3 Third set of computations from solving the Production-Planning Problem

State	Optimal Action	Total Hours Remaining
4	G	12
5	G	10
6	F	11

The next two sets of calculations are first for states 2 and 3 and then finally for state 1. We do not go through these explicitly. Instead, we leave it to the reader to verify that the values that are reported in Tables 6.4 and 6.5 are correct. Note that when conducting calculations for states 2 and 3, we use the time-remaining calculations that are summarized in Table 6.3 *only*. Similarly, when conducting calculations for state 1, we *only* use the time-remaining information that is recorded in Table 6.4.

Table 6.4 Fourth set of computations from solving the Production-Planning Problem

State	Optimal Action	Total Hours Remaining
2	D	12
3	D	12

Having completed the computations in Tables 6.1 through 6.5, the next question is how to determine an optimal solution and the associated objective-function value. First, note that the third column of Table 6.5 gives us the optimal objective-function value. The third column of this table tells us how many hours are remaining to

Table 6.5 Fifth set of computations from solving the Production-Planning Problem

State	Optimal Action	Total Hours Remaining
1	B	16

reach the finished product if we are at state 1. Note, however, that state 1 is the raw material that we start with. Thus, we know that it takes 16 hours to produce the finished product.

Next, note that the second column of Table 6.5 tells us what optimal action to follow if we are in state 1 (*i.e.*, when we have raw material). This optimal action is process 'B.' Knowing this, the next equation is how to determine the subsequent optimal action? This information is recorded in the second column of Table 6.4, which tells us what to do in each of states 2 and 3. Note that if we start by using process 'B,' the raw material is converted from state 1 to state 3 (we determine this from Figure 6.1). The associated row of Table 6.4 tells us that the optimal process to choose after process 'B' is process 'D.' We know from Figure 6.1 that if the material is in state 3 and we use process 'D,' the material is next converted to state 5. We next examine Table 6.3, which tells us what process to choose if the material is in any of states 4, 5, or 6. The second column of the associated row of Table 6.3 tells us that if the material is in state 5, we should choose process 'G.' Continuing this logic through the remaining tables, we find that our optimal process sequence is:

$$B \implies D \implies G \implies I \implies K.$$

6.1.3 Computational Benefits of Dynamic Optimization

Before we generalize the method that is employed in this section to decompose and solve the Production-Planning Problem, we can briefly examine the computational complexity of the solution method just used. One of the benefits of using dynamic optimization techniques is that computations can be reused. For instance, when deciding which of processes 'C,' 'D,' or 'E' to use if the material is in state 3, we do not need to do an 'on-the-fly' computation of what sequence of processes would be used after that. Rather, we use the values that are recorded in Table 6.3, which are already computed at the point that we are examining state 3, to determine what the total time associated with each of processes 'C,' 'D,' and 'E' are.

The second computational savings of dynamic optimization is that it allows us to eliminate combinations of decisions that are suboptimal. To understand this, recall that the brute-force technique that is first outlined in Section 6.1.2 requires us to examine 22 different combinations of processes and compare their total processing times. Note that after we examine state 9 in the dynamic optimization technique, we determine that 'I' is the optimal process to follow. In making this observation, we eliminate from further consideration all of the process combinations that employ process 'J' at state 9 (of which there are three). Indeed, we further eliminate more process combinations after examining each of states 1 through 6 and 8. Examining state 7 does not eliminate any process combinations, because there are no decisions to be made at this point.

6.2 Formulating Dynamic Optimization Problems

As the solution method that is applied in Section 6.1 to the Production-Planning Problem suggests, dynamic optimization problems are approached in a substantively different manner than linear, mixed-integer linear, and nonlinear optimization problems are. For this reason, they are formulated in a vastly different manner as well. When formulating a dynamic optimization problem there are six problem elements that must be identified: the (i) stages, (ii) state variables, (iii) decision variables, (iv) state-transition functions, (v) constraints, and (vi) objective-contribution functions. These six problem elements 'encode' the structure of the system being modeled and allow us to solve it using the efficient recursive method that is applied to the Production-Planning Problem in Section 6.1. Bellman [1] outlines this generic means of structuring and solving dynamic optimization problems.

We proceed by first defining these six elements of a dynamic optimization problem abstractly and then identifying these elements for the problem that is introduced in Section 6.1. We then give some further formulation examples in Section 6.3.

6.2.1 Stages, State Variables, and Decision Variables

We begin by defining the stages, state variables, and decision variables in a dynamic optimization problem, because these three often go hand-in-hand when formulating the problem. The **stages** of the problem represent points at which decisions are made, which are represented by the **decision variables**. Decision variables are also sometimes referred to as action variables. The **state variables** summarize all of the information needed at each stage to determine an optimal action and how the problem progresses from one stage to the next.

The Production Planning Problem, which is discussed in Section 6.1 has five stages, which are shown in Figure 6.2. These stages correspond to the fact that the production process requires a sequence of five processes to convert the raw material into the finished product. Each stage corresponds to the point at which a production process is chosen. There is a single set of state variables, which is the state that the material is in at the beginning of each stage. The nodes in Figure 6.2 (which are the same as those in Figure 6.1) represent the different possible values that the state variables can take on. There is a single set of decision variables in the Production-

Planning Problem, which represents the process chosen in each stage. The arcs in Figures 6.2 (and 6.1) represent the different possible decisions that can be chosen.

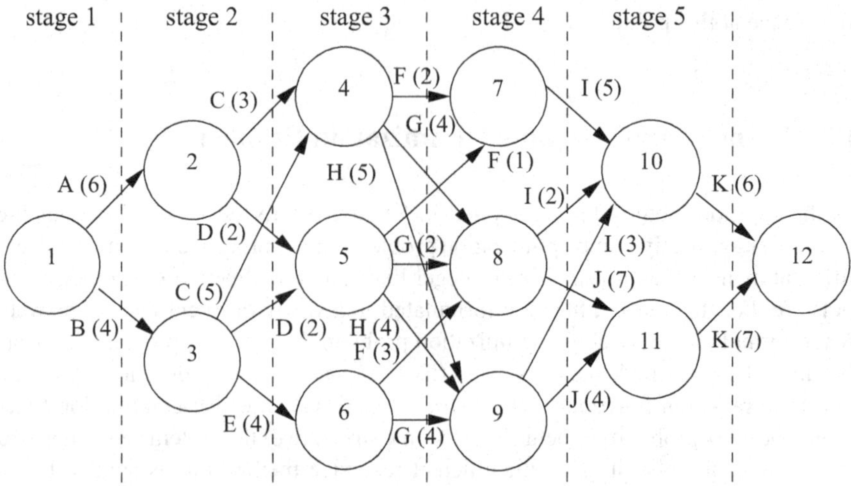

Fig. 6.2   Stages in the Production-Planning Problem

Table 6.6 summarizes the possible states and decisions in each stage of the problem. There is one possible state in stage 1, which is state 1 (*i.e.*, raw material). Figures 6.2 (and 6.1) show that there are two alternatives that can be chosen in stage 1, which are 'A' and 'B.' Thus, the stage-1 decision variable can only take on a value of 'A' or 'B.' This restriction on the value that the stage-1 decision can take is an example of a constraint. Constraints are discussed in Section 6.2.3. Stage 2 has two possible states—2 and 3—and the stage-2 decision variable can take on three values—'C,' 'D,' and 'E.' Note, however, that process 'E' is only available in state 3. This is yet another example of a constraint.

Table 6.6   Possible states and decisions in each stage of the Production-Planning Problem

Stage	States	Decisions
1	1	A, B
2	2, 3	C, D, E
3	4, 5, 6	F, G, H
4	7, 8, 9	I, J
5	10, 11	K

There is not a sixth stage, consisting of state 12. The reason for this is that once the material is in state 12, we have the finished product and there are no decisions remaining to be made. Recall that stages are points in the problem at which decisions are made, and there are no decisions to be made once the product is finished.

The rationale behind the stages of the problem stems from the fact that producing the finished product requires five successive process. The points at which we decide which process to use correspond to the five stages that are identified. Moreover, to make a decision at each stage regarding what process to next use, we must know which specific node the material is at. For instance, in stage 2, which of nodes 2 or 3 the material is in is vitally important information. Knowing which of these two states the material is in determines what decisions are available—if the material is at node 2 process 'E' is unavailable. Moreover, the amount of time that the stage-2 processes take depend upon the node that the material is at. For instance, process 'C' takes three hours if the material is at node 2 as opposed to five hours if it is at node 3. Because the nodes encode this important information, they correspond to the problem states.

Determining the stages and state variables of a DPP can often be the most challenging part of formulating a problem. Many dynamic optimization problems have a natural (usually temporal) sequence of decisions being made. In such cases, the stages typically correspond to this sequence. There are cases of problems that do not have such a natural sequence, which can make determining the stages more difficult. Section 6.3.4 provides one such example. Oftentimes if the state variables are not defined properly, it is difficult to formulate the other elements of the DPP. If this occurs, it may indicate that there are important state variables that are not identified or that the state variables identified are not defined properly. Examples in which this occurs are provided in Sections 6.3.2 and 6.3.3.

6.2.2 State-Transition Functions

The **state-transition functions** define how the state variables transition or change from one stage to the next. More specifically, the stage-t state-transition function takes the stage-t state and decision variables as inputs and defines the values of the stage-$(t + 1)$ state variables as outputs.

The terminal node of each arc in Figures 6.2 (and 6.1) defines the state transitions for the Production-Planning Problem that is introduced in Section 6.1. That is because the arcs tell us what state the material is transformed into after a particular production process is chosen. We can also write the state-transition functions as analytical expressions. To do so, let s_t denote the material's state at the beginning of stage t and let d_t denote the stage-t decision variable (*i.e.*, the production process chosen in stage t). We can write the stage-1 state-transition function as:

$$s_2 = f_1(s_1, d_1) = \begin{cases} 2, & \text{if } d_1 = \text{A}, \\ 3, & \text{if } d_1 = \text{B}. \end{cases} \tag{6.3}$$

This function tells us what state, s_2, the material will be in after the stage-1 production process is chosen.

Note that the function, $f_1(s_1, d_1)$, in (6.3) is written as depending on both s_1 and d_1. This is despite the fact that s_2 is determined solely by d_1 (*i.e.*, s_2 does not depend directly on s_1, as seen in the conditional expressions to the right of the second equality in (6.3)). This is the standard way of writing a state-transition function, to make it clear that generically speaking, the stage-$(t + 1)$ state variable can depend on the stage-t state and decision variables. Although it is less generic, we could also write this state-transition function as:

$$s_2 = f_1(d_1) = \begin{cases} 2, & \text{if } d_1 = \text{A}, \\ 3, & \text{if } d_1 = \text{B}. \end{cases}$$

We can similarly define the stage-2 state-transition function (in the more generic form) as:

$$s_3 = f_2(s_2, d_2) = \begin{cases} 4, & \text{if } d_2 = \text{C}, \\ 5, & \text{if } d_2 = \text{D}, \\ 6, & \text{if } d_2 = \text{E}. \end{cases} \tag{6.4}$$

We leave it to the reader to derive analytical expressions for the state transitions of the remaining stages.

Before proceeding, there are several important details regarding state-transition functions that should be noted. First, state-transition functions should always be written as transitions from the current stage to the next. That is to say, the state variable in the next stage, s_{t+1}, should be dependent solely on the state and decision variables in the current stage, s_t and d_t, respectively. There should not be any dependence on state or decision variables from previous stages. Similarly, there should not be any dependence on state or decision variables in subsequent stages. If the state-transition functions cannot be written in a way to have a single-stage transition, this may indicate that additional state variables must be included in the formulation. Examples of problems in which this occurs are given in Sections 6.3.2 and 6.3.3.

Second, we may use conditional statements, such as those in Equations (6.3) and (6.4), when defining state-transition functions. That is to say, we do not have to employ binary variables and the related techniques that are introduced in Section 3.3 to define state-transition functions for dynamic optimization problems.

Third, the stage-2 state-transition function given by Equation (6.4) defines s_3 to be 6 if $d_2 = \text{E}$. However, the decision 'E' is only available in stage 2 if the material is in state 3 (*i.e.*, only if $s_2 = 3$). Thus, one could alternately define the stage-2 state-transition function as:

$$s_3 = f_2(s_2, d_2) = \begin{cases} 4, & \text{if } d_2 = \text{C}, \\ 5, & \text{if } d_2 = \text{D}, \\ 6, & \text{if } d_2 = \text{E and } s_2 = 3. \end{cases}$$

We do not typically do this, however, because the restriction on the stage-2 decision variable is an example of a constraint, which is formulated separately from the state-transition function. Constraints are discussed in more detail in Section 6.2.3.

Finally, the Production-Planning Problem is an example of a problem in which the state-transition functions are **state-independent** or **state-invariant**. That is to say, at each stage of the problem the material's state in the next stage depends solely on the decision made in the current stage, and not on the current state of the material. We see examples of dynamic optimization problems that have **state-dependent** state-transition functions in Section 6.3. In these problems the state-variable value in each stage depends on the state-variable value in the previous stage.

6.2.3 Constraints

As in linear, mixed-integer linear, and nonlinear optimization problems, the **constraints** indicate what decisions can be feasibly made. More specifically, the constraints of a DPP indicate what decision-variable values are feasible at each stage of the problem. Depending on the particulars of a DPP, the constraints can be stage- and state-dependent. This means that the feasible decisions vary depending on what stage of the problem we are in and what values the state variables take on. In the Production-Planning Problem the constraints are given by which arcs originate from each node in Figures 6.2 (and 6.1). For instance, in stage 1 we know that processes 'A' and 'B' are available. Thus, the stage-1 constraints can be written as:

$$d_1 \in \{A, B\}.$$

The feasible stage-2 decisions depend upon the state that the material is in. Thus, we can write the stage-2 constraints as:

$$d_2 \in \begin{cases} \{C, D\}, & \text{if } s_2 = 2, \\ \{C, D, E\}, & \text{if } s_2 = 3. \end{cases}$$

We see that in the Production-Planning Problem the constraints are both stage- and state-dependent. The stage-dependence is exemplified by 'A' and 'B' only being feasible in stage 1 and 'C,' 'D,' and 'E' only being feasible in stage 2. The state-dependence in exemplified by 'E' only being available in stage 2 if $s_2 = 3$. We also see that as with state-transition functions, the constraints of a dynamic optimization problem can have conditional statements, such as in the stage-2 constraint that is given above.

It is important to stress that the constraints should always be defined on decision variables and not on state variables. This is because the constraints are intended to tell us what decisions are feasible at each stage. In some problems, it may be more natural to write a restriction on a state variable. In such instances, we can use the state-transition function to transform state-variable restrictions into decision-variable constraints.

It should also be noted that the problem constraints should always depend on the current state variables *only*. That is to say, the constraints on the stage-t decision variables should depend solely on the stage-t state variables. If a constraint involves decision or state variables from previous stages, this may indicate that the state variable definitions need to be refined or that additional state variables are needed to properly formulate a DPP. The examples that are discussed in Sections 6.3.1 and 6.3.2 illustrate some of these issues in defining problem constraints.

6.2.4 Objective-Contribution Functions

The **objective-contribution functions** indicate how much is contributed toward the problem's overall objective in each stage of the problem. The objective-contribution function can generally depend on the decision and state variables in each stage. As with the state-transition functions and constraints, the stage-t objective-contribution function should depend *solely* on the stage-t state and decision variables. If state or decision variables from previous stages must be included in the objective-contribution function, this may indicate that additional state variables are needed to properly formulate the DPP. The example that is given in Section 6.3.3 illustrates this.

The overall objective of the Production-Planning Problem is the total processing time (which we seek to minimize). Thus, the amount contributed toward this overall objective is the amount of processing time that the process chosen in each stage involves. These processing times are defined by the numbers next to each arc in Figures 6.2 (and 6.1).

We can also write analytical expressions for the objective-contribution functions. For instance, the stage-1 objective-contribution function is:

$$c_1(s_1, d_1) = \begin{cases} 6, & \text{if } d_1 = \text{A}, \\ 4, & \text{if } d_1 = \text{B}. \end{cases} \tag{6.5}$$

Note that we write the objective-contribution function, $c_1(s_1, d_1)$, as depending on both s_1 and d_1. As with writing state-transition functions, this is the most general manner in which to express an objective-contribution function. However, because the amount contributed toward the overall objective function in stage 1 does not depend directly on s_1, we could write this function less generically as:

$$c_1(d_1) = \begin{cases} 6, & \text{if } d_1 = \text{A}, \\ 4, & \text{if } d_1 = \text{B}. \end{cases}$$

We opt to write objective-contribution functions in the former more generic manner, in line with (6.5), however.

The stage-2 objective-contribution function is:

$$c_2(s_2, d_2) = \begin{cases} 3, & \text{if } s_2 = 2 \text{ and } d_2 = \text{C}, \\ 2, & \text{if } s_2 = 2 \text{ and } d_2 = \text{D}, \\ 5, & \text{if } s_2 = 3 \text{ and } d_2 = \text{C}, \\ 2, & \text{if } s_2 = 3 \text{ and } d_2 = \text{D}, \\ 4, & \text{if } s_2 = 3 \text{ and } d_2 = \text{E}. \end{cases} \tag{6.6}$$

As with state-transition functions and constraints, the objective-contribution function can include conditional statements, such as those used in Equations (6.5) and (6.6).

6.3 Formulation Examples

With the six fundamental elements of a dynamic optimization problem defined, we now turn to a series of additional formulation examples. For each we identify the six elements that must be formulated and, as appropriate, comment on some of the details regarding the definition of the six elements that are introduced in Section 6.2.

6.3.1 Energy-Storage Problem

A company owns a pumped hydroelectric storage plant. This plant consists of two reservoirs, one of which is above the other. The two reservoirs are connected by a turbine and pump. In each hour, the company can choose to do exactly one of either: (i) operate the turbine, (ii) operate the pump, or (iii) nothing. If it chooses to operate the turbine, the plant releases one unit of water from the upper reservoir into the lower reservoir and produces 80 MWh of electricity, which can be sold on the wholesale electricity market. If it chooses to operate the pump, it consumes 100 MWh of electricity, which is purchased from the wholesale market and pumps one unit of water from the lower reservoir into the upper reservoir. If it chooses to do nothing, the water levels in the two reservoirs remain the same and it does not earn money in or pay money to the wholesale market.

The upper reservoir can hold at most four units of water and has one unit of water in it at the beginning of hour 1. There are effectively no limits on the water level of the lower reservoir, as it is too big for the pumped hydroelectric storage plant to have any effect on its water level. Table 6.7 lists the wholesale prices over the following 24 hours of the day. The company would like to determine an operating schedule for the pumped hydroelectric storage plant to maximize the total profit that it earns over the following 24 hours.

The stages of this problem are easy to determine, because there is a natural time sequence. The company must determine what to do with the pumped hydroelectric storage plant in each of the next 24 hours. Thus, the problem has 24 stages and each of the next 24 hours represents one of those stages.

The decisions that are being made in each stage of the problem are easy to identify as well—the company must determine whether to operate the turbine, operate the pump, or do nothing. There are at least two ways that the decision variables can be defined, and we formulate the problem using both. The first is to define three separate decision variables, $r_t, u_t, n_t, \forall t = 1, \ldots, 24$, for each of the three options available in each stage. These variables are defined as:

$$r_t = \begin{cases} 1, & \text{if the turbine is operated in stage } t, \\ 0, & \text{otherwise,} \end{cases}$$

$$u_t = \begin{cases} 1, & \text{if the pump is operated in stage } t, \\ 0, & \text{otherwise,} \end{cases}$$

Table 6.7 Wholesale electricity prices over the next 24 hours in the Energy-Storage Problem

Hour	Price [\$/MWh]
1	21.78
2	19.81
3	17.10
4	16.20
5	16.28
6	16.82
7	17.77
8	21.07
9	24.13
10	25.86
11	26.17
12	25.97
13	25.91
14	24.64
15	23.68
16	23.81
17	27.82
18	36.15
19	37.26
20	34.64
21	32.16
22	30.01
23	25.26
24	21.63

and:

$$n_t = \begin{cases} 1, & \text{if the company does nothing in stage } t, \\ 0, & \text{otherwise.} \end{cases}$$

The second uses one variable in each stage, $a_t, \forall t = 1, \ldots, 24$, which compactly gives the three options. This variable is defined as:

$$d_t = \begin{cases} -1, & \text{if the turbine is operated in stage } t, \\ 1, & \text{if the pump is operated in stage } t; \\ 0, & \text{otherwise.} \end{cases}$$

These are not the only ways to define the decision variables. Moreover, the choice of interpreting the values -1 and 1 in defining d_t is arbitrary and their interpretations can be interchanged (so long as the rest of the model formulation is changed accordingly).

To define the state variables, we first note that we must know how much water is in the upper reservoir at the beginning of each stage. Thus, we define a set of state variables, $s_t, \forall t = 1, \ldots, 24$, where s_t represents the units of water in the upper reservoir at the beginning of stage t. It is also useful, when writing the objective-contribution functions, to have a second set of state variables representing the wholesale electricity prices. We can define the variables, $p_t, \forall t = 1, \ldots, 24$, where p_t represents the stage-t wholesale electricity price in \$/MWh.

Before proceeding, let us draw an important distinction between s_t and p_t. The former, s_t, is an example of an **endogenous state variable**. An endogenous state variable is one that evolves based on decisions that are made in the problem (*i.e.*, based on the decision variables). For instance, if the company chooses to operate the pump in stage t, the water level of the upper reservoir will be one unit higher at the beginning of stage $(t + 1)$. Endogenous state variables *must* be included as state variables when formulating a dynamic optimization problem. This is because solving a dynamic optimization properly requires that we capture how endogenous state variables evolve from one stage to the next based on the decisions made. If an endogenous state variable is excluded from the formulation, this dynamic is not captured and the problem cannot be solved properly.

The second state variable, p_t, is an example of an **exogenous state variable**. Exogenous state variables are not affected by decisions made in the problem. Rather, they are fixed data or parameters that are given in the problem description. Exogenous state variables *do not* have to be explicitly included as state variables when formulating a dynamic optimization problem. It is sometimes convenient for notational purposes to give exogenous data a variable name. This is what we are doing by calling the stage-t price p_t. Doing so makes the objective-contribution functions easy to write in a compact form.

Almost all dynamic optimization problems have fixed data that can be listed as exogenous state variables. For instance, in the Production-Planning Problem that is introduced in Section 6.1 we could define state variables representing the amount of time that different production processes take. These variables would represent the values that are next to the arcs in Figure 6.1.

We define two versions of the state-transition functions for the endogenous state variables, s_t. These correspond to the two different definitions of the decision variables. If we define three decisions variables, r_t, u_t, and n_t, in each stage, the state-transitions are given by:

$$s_{t+1} = f_t^s(s_t, r_t, u_t, n_t) = s_t - r_t + u_t, \forall t = 1, \ldots, 24. \tag{6.7}$$

This function defines the amount of water in the upper reservoir at the beginning of stage $(t+1)$ as the amount in the reservoir at the beginning of stage t, minus the amount (one unit) removed if the turbine is operated, plus the amount (one unit) if the pump is operated. We place the superscript 's' on the state-transition function in (6.7) to make it explicitly clear that this is the state-transition function for the s_t state variables. If we have a single decision variable in each stage, the state-transition functions are given by:

$$s_{t+1} = f_t^s(s_t, d_t) = s_t + d_t, \forall t = 1, \ldots, 24.$$

This has the exact same interpretation as (6.7). Note that the state-transition functions for this problem are state-dependent, unlike the state-transition functions in the Production-Planning Problem that is introduced in Section 6.1. That is to say the endogenous state variable in stage $(t+1)$ is a function of the stage-t state-variable value. The Production-Planning Problem does not have this property.

If we define the prices as exogenous state variables, then we must define state-transition functions for these variables as well. These state-transition functions are somewhat trivial, however, as they simply specify that the prices in the different stages are equal to the values that are given in Table 6.7. Thus, the state-transition functions for the prices are:

$$p_1 = 21.78,$$

$$p_2 = 19.81,$$

$$\vdots$$

$$p_{24} = 21.63.$$

Because the p_t state variables represent exogenously given problem data, it is common to not explicitly write such state-transition functions for sake of brevity.

This problem has either two or three sets of constraints, depending on which of the two definitions of the decision variables is used. The first set requires that the decision variables take on the discrete values that are given in their definitions. If the problem is formulated using three decision variables in each stage, these constraints take the form:

$$r_t, u_t, n_t \in \{0, 1\}, \forall t = 1, \ldots, 24.$$

Otherwise, with a single decision variable in each stage, the constraints take the form:

$$d_t \in \{-1, 0, 1\}, \forall t = 1, \ldots, 24.$$

The second set of constraints ensure that the water level of the upper reservoir does not go below 0 or above 4 after each decision is made. We require the former restriction because it is physically impossible for the upper reservoir to have a negative amount of water. The latter restriction is imposed to ensure that the upper reservoir is not flooded. These restrictions have the form:

$$0 \leq s_{t+1} \leq 4, \forall t = 1, \ldots, 24. \tag{6.8}$$

Note that (6.8) defines a restriction on a state variable. As discussed in Section 6.2.3, constraints should always be written as restrictions on decision variables. We convert (6.8) into constraints on the decision variables by using the state-transition functions. Recall that the state-transition functions are:

$$s_{t+1} = s_t - r_t + u_t, \forall t = 1, \ldots, 24;$$

in the three-decision-variable formulation and:

$$s_{t+1} = s_t + d_t, \forall t = 1, \ldots, 24;$$

in the single-decision-variable formulation. If we substitute the state-transition functions into (6.8) we have:

$$0 \leq s_t - r_t + u_t \leq 4, \forall t = 1, \ldots, 24;$$

in the three-decision-variable formulation and:

$$0 \leq s_t + d_t \leq 4, \forall t = 1, \ldots, 24;$$

in the single-decision-variable formulation. We can further manipulate these two expressions (by subtracting s_t from both sides of the double-sided inequalities) to arrive at the constraints:

$$- s_t \leq -r_t + u_t \leq 4 - s_t, \forall t = 1, \ldots, 24; \tag{6.9}$$

in the three-decision-variable formulation and:

$$- s_t \leq d_t \leq 4 - s_t, \forall t = 1, \ldots, 24; \tag{6.10}$$

in the single-decision-variable formulation. The restrictions in (6.9) and (6.10) are of the form that constraints in dynamic optimization problems must be. First, the constraints are on the decision variables (r_t and u_t in (6.9) and d_t in (6.10)). Secondly, the stage-t constraints only depend on stage-t information. That is to say, the restrictions on what values the stage-t decision variables can take depends on the stage-t state variable, s_t. This is because s_t appears on the two sides of the double-sided inequalities.

We have one final set of constraints, which ensures that exactly one of the three options available is chosen in each stage. In the three-decision-variable formulation these constraints are written as:

$$r_t + u_t + n_t = 1, \forall t = 1, \ldots, 24.$$

No such constraint is needed in the single-decision-variable formulation of the problem. This is because the way that the decision variable, d_t, is defined ensures that the plant does exactly one of operating the pump, operating the turbine, or nothing in each stage.

Putting these constraint sets together, we can write the constraints of the three-decision-variable formulation as:

$$r_t, u_t, n_t \in \{0, 1\}, \forall t = 1, \ldots, 24;$$

$$-s_t \le -r_t + u_t \le 4 - s_t, \forall t = 1, \ldots, 24;$$

and:

$$r_t + u_t + n_t = 1, \forall t = 1, \ldots, 24.$$

The constraints of the single-decision-variable formulation are:

$$d_t \in \{-1, 0, 1\}, \forall t = 1, \ldots, 24;$$

and:

$$-s_t \le d_t \le 4 - s_t, \forall t = 1, \ldots, 24.$$

The objective-contribution functions can be written in two ways, depending on whether we formulate the problem with one or three sets of decision variables. With three decision variables the objective-contribution functions are:

$$c_t(s_t, p_t, r_t, u_t, n_t) = \begin{cases} -100p_t, & \text{if } u_t = 1, \\ 80p_t, & \text{if } r_t = 1, \\ 0, & \text{otherwise,} \end{cases} \tag{6.11}$$

for each $t = 1, \ldots, 24$. This function says that the company pays $100p_t$ if the pump operates (as it purchases 100 MWh at a price of $\$p_t$/MWh) and earns $80p_t$ if the turbine operates in stage t. Otherwise, the company does not pay or earn anything if it chooses to do nothing in stage t. Note that defining p_t as an exogenous state variable makes the objective-contribution functions more compact, as they can all be written in the same generic form in terms of p_t.

If the problem if formulated with a single decision variable in each stage, the objective-contribution functions are defined as:

$$c_t(s_t, p_t, d_t) = \begin{cases} -100p_t, & \text{if } d_t = 1, \\ 80p_t, & \text{if } d_t = -1, \\ 0, & \text{otherwise,} \end{cases}$$

for each $t = 1, \ldots, 24$. These functions have the same interpretation that (6.11) does.

6.3.2 *Energy-Production Problem*

A company owns an electricity-generating plant. In each hour the company can choose to either operate the plant or not. If the plant is operated in a given hour, it can produce either 100 MW or 350 MW in that hour (*i.e.*, these are the only output levels possible). If the plant is not operated in a given hour, then it cannot produce any power in that hour.

If the plant is operated in a given hour, the company incurs a fixed cost of $400, regardless of how much power the plant produces. The plant also incurs a variable cost of $45/MWh produced. Finally, if the plant is switched on in a given hour (*i.e.*, it is operated in hour t but was not operated in hour $(t - 1)$), the company incurs a startup cost of $600 (which is in addition to the $400 fixed cost of operating the plant and its variable operation cost). The plant has a further ramping constraint on its operation. If the plant is operating in hour t it has to be producing 100 MW for the company to be able to switch the plant off in hour $(t + 1)$. Otherwise, if the plant is producing 350 MW in hour t, the company cannot switch the plant off in hour $(t+1)$.

Energy produced by the generating plant can be sold in the wholesale electricity market. Table 6.8 summarizes wholesale electricity prices over the next 24 hours. The plant was online and producing 100 MW in the previous hour. The company would like to determine how to operate its plant over the next 24 hours to maximize profits earned from energy sales.

This is another example of a problem in which the stages are easy to determine. This is because of the natural time sequence of the decisions being made. The company must determine what to do with its generating plant in each of the next 24 hours. As such, the problem has 24 stages and each of the next 24 hours represents one of those stages.

The problem has two sets of decisions, both of which are relatively easy to determine. The first is whether to operate the plant in each stage or not. We represent this using binary decision variables, u_t, $\forall t = 1, \ldots, 24$, with u_t defined as:

$$u_t = \begin{cases} 1, & \text{if the plant is operated in stage } t, \\ 0, & \text{otherwise.} \end{cases}$$

The other set of decisions is how much power to produce in each stage. We define an additional set of decision variables, q_t, $\forall t = 1, \ldots, 24$, with q_t representing the MW produced by the plant in stage t.

Defining the endogenous state variables (and, as such, the state-transition functions) of this problem is more difficult than in the other examples seen thus far. It is easier to determine what state variables are needed when we attempt to write the constraints and objective-contribution functions. Thus, we tackle these first. We can, however, define a set of exogenous state variables, p_t, $\forall t = 1, \ldots, 24$, with p_t representing the stage-t wholesale electricity price. As in the Energy-Storage Problem, which is discussed in Section 6.3.1, this set of exogenous state variables makes it easier to write the objective-contribution functions compactly. We can also define

Hour	Price [\$/MWh]
1	65
2	61
3	42
4	35
5	30
6	32
7	38
8	42
9	53
10	57
11	59
12	64
13	72
14	77
15	64
16	62
17	40
18	64
19	55
20	43
21	40
22	55
23	30
24	21

Table 6.8 Wholesale electricity prices over the next 24 hours in the Energy-Production Problem

the state-transition functions for the prices as:

$$p_1 = 65,$$

$$p_2 = 61,$$

$$\vdots$$

$$p_{24} = 21.$$

These state-transition functions can be excluded, for sake of brevity, as they simply repeat the fixed price data that are in Table 6.8 (as in the case of the Energy-Storage Problem).

There are three sets of constraints in this problem. The first requires that the operation variables take on the discrete values given in their definition:

$$u_t \in \{0, 1\}, \forall t = 1, \ldots, 24.$$

The second set of constraints restricts the values that the production variables take on. There are two types of restrictions on the production variables. The first is that the plant can only produce 0, 100, or 350 MW. The other is that the stage-t production of the plant must be positive if the plant is operated in stage t. Otherwise, if the plant is not operated in stage t, its stage-t production must be zero. The simplest way to express these two types of restrictions is through the conditional constraints:

$$q_t \in \begin{cases} \{0\}, & \text{if } u_t = 0, \\ \{100, 350\}, & \text{if } u_t = 1, \end{cases}$$

for all $t = 1, \ldots, 24$. Recall that such conditional constraints can be used when formulating a dynamic optimization model. A less compact way to write the constraints on the production variables is to first restrict the production variable to take on one of the three possible values:

$$q_t \in \{0, 100, 350\}, \forall t = 1, \ldots, 24;$$

and to then relate the q_t and u_t variables as:

$$100u_t \leq q_t \leq 350u_t, \forall t = 1, \ldots, 24. \tag{6.12}$$

Constraint set (6.12) is an example of a variable-discontinuity constraint (*cf.* Section 3.3.1 for further details and examples). Either of these two approaches to modeling the constraints is equally valid.

The third set of constraints impose the ramping restriction on shutting the plant down. One way of writing these constraints is:

$$u_t = 1 \text{ if } q_{t-1} = 350, \forall t = 1, \ldots, 24. \tag{6.13}$$

In words, this constraint says that if the plant's stage-$(t-1)$ production level is 350 MW, then it must be operating in stage t. This is not a valid constraint, however, as the restriction on a stage-t decision variable is written in terms of a stage $(t-1)$ decision variable.

As suggested in Section 6.2.3, we address this issue by defining a set of endogenous state variables, $x_t, \forall t = 1, \ldots, 24$, where x_t represents how much power is produced by the plant in stage $(t-1)$. Having defined this new set of state variables, we write constraints (6.13) as:

$$u_t = 1 \text{ if } x_t = 350, \forall t = 1, \ldots, 24;$$

which are of the form needed when formulating a dynamic optimization problem.

We next turn to writing the objective-contribution functions. We can write these as:

$$\begin{cases} (p_t - 45)q_t - 400u_t - 600, & \text{if } u_t = 1 \text{ and } u_{t-1} = 0, \\ (p_t - 45)q_t - 400u_t, & \text{otherwise,} \end{cases} \tag{6.14}$$

for all $t = 1, \ldots, 24$. Each of these functions consists of three terms. The first, $(p_t - 45)q_t$, represents revenues earned from selling energy less the variable production cost. The second, $-400u_t$, represents the fixed cost of operating the plant in stage t, which is only incurred if the plant is operating (*i.e.*, only if $u_t = 1$). The last term, -600, represents the fixed cost of starting up the plant. This cost is only incurred if the plant is operating in stage t (*i.e.*, if $u_t = 1$) but is not operating in the previous stage (*i.e.*, if $u_{t-1} = 0$).

This is an invalid objective-contribution function, however, because the stage-t objective depends on a decision from the previous stage. As with the ramping constraint, we address this problem by introducing a new set of endogenous state variables, y_t, $\forall t = 1, \ldots, 24$, where we define y_t as:

$$
y_t = \begin{cases} 1, & \text{if the plant is operated in stage } (t-1), \\ 0, & \text{otherwise.} \end{cases}
$$

With this new set of state variables we can define the objective-contribution functions as:

$$
c_t(x_t, y_t, p_t, u_t, q_t) = \begin{cases} (p_t - 45)q_t - 400u_t - 600, & \text{if } u_t = 1 \text{ and } y_t = 0, \\ (p_t - 45)q_t - 400u_t, & \text{otherwise,} \end{cases}
$$

for all $t = 1, \ldots, 24$.

We must finally define state-transition functions for the two sets of endogenous state variables that we define to properly formulate the constraints and objective-contribution functions. The state-transition functions for the x_t state variables are:

$$
x_{t+1} = f_t^x(x_t, y_t, p_t, u_t, q_t) = q_t, \forall t = 1, \ldots, 24.
$$

The state-transition function for the y_t state variables are:

$$
y_{t+1} = f_t^y(x_t, y_t, p_t, u_t, q_t) = u_t, \forall t = 1, \ldots, 24.
$$

We summarize the problem formulation as follows. The problem has 24 stages, each of which corresponds to one of the following 24 hours. There are two decision variables in each stage: $u_t, q_t, \forall t = 1, \ldots, 24$. We define u_t as:

$$
u_t = \begin{cases} 1, & \text{if the plant is operated in stage } t, \\ 0, & \text{otherwise,} \end{cases}
$$

and q_t as MW produced by the plant in stage t. There are three sets of state variables, $p_t, x_t, y_t, \forall t = 1, \ldots, 24$. p_t is an exogenous state variable, which represents the stage-t wholesale electricity price (measured in \$/MWh). x_t is an endogenous state variable that represents how many MW are produced by the plant in stage $(t-1)$. y_t is an endogenous state variable, which is defined as:

$$y_t = \begin{cases} 1, & \text{if the plant is operated in stage } (t-1), \\ 0, & \text{otherwise.} \end{cases}$$

The state-transition functions for the two endogenous state variables are:

$$x_{t+1} = f_t^x(x_t, y_t, p_t, u_t, q_t) = q_t, \forall t = 1, \ldots, 24;$$

and:

$$y_{t+1} = f_t^y(x_t, y_t, p_t, u_t, q_t) = u_t, \forall t = 1, \ldots, 24.$$

There are three sets of constraints. The first restricts u_t to the discrete values given in its definition:

$$u_t \in \{0, 1\}, \forall t = 1, \ldots, 24.$$

The second restricts q_t to the discrete values that it can take and relates it to u_t:

$$q_t \in \begin{cases} \{0\}, & \text{if } u_t = 0, \\ \{100, 350\}, & \text{if } u_t = 1, \end{cases}$$

for all $t = 1, \ldots, 24$. The third:

$$u_t = 1 \text{ if } x_t = 350, \forall t = 1, \ldots, 24;$$

imposes the ramping constraint on shutting down the plant. Finally, the objective-contribution functions are:

$$c_t(x_t, y_t, p_t, u_t, q_t) = \begin{cases} (p_t - 45)q_t - 400u_t - 600, & \text{if } u_t = 1 \text{ and } y_t = 0, \\ (p_t - 45)q_t - 400u_t, & \text{otherwise,} \end{cases}$$

for all $t = 1, \ldots, 24$.

6.3.3 Staffing Problem

A company needs to schedule its pool of maintenance staff over the next 10 work shifts. It wishes to do so to minimize its total staffing costs. The company needs to have a minimum number of staffers working during each of the next 10 shifts. Table 6.9 lists these minimum staffing levels. Moreover, the company can never have more than 20 people working during any shift.

At the beginning of each work shift, the company decides how many staffers to schedule to begin working in that shift. When doing so, it has the option to schedule an employee for one shift or for two consecutive shifts and can schedule any combination of single- and double-shift employees at any time. The only exception is in shift 10, when it can only hire single-shift employees. An employee that is scheduled for a

Table 6.9 Minimum number of employees that must be working in the maintenance center over the next 10 shift periods in the Staffing Problem

Shift Period	Minimum Number of Employees
1	10
2	15
3	9
4	16
5	16
6	10
7	15
8	13
9	14
10	8

single shift costs $100, which is paid immediately. An employee that is scheduled for a double shift costs $80/shift, which is paid in each of the two shifts. The company must hire an integer number of employees. The company has three employees available to work a single shift at the beginning of shift period 1, who are carried over from the previous shift period.

This problem has a natural sequence of decisions being made. Specifically, at the beginning of each of the 10 shifts the company must decide how many employees to schedule. Thus, there are 10 stages and each stage corresponds to one of the 10 shifts.

There are two decisions being made in each stage—the number of employees to schedule to work a single shift and the number of employees to schedule to work a double shift. Thus, we define two sets of decision variables $a_t, b_t, \forall t = 1, \ldots, 10$. We define a_t as the number of employees scheduled to begin a single shift in stage t and b_t as the number of employees scheduled to begin a double shift in stage t.

The state variables for this problem may not be immediately apparent. However, if we attempt to find the objective-contribution functions, it becomes more clear what our endogenous state variables are. The amount that is contributed toward our overall objective in stage t is:

$$100a_t + 80b_t + 80b_{t-1}. \tag{6.15}$$

Equation (6.15) says that we pay $100 and $80, respectively, for the single- and double-shift employees scheduled to begin work in stage t. We also pay $80 for the double-shift employees that are scheduled to begin work in stage $(t - 1)$. This is not, however, a proper objective-contribution function. This is because the stage-t objective-contribution may only depend on the stage-t state and decision variables. As in the Electricity-Production Problem, which is introduced in Section 6.3.2, we rectify this problem by adding a set of endogenous state variables, $s_t, \forall t = 1, \ldots, 10$. We define s_t as the number of double-shift employees who are scheduled to start work in stage $(t - 1)$ (and are scheduled to still be working in stage t).

We can also define a set of exogenous state variables, $m_t, \forall t = 1, \ldots, 10$, where m_t represents the minimum number of employees that must be staffing the maintenance center in stage t. These state variables store the values that are summarized in Table 6.9 and allow us to write the problem constraints more compactly.

Given these state variable definitions, we can write the state-transition functions for the s_t state variables as:

$$s_{t+1} = f_t^s(s_t, a_t, b_t) = b_t, \forall t = 1, \ldots, 10.$$

This function says that $s_{t+1} = b_t$, meaning that the number of double-shift employees carried into stage $(t + 1)$ is equal to the number of double-shift employees that are scheduled to start working in stage t. The state-transition functions for the m_t state variables are:

$$m_1 = 10,$$

$$m_2 = 15,$$

$$\vdots$$

$$m_{10} = 8.$$

The problem constraints include non-negativity and integrality of the decision variables:

$$a_t, b_t \in \mathbb{Z}^+, \forall t = 1, \ldots, 10;$$

and the minimum and maximum number of employees that must be staffing the maintenance center in each stage. This second set of constraints can be written as:

$$m_t \leq s_t + a_t + b_t \leq 20, \forall t = 1, \ldots, 10. \tag{6.16}$$

The term in the middle of (6.16) gives the total number of employees working in stage t—s_t employees are carried over from stage $(t - 1)$ and $(a_t + b_t)$ employees are scheduled to begin a single or double shift in stage t. We also have a constraint that double-shift employees cannot start work in shift 10:

$$b_{10} = 0.$$

Finally, we can write the objective-contribution functions as:

$$c_t(s_t, a_t, b_t) = 100a_t + 80(s_t + b_t), \forall t = 1, \ldots, 10.$$

This expression is analogous to (6.15). Unlike (6.15) it is a properly structured objective-contribution function because the stage-t objective-contribution function depends solely on the stage-t state and decision variables.

6.3.4　Investment Problem

A dapper youngster has \$7000 to invest in any combination of three assets. If she chooses to invest in an asset, she must invest an integer multiple of \$1000. If she chooses to invest in the first asset she earns a \$1000 fixed return on her investment plus the amount invested is doubled. If she chooses to invest in the second asset, she earns a \$5000 fixed return on her investment plus the amount invested is quadrupled. If she chooses to invest in the third asset, she earns a \$2000 fixed return on her investment plus the amount invested is tripled. These returns on her investments are earned after five years. The investor would like to determine how to invest the \$7000 that she has on-hand today to maximize the total return on her investment after five years.

This is an example of a problem in which determining the problem stages can be difficult, because there is not a natural time sequence. Instead, the easiest way to determine the stages is to first identify the decisions that are being made. In this problem, the decisions are how much to invest in each of the three assets, which we denote d_1, d_2, and d_3, respectively. Knowing that these are the three decisions being made, we can next identify the problem as having three stages. The stages correspond to when the investor decides how much to invest in each of assets 1 through 3. Note that the problem description implies that these investment decisions are made simultaneously. However, we represent this as a sequential decision to decompose the problem as a DPP.

We next need to determine the problem states. The information that is critically important when making each of the three investment decisions in how much cash is on hand for investing at the beginning of each stage (*i.e.*, how much of the \$7000 is uninvested when making each of the decisions in the three stages). We let s_1, s_2, and s_3 denote these three state variables. We could also define exogenous state variables that represent the fixed return on investment that each asset provides or the factor by which the amount invested in each asset grows. We do not do so, however, because the objective-contribution functions of this problem are easy to write without defining these exogenous state variables.

Next, we write the state-transition functions as:

$$s_{t+1} = f_t(s_t, d_t) = s_t - d_t, \forall t = 1, 2, 3. \tag{6.17}$$

These state-transition functions say that the amount of cash that is on hand at the beginning of stage $(t+1)$ (*i.e.*, when determining how much to invest in asset $(t+1)$) equals the amount of cash that is on hand at the beginning of stage t less the amount that is invested in stage t (*i.e.*, the amount that is invested in asset t).

Before proceeding, note that one may be tempted to write a state-transition function of the form:

$$s_3 = s_1 - d_1 - d_2. \tag{6.18}$$

This equation says that the amount of cash on hand at the beginning of stage 3 equals the amount of cash that is on hand at the beginning of stage 1, less the amounts invested in stages 1 and 2. This is an invalid state-transition function, however. The reason that this is invalid is because the change in the state variable between stages 2 and 3 should *only* depend on the stage-2 state and decision variables. Equation (6.18) gives the stage-3 state variable in terms of the stage-1 decision variable, which is invalid. As discussed in Section 6.2.2, state-transition functions should always define single-state transitions. Equation (6.17) does this, whereas (6.18) does not.

To determine the problem constraints, we first note that the amount invested in each asset must be non-negative and an integer multiple of $1000. These constraints can be written as:

$$d_t \in \{0, 1000, 2000, \ldots\}, \forall t = 1, 2, 3. \tag{6.19}$$

We must also ensure that we not invest more money than we have available in any stage. These constraints can be written as:

$$d_t \leq s_t, \forall t = 1, 2, 3. \tag{6.20}$$

We finally need to define the objective-contribution functions, which are the returns on investment earned in each of the three stages. Based on the problem information, these are given by:

$$c_1(s_1, d_1) = \begin{cases} 2d_1 + 1000, & \text{if } d_1 > 0, \\ 0, & \text{otherwise,} \end{cases}$$

$$c_2(s_2, d_2) = \begin{cases} 4d_2 + 5000, & \text{if } d_2 > 0, \\ 0, & \text{otherwise,} \end{cases}$$

and:

$$c_3(s_3, d_3) = \begin{cases} 3d_3 + 2000, & \text{if } d_3 > 0, \\ 0, & \text{otherwise.} \end{cases} \tag{6.21}$$

6.4 Iterative Solution Algorithm for Dynamic Optimization Problems

We now turn to solving dynamic optimization problems. More specifically, we discuss how to generalize the solution technique that we apply to the Production-Planning Problem in Section 6.1.2 to solve any dynamic optimization problem. This solution technique is known as the **Dynamic Programming Algorithm**. Although it can be tedious, the Dynamic Programming Algorithm is trivially easy to apply to any problem, so long as the six elements that are required to formally formulate the problem are properly identified.

Let us begin by recalling how the Production-Planning Problem is decomposed and explain the steps that we go through to solve it in Section 6.1.2. We explain these steps in terms of the elements of a generic dynamic optimization problem (*i.e.*, in terms of the stages, state and decision variables, state-transition functions, constraints, and objective-contribution functions). Explaining the steps that are used to solve the Production-Planning Problem in terms of these six elements allows us to apply the same methodology to any DPP.

In the Production-Planning Problem, we first start by examining states 10 and 11. For each of these states we determine the optimal action to choose and record this and the associated amount of time that it takes to finish the product in Table 6.1.

We then examine states 7 through 9, repeating the same process. Recall that when determining the amount of time that it takes to finish the product from each of states 7 through 9, we add two separate terms together. The first accounts for the amount of time associated with the immediately chosen process (*i.e.*, process 'I' or 'J'). The second accounts for what happens after this immediate process is completed and the remaining amount of time to complete the product thereafter. Equations (6.1) and (6.2) show these two terms being added together. Recall, also, that the second term, which is the amount of time that it takes to complete the final product from one of states 10 or 11, is taken from Table 6.1. Finally, note that when examining states 8 and 9, there are two feasible alternatives available (states 7, 10, and 11 have only one feasible decision available). For states 8 and 9 we compute the total time associated with each of the two alternatives available and select the optimal (minimal-time) one.

From there, we move further back through the problem, examining states 4, 5, and 6, followed by states 2 and 3, and then finally state 1, at which point we have worked backward through the entire problem.

To generalize this process, first note that we decompose the Production-Planning Problem by considering sets of states together. More specifically, what we are doing is working backward through the *stages* of the problem. In each stage we consider each possible state that the system can be in (*i.e.*, different possible values that the state variables can take on). For each possible state, what we do is enumerate all of the feasible decision-variable values that are available to us. We then determine what the total objective-function value over the current and subsequent stages of the problem is if that decision is chosen. We must stress that when computing this total objective-function value, we do not directly sum the objective-contribution functions for all of the subsequent stages of the problem. Rather, we consider the sum of how much is contributed toward the overall objective in the current stage (this is given by the current-stage objective-contribution function) and what is contributed in the remaining stages of the problem (this comes from information that is recorded in the third column of Tables 6.1 through 6.5 in the Production-Planning Problem). The fact that we do not compute the objective-contribution function for each subsequent stage when computing the total objective-function value is an important and time-saving benefit of the Dynamic Programming Algorithm.

6.4.1 Dynamic Programming Algorithm Example

To further illustrate the Dynamic Programming Algorithm, we apply it to solve the Investment Problem, which is introduced in Section 6.3.4. Following the process used to solve the Production-Planning Problem in Section 6.1.2, we begin by first examining the final stage. The Investment Problem has three stages, thus we begin by examining stage 3. To examine stage 3 we first determine what possible values the state variables can take in stage 3. The endogenous state variable, s_3, represents the amount of cash that is on hand to invest at the beginning of stage 3. We know that the investor starts with \$7000 and can only invest integer multiples of \$1000 in the assets. Thus, there are only eight possible values that s_3 can take on: 0, 1000, 2000, 3000, 4000, 5000, 6000, and 7000.

For each of these possible state-variable values, we next determine what decisions are feasible. This is done using the constraints. Let us begin by considering the first possible state-variable value, $s_3 = 0$. Constraints (6.19) and (6.20) for stage 3 are:

$$d_3 \in \{0, 1000, 2000, \ldots\},$$

and:

$$d_3 \leq s_3.$$

If we substitute $s_3 = 0$ (because 0 is the state-variable value that we are currently examining) into these constraints they become:

$$d_3 \in \{0, 1000, 2000, \ldots\},$$

and:

$$d_3 \leq 0.$$

Taken together, these constraints simplify to:

$$d_3 = 0.$$

Thus, there is no decision to be optimized in the case that $s_3 = 0$, because there is only one feasible decision. Nevertheless, we do need to determine what the overall objective function value from stage 3 through all of the subsequent stages is. Remember from the Production-Planning Problem that this information is used later when examining other stages of the problem. We compute the objective-function value using the stage-3 objective-contribution function. Because stage 3 is the final stage of the problem, we do not have a second term to include in computing the overall objective-function value. This is because we are at the final stage of the problem and there are no further decisions to be made that add to the total objective after the stage-3 decision is made. We compute the objective-function value using objective-contribution function (6.21), which is:

$$c_3(s_3, d_3) = \begin{cases} 3d_3 + 2000, & \text{if } d_3 > 0, \\ 0, & \text{otherwise.} \end{cases}$$

Substituting $s_3 = 0$ (because 0 is the case that we are examining) and $d_3 = 0$ (because 0 is the feasible decision that we are considering) into this function gives:

$$c_3(0, 0) = 0.$$

We summarize the results of examining the case in which $s_3 = 0$ in the first row of Table 6.10. The first column of Table 6.10, which is labeled s_3, lists the different possible stage-3 state-variable values. The second column, which is labeled $d_3^*(s_3)$, indicates the optimal decision to choose, as a function of the stage-3 state variable. The second column of the table is commonly referred to as the **decision policy**. This is because the second column gives the optimal decision to make in stage 3 for each possible stage-3 state. We enter a value of 0 in the first row of this column, because we have that $d_3 = 0$ is the optimal (because it is the only feasible) decision if $s_3 = 0$.

Table 6.10 Stage-3 calculations for applying the Dynamic Programming Algorithm to the Investment Problem

s_3	$d_3^*(s_3)$	$J_3(s_3)$
0	0	0
1000	1000	5000
2000	2000	8000
3000	3000	11000
4000	4000	14000
5000	5000	17000
6000	6000	20000
7000	7000	23000

The final column of the table, which is labeled $J_3(s_3)$, holds what is called the **cost-to-go function** or **value function** for stage 3. The value function gives the overall objective-function value from the current stage (in the case of Table 6.10, stage 3) to the end of the problem (*i.e.*, including all subsequent stages of the problem). The stage-3 value function is given as a function of the stage-3 state. We enter a value of 0 in the first row of this column, because we have that if $s_3 = 0$ and we optimally choose $d_3 = 0$, the resulting objective-function value from stage 3 and all subsequent stages is 0. Note, finally, that the three columns of Table 6.10 have the same interpretations as the columns of Tables 6.1 through 6.5, which we use to organize our calculations when solving the Production-Planning Problem in Section 6.1.2.

We repeat the same steps that are used for $s_3 = 0$ for each of the remaining seven possible stage-3 state-variable values. We begin with $s_3 = 1000$. We must first examine constraints (6.19) and (6.20) for stage 3 to determine what decisions are feasible if the investor has \$1000 available at the beginning of stage 3. Substituting $s_3 = 1000$ (because 1000 is the state that we are now examining) into the constraints, gives:

$$d_3 \in \{0, 1000\}. \tag{6.22}$$

Because there are now two feasible decisions, there is a decision to be optimized. We select among the two alternatives by first determining what the overall objective-function values from stage 3 and all subsequent stages is if each alternative is chosen. We then select the alternative that gives the highest objective-function value (because our objective in this problem is to maximize overall return on investment). Remember, however, that because we are examining stage 3, which is the final stage of the problem, we only use the stage-3 objective-contribution function to make this determination. Using (6.21) we have:

$$c_3(1000, 0) = 0, \tag{6.23}$$

and:

$$c_3(1000, 1000) = 5000. \tag{6.24}$$

Thus, we choose $d_3 = 1000$ as the optimal stage-3 action. The results of these calculations are recorded in the second row of Table 6.10—our optimal choice of $d_3 = 1000$ is in the second column and the associated value-function value of 5000 is recorded in the third.

We do not explicitly run through the remaining six cases for stage 3. Rather, we leave it to the reader to repeat the exact steps used to analyze $s_3 = 0$ and $s_3 = 1000$ to verify the values in the remaining rows of Table 6.10.

After the remaining stage-3 calculations are conducted and recorded in Table 6.10, we next move back one stage in the problem to stage 2. We repeat the same process to analyze stage 2 by first determining what are possible stage-2 state-variable values. The endogenous state variable, s_2, represents the amount of cash on hand to invest at the beginning of stage 2. Moreover, we know that the investor starts with $7000 and can only invest integer multiples of $1000 in the assets. Thus, there are the same eight possible stage-2 states as in stage 3—s_2 can equal any of 0, 1000, 2000, 3000, 4000, 5000, 6000, or 7000.

Starting with the first case of $s_2 = 0$, we substitute this into constraints (6.19) and (6.20) for stage 2 to determine what decision-variable values are feasible. These constraints reduce to:

$$d_2 = 0.$$

Thus, just as in the case of $s_3 = 0$ when analyzing stage 3, there is only one feasible decision available. By definition, this one feasible decision is optimal. Before proceeding to the next state, we must determine what the total objective-function value from stage 2 through the remaining stages of the problem is. Substituting $s_2 = 0$ and $d_2 = 0$ (because those are the state and decision, respectively, that we are examining) into the stage-2 objective contribution function gives us:

$$c_2(0, 0) = 0,$$

meaning that we earn a \$0 return on investment in stage 2 if $s_2 = 0$ and $d_2 = 0$.

This only tells us what we earn in stage 2, however. We need to add to $c_2(0, 0)$ what is earned over the remaining stages until the end of the problem. Recall from solving the Production-Planning Problem that when examining stage 4 in Equations (6.1) and (6.2) we use the stage-5 value-function information that is recorded in Table 6.1 (*i.e.*, the value function from the next stage). We do the same here, using the stage-3 value-function information, which is recorded in Table 6.10. The only question that remains is which row of Table 6.10 to take the value-function information from. In the Production-Planning Problem this is determined by which state the material is converted into after the stage-4 production process is completed. This information is given by the terminal node of the arc in Figure 6.1 that is associated with the production process that is chosen in stage 4. For instance, if process 'I' is chosen in stage 4, the material transitions to state 10 in stage 5. Thus, the value-function value from the first row of Table 6.1, corresponding to state 10, is used. The terminal nodes of the arcs in Figure 6.1 represent the state-transition functions. Thus, we likewise use the state-transition function to determine which row of Table 6.10 to take the value-function information from in solving the Investment Problem.

The stage-2 state-transition function is:

$$s_3 = f_2(s_2, d_2) = s_2 - d_2.$$

Substituting $s_2 = 0$ and $d_2 = 0$ into the state-transition function gives:

$$s_3 = f_2(0, 0) = 0,$$

meaning that if $s_2 = 0$ and the investor chooses $d_2 = 0$ as the stage-2 decision, the stage-3 state variable equals $s_3 = 0$. Thus, we use $J_3(s_3) = J_3(0)$ to represent the objective-function value from stage 3 through the remainder of the problem. Thus, if $s_2 = 0$ and $d_2 = 0$ the total return on investment from stage 2 through the end of the problem is:

$$c_2(s_2, d_2) + J_3(s_3) = c_2(0, 0) + J_3(0) = 0.$$

The results of analyzing the case with $s_2 = 0$ are summarized in the first row of Table 6.11.

We next examine the case of $s_2 = 1000$. Substituting this into stage-2 constraints (6.19) and (6.20) gives us:

$$d_2 \in \{0, 1000\},$$

as feasible decision-variable values. Thus, we have a decision to optimize in this case. For each of the two feasible decisions, we need to compute how much is contributed toward the overall objective in stage 2 *plus* the amount earned over the subsequent stages of the problem. As in the case of $s_2 = 0$, the first term is given by the stage-

Table 6.11 Stage-2 calculations for applying the Dynamic Programming Algorithm to the Investment Problem

s_2	$d_2^*(s_2)$	$J_2(s_2)$
0	0	0
1000	1000	9000
2000	1000	14000
3000	2000	18000
4000	3000	22000
5000	4000	26000
6000	5000	30000
7000	6000	34000

2 objective-contribution function. The second term is given by the value-function information that is recorded in Table 6.10. As in the case of $s_2 = 0$, we use the stage-2 state-transition function to determine which row of Table 6.10 to take the value-function information from.

For the decision-variable value of $d_2 = 0$ we compute the total objective-function value from stage 2 forward as:

$$\left[\begin{array}{c} \text{immediate stage-2 return on} \\ \text{investment from } d_2 = 0 \end{array}\right] + \left[\begin{array}{c} \text{return on investment} \\ \text{in subsequent stages} \end{array}\right]$$

$$= c_2(s_2, d_2) + J_3(s_3) \tag{6.25}$$

$$= c_2(1000, 0) + J_3(f_2(s_2, d_2)) \tag{6.26}$$

$$= 0 + J_3(f_2(1000, 0)) \tag{6.27}$$

$$= J_3(1000) \tag{6.28}$$

$$= 5000. \tag{6.29}$$

Equation (6.25) defines the first of these two terms as the stage-2 objective-contribution function and the second as the stage-3 value function, which is evaluated at the stage-3 state, s_3. We substitute $s_2 = 1000$ and $d_2 = 0$ into the first-term of (6.26), because we are examining the case in which the state is 1000 and the decision is 0. Furthermore, we know that the stage-3 state, s_3, is determined by the stage-2 state-transition function as:

$$s_3 = f_2(s_2, d_2),$$

and make this substitution into the second term in (6.26). Next, we know that $c_2(1000, 0) = 0$, and substitute this for the first term in (6.27). We also substitute $s_2 = 1000$ and $d_2 = 0$ into the second term in (6.27). We next know that $f_2(1000, 0) = 1000$, and substitute this into (6.28). Using the information that is recorded in Table 6.10, we know that $J_3(1000) = 5000$, which gives us (6.29).

A similar calculation for $d_2 = 1000$ gives:

$$c_2(s_2, d_2) + J_3(s_3)$$
$$= c_2(1000, 1000) + J_3(f_2(s_2, d_2))$$
$$= 9000 + J_3(f_2(1000, 1000))$$
$$= 9000 + J_3(0)$$
$$= 9000 + 0$$
$$= 9000.$$

These two calculations tell us that if the investor enters stage 2 with \$1000 on hand, there are two options available, which are to either invest \$0 or \$1000 in stage 2. Investing \$0 in stage 2 earns the investor \$5000 from stage 2 through the subsequent stages of the problem, whereas investing \$1000 in stage 2 earns \$9000. Because the investor is looking to maximize return on investment, the optimal stage-2 decision if the investor has \$1000 on hand in stage 2 is to invest \$1000 in stage 2, which earns \$9000 from stage 2 to the end of the problem. These findings are recorded in the second row of Table 6.11.

We do one further set of stage-2 calculations for the case of $s_2 = 2000$. Repeating the same steps as before, we first substitute the state-variable value into stage-2 constraints (6.19) and (6.20) and find that:

$$d_2 \in \{0, 1000, 2000\},$$

are the feasible decision-variable values available. For $d_2 = 0$ we compute the overall objective-function value for stage 2 and all subsequent stages as:

$$c_2(2000, 0) + J_3(f_2(2000, 0)) = 0 + J_3(2000) = 0 + 8000 = 8000,$$

for $d_2 = 1000$ we have:

$$c_2(2000, 1000) + J_3(f_2(2000, 1000)) = 9000 + J_3(1000) = 9000 + 5000 = 14000,$$

and for $d_2 = 2000$ we have:

$$c_2(2000, 2000) + J_3(f_2(2000, 2000)) = 13000 + J_3(0) = 13000 + 0 = 13000.$$

Thus, if the investor's stage-2 state is $s_2 = 2000$ the optimal stage-2 decision is $d_2 = 1000$ with an overall objective-function value from stage-2 forward to the end of the problem of \$14000. These findings are recorded in the third row of Table 6.11.

We leave it to the reader to confirm that the remaining entries in Table 6.11 are correct. After doing so, we move back yet another stage to stage 1 of the problem. Because it is the first stage of the problem, stage 1 differs slightly from the others in terms of what state-variable values are possible. We are told that the investor starts with \$7000 to invest. Thus, we know that the stage-1 state variable is $s_1 = 7000$. Substituting this into the stage-1 constraints tells us that:

$$d_1 \in \{0, 1000, 2000, \ldots, 7000\},$$

are the feasible decision-variable values available. As before, we must first compute the overall objective-function value from stage 1 through all of the subsequent stages for each of these eight alternatives. We then choose an alternative, which is the one that maximizes this overall objective-function value. The overall objective-function value again consists of two terms, the stage-1 objective contribution and the stage-2 value function. Note that we do not explicitly include the stage-3 value function (or value functions from any subsequent stages, if the problem has more than three stages) in this calculation. This is because the stage-2 value function captures the total amount earned from stage-2 to the end of the problem. We can see this by recalling that the stage-2 objective-function calculation conducted in (6.25) (and the other calculations for other values of s_2) includes stage-3 value-function information. Thus, the stage-2 value-function information in Table 6.11 implicitly includes all of the value-function information from subsequent stages.

For $d_1 = 0$ the overall objective-function value is:

$$c_1(7000, 0) + J_2(f_1(7000, 0)) = 0 + J_2(7000) = 0 + 34000 = 34000,$$

for $d_1 = 1000$ we have:

$$c_1(7000, 1000) + J_2(f_1(7000, 1000)) = 3000 + J_2(6000) = 3000 + 30000 = 33000,$$

for $d_1 = 2000$ we have:

$$c_1(7000, 2000) + J_2(f_1(7000, 2000)) = 5000 + J_2(5000) = 5000 + 26000 = 31000,$$

for $d_1 = 3000$ we have:

$$c_1(7000, 3000) + J_2(f_1(7000, 3000)) = 7000 + J_2(4000) = 7000 + 22000 = 29000,$$

for $d_1 = 4000$ we have:

$$c_1(7000, 4000) + J_2(f_1(7000, 4000)) = 9000 + J_2(3000) = 9000 + 18000 = 27000,$$

for $d_1 = 5000$ we have:

$$c_1(7000, 5000) + J_2(f_1(7000, 5000)) = 11000 + J_2(2000) = 11000 + 14000 = 25000,$$

for $d_1 = 6000$ we have:

$$c_1(7000, 6000) + J_2(f_1(7000, 6000)) = 13000 + J_2(1000) = 13000 + 9000 = 24000,$$

and for $d_1 = 7000$ we have:

$$c_1(7000, 7000) + J_2(f_1(7000, 7000)) = 15000 + J_2(0) = 15000 + 0 = 15000.$$

Thus, the optimal stage-1 decision is $d_1 = 0$, which gives an objective-function value over stage 1 and all subsequent stages of 34000. These findings are recorded in Table 6.12.

Table 6.12 Stage-1 calculations for applying the Dynamic Programming Algorithm to the Investment Problem

s_1	$d_1^*(s_1)$	$J_1(s_1)$
7000	0	34000

Having worked through all of the problem stages, we finally want to recover optimal-solution information. There are typically two important pieces of information that we want. The first is the optimal objective-function value. The second is an optimal sequence of decisions. Just as with the Production-Planning Problem, the stage-1 value-function information, which is in Table 6.12, tells us the overall objective-function value, which is \$34000. This is because $J_1(7000)$ tells us what the objective-function value is from stage 1 to the end of the problem if the investor enters stage 1 in the state $s_1 = 7000$. The objective-function value from stage 1 to the end of the problem is exactly equal to the optimal overall objective-function value.

We can also reconstruct an optimal sequence of decisions to make by following the same approach that is used in the Production-Planning Problem. This is done using the decision-policy information that is recorded in the second column of Tables 6.10 through 6.12 and the state-transition functions. Specifically, we work backward through Tables 6.10 through 6.12 (meaning that we work forward through the stages of the problem). First, Table 6.12 tells us that the optimal stage-1 decision if $s_1 = 7000$ is $d_1 = 0$. We next use the stage-1 state-transition function to determine that if this decision is chosen in stage 1, the stage-2 state variable is:

$$s_2 = f_1(s_1, d_1) = f_1(7000, 0) = 7000. \tag{6.30}$$

We then use the last row of Table 6.11, which corresponds to $s_2 = 7000$, to determine that if the stage-2 state is $s_2 = 7000$ the optimal stage-2 decision is $d_2 = 6000$. We then use the stage-2 state-transition function to determine the resulting stage-3 state, which is:

$$s_3 = f_2(s_2, d_2) = f_2(7000, 6000) = 1000. \tag{6.31}$$

The second row of Table 6.10, which corresponds to $s_3 = 1000$, then tells us that the optimal stage-3 decision is $d_3 = 1000$. Thus, our optimal sequence of decisions is to invest \$0 in stage 1 (asset 1), \$6000 in stage 2 (asset 2), and \$1000 in stage 3 (asset 3).

6.4.2 Dynamic Programming Algorithm

We now provide a more general outline of how to apply the Dynamic Programming Algorithm to solve any dynamic optimization problem. In doing so, we use the same notational conventions that are developed in the previous sections. Specifically, we let T denote the number of problem stages and $t = 1, \ldots, T$ denote a particular stage. Any variable or function subscripted by t refers to the stage-t value of that variable or function. More specifically, we let s_t and d_t denote the stage-t state and decision variables, respectively. Depending on whether the problem that we are solving has one or multiple state or decision variables in each stage, s_t and d_t can be scalars or vectors. We also define $f_t(s_t, d_t)$ to be the stage-t state-transition function and $c_t(s_t, d_t)$ the stage-t objective-contribution function.

We next define $\mathscr{S}_t$ to be the set of different possible stage-t state-variable values. For instance, when solving the Investment Problem in Section 6.4.1 we conclude that the investor could have any of \$0, \$1000, \$2000, \$3000, \$4000, \$5000, \$6000, or \$7000 available to invest in stage 3. Because these are the eight possible values that s_3 can take on, we would say that:

$$\mathscr{S}_3 = \{0, 1000, 2000, 3000, 4000, 5000, 6000, 7000\}.$$

We would similarly define:

$$\mathscr{S}_2 = \{0, 1000, 2000, 3000, 4000, 5000, 6000, 7000\},$$

and:

$$\mathscr{S}_1 = \{7000\},$$

based on what we conclude about possible state-variable values in each of stages 2 and 1, respectively.

We define $\mathscr{D}_t(s_t)$ to be the set of stage-t decisions that are feasible as a function of the stage-t state variable. In essence, $\mathscr{D}_t(s_t)$ represents what decisions the constraints tell us are feasible when we substitute particular state-variable values into them. For instance, when conducting the stage-3 calculations to solve the Investment Problem in Section 6.4.1, we find that if $s_3 = 0$ the constraints reduce to:

$$d_3 \in \{0\}.$$

We could express this constraint, using our newly defined notation, as:

$$\mathscr{D}_3(0) = \{0\}.$$

The other cases that are explicitly examined in Section 6.4.1 give us that:

$$\mathscr{D}_3(1000) = \{0, 1000\},$$

$$\mathscr{D}_2(0) = \{0\},$$

$$\mathscr{D}_2(1000) = \{0, 1000\},$$

$$\mathscr{D}_2(2000) = \{0, 1000, 2000\},$$

and:

$$\mathscr{D}_1(7000) = \{0, 1000, 2000, 3000, 4000, 5000, 6000, 7000\}.$$

We finally let $J_t(s_t)$ denote the stage-t value function and $d_t^*(s_t)$ the stage-t decision policy. The values of these functions are unknown when we begin solving the problem and are found as we conduct our calculations. For instance, as we solve the Production-Planning Problem in Section 6.1.2, we record value-function and decision-policy information in Tables 6.1 through 6.5. We could summarize the value-function information for the Production-Planning Problem as:

$$J_5(11) = 7,$$

$$J_5(10) = 6,$$

$$\vdots$$

$$J_1(1) = 16,$$

and the corresponding decision-policy information as:

$$d_5^*(11) = \text{'K'},$$

$$d_5^*(10) = \text{'K'},$$

$$\vdots$$

$$d_1^*(1) = \text{`B'}.$$

Likewise, when solving the Investment Problem in Section 6.4.1, value-function and decision-policy information are recorded in Tables 6.10 through 6.12. We could summarize the value-function information for the Investment Problem as:

$$J_3(7000) = 23000,$$

$$J_3(6000) = 20000,$$

$$\vdots$$

$$J_1(7000) = 34000,$$

and the corresponding decision-policy information as:

$$d_3^*(7000) = 7000,$$

$$d_3^*(6000) = 6000,$$

$$\vdots$$

$$d_1^*(7000) = 0.$$

The value function is used when solving the dynamic optimization problem whereas the decision policy is used to determine an optimal sequence of decision-variable values.

Below we give a high-level outline of the first part of the Dynamic Programming Algorithm, which we term the Backward Recursion. This part of the algorithm works backwards through the problem stages to determine optimal decisions at each stage for each possible state. The Backward Recursion is really three nested **for** loops. The outermost loop, on Line 2, works backwards through the problem stages. The next innermost loop, on Line 3, works through each possible value that the state variable can take in each stage. The third loop, on Line 4, works through each decision-variable value that is feasible for each possible state.

Dynamic Programming Algorithm: Backward Recursion

1: **procedure** BACKWARD RECURSION
2:      **for** $t \leftarrow T, \ldots, 1$ **do**   ▷ Step backward through problem stages
3:          **for** $s_t \in \mathscr{S}_t$ **do**   ▷ Step through each possible stage-t state
4:              **for** $d_t \in \mathscr{D}_t(s_t)$ **do**   ▷ Step through each feasible stage-t decision for state s_t
5:                  **if** $t = T$ **then**
6:                      $b(d_t) \leftarrow c_t(s_t, d_t)$ ▷ Compute objective value
7:                  **else**
8:                      $b(d_t) \leftarrow c_t(s_t, d_t) + J_{t+1}(f_t(s_t, d_t))$ ▷ Compute objective value
9:                  **end if**
10:              **end for**
11:              $J_t(s_t) \leftarrow \min_{d_t \in \mathscr{D}_t(s_t)} b(d_t)$
12:              $d_t^*(s_t) \leftarrow \arg\min_{d_t \in \mathscr{D}_t(s_t)} b(d_t)$
13:          **end for**
14:      **end for**
15: **end procedure**

To further illustrate these loops, recall how the Investment Problem is analyzed in Section 6.4.1. Moreover, take the particular case of $t = 3$ (*i.e.*, the final problem stage). We analyze stage 3 by first enumerating all of the possible stage-3 states. The second **for** loop on Line 3 exactly does this by looping through all of the possible stage-t states (in the case of $t = 3$ in the Investment Problem, Line 3 loops through the possible stage-3 states of $s_3 = 0, 1000, 2000, 3000, 4000, 5000, 6000$, and 7000).

Next, take as an example s_3 set equal to the particular value of 1000 in Line 3 (this is one of the eight values that s_3 can potentially take). The third loop, on Line 4, works through all of the feasible stage-3 decisions that are available if the stage-3 state variable is $s_3 = 1000$. In the case of the Investment Problem, this amounts to looping through the feasible decisions of $d_3 = 0$ and $d_3 = 1000$. These are exactly the two feasible decisions that are identified in (6.22) in the case that $s_3 = 1000$.

Next, Lines 5 through 9 do the actual computations of the objective-function values. These calculations differ depending on whether we are examining the final problem stage or not (*i.e.*, depending on whether $t = T$ or $t < T$). If we are examining the final stage, then the objective function is computed as shown on Line 6, using the objective-contribution function *only*. Recall that this is how the objective function is computed when examining stage 5 in the Production-Planning Problem and when examining stage 3 in the Investment Problem. Otherwise, if we are examining any other problem stage, we include both the objective-contribution and value functions, as shown on Line 8. The value function is evaluated at whatever the new state-variable value will be in the subsequent problem stage. This new state-variable value in the subsequent stage is determined by the state-transition function, as noted in Line 8. Again, the objective-function calculations on Lines 6 and 8 are simply generalizations of how the Production-Planning and Investment Problems are solved in Sections 6.1.2 and 6.4.1. The term, $b(d_t)$, is a placeholder, used in Lines 6 and 8, to which the objective-function value associated with choosing the decision, d_t, is assigned.

After the objective-function values for each feasible decision (associated with a particular possible state) are computed, Lines 11 and 12 determine an optimal decision to make and the associated value-function value for that state. More specifically, Line 11 determines the value-function value, by setting $J_t(s_t)$ (*i.e.*, the value of the value function associated with the state-variable value being considered) equal to the minimum objective-function value among all feasible decisions. This is done by choosing the feasible decision that minimizes $b(d_t)$, which is the placeholder that holds the objective-function values that are computed in Lines 6 or 8. Note that if we are solving a dynamic optimization problem that is a maximization, then the 'min' operator that is in Line 11 is changed to a 'max.' Line 12 determines the associated optimal decision, by using an 'arg min' operator. The 'arg' operator returns the value of d_t that minimizes the function (*i.e.*, the optimal choice of d_t). Again, if solving a problem that is a maximization, the 'arg min' operator that is in Line 12 is replaced with an 'arg max.' The computations that are on Lines 11 and 12 are analogous to the comparisons between the objective values of choosing $d_3 = 0$ or $d_3 = 1000$ if the stage-3 state variable is $s_3 = 1000$ that are in (6.23) and (6.24).

It is also possible that there are multiple feasible decisions that give the same minimum (or in the case of a maximization problem, maximum) objective value in Lines 11 and 12. If so, such ties can be broken arbitrarily and the decision policy, *i.e.*, $d_t^*(s_t)$, can be set equal to any of the optimal decisions that give the lowest objective-function value.

We now outline the second part of the Dynamic Programming Algorithm, which we term the Forward Recursion. After the Backward Recursion is used to determine the decision policies in each stage, the Forward Recursion works forward through the problem stages to reconstruct an optimal sequence of decisions to make at each stage. In the first stage (*i.e.*, $t = 1$) we are given the starting state-variable value. This is fixed in Line 4. Otherwise, in the other problem stages, we determine the state-variable value using the state-transition function and the state- and decision-variable values from the previous stage, as shown in Line 6. The optimal decision in the current stage is then determined in Line 8 using the decision policy (which is determined in the Backward Recursion part of the Dynamic Programming Algorithm) and the current stage's state-variable value (which is determined in either Line 4 or 6).

Dynamic Programming Algorithm: Forward Recursion

```
 1: procedure FORWARD RECURSION
 2:     for t ← 1, ..., T do                      ▷ Step forward through problem stages
 3:         if t = 1 then
 4:             Fix s₁                            ▷ Fix state variable value
 5:         else
 6:             sₜ ← fₜ₋₁(sₜ₋₁, dₜ₋₁)             ▷ Fix state variable value
 7:         end if
 8:         dₜ ← dₜ*(sₜ)                          ▷ Determine optimal decision
 9:     end for
10: end procedure
```

To understand how the Forward Recursion works, recall how we determine an optimal sequence of decisions after solving the Investment Problem in Section 6.4.1. We begin in stage 1 and determine that $s_1 = 7000$, based on the information that is given in the problem description. This is akin to fixing $s_1 = 7000$ in Line 4 of the Forward Recursion. We next use the decision-policy information that is recorded in Table 6.12 to determine that if $s_1 = 7000$, then $d_1 = d_1^*(7000) = 0$ is an optimal stage-1 decision. This is Line 8 of the Forward Recursion.

We next move forward to stage 2. The first step in stage 2 is to determine what the stage-2 state-variable value will be. We use the state-transition function in (6.30) to determine that if $s_1 = 7000$ and $d_1 = 0$ then $s_2 = 7000$. This calculation is Line 6 of the Forward Recursion. Knowing that $s_2 = 7000$, we next use the decision-policy information that is recorded in Table 6.11 to determine that if $s_2 = 7000$ the optimal stage-2 decision is $d_2 = d_2^*(7000) = 6000$. This determination of the optimal stage-2 decision is Line 8 of the Forward Recursion.

We finally move to stage 3 and use the state-transition function, which is in (6.31), to determine that if $s_2 = 7000$ and $d_2 = 6000$ then $s_3 = 1000$. We then use the decision-policy information that is in Table 6.10 to determine that if $s_3 = 1000$ then $d_3 = d_3^*(1000) = 1000$ is an optimal decision. These two determinations are Lines 6 and 8 of the Forward Recursion.

6.4.3   *Optimality of the Dynamic Programming Algorithm*

The Dynamic Programming Algorithm provides an efficient way to decompose and solve an optimization problem. An important question, however, is whether the algorithm provides an optimal solution to the overall problem. Here we provide an intuitive explanation of why the Dynamic Programming Algorithm is guaranteed to yield an optimal solution [1, 2]. We demonstrate this result for any generic dynamic optimization problem, but show how the derivations involving a generic problem can be applied to the Investment Problem that is introduced in Section 6.3.4. We then use the example of the Production-Planning Problem from Section 6.1 to provide a more intuitive explanation of the optimality of the Dynamic Programming Algorithm.

We begin by noting that a generic dynamic optimization problem that is a minimization can be written as:

$$\min_{d_1,\ldots,d_T,s_2,\ldots,s_T} \sum_{t=1}^{T} c_t(s_t, d_t) \tag{6.32}$$

$$\text{s.t.} \quad d_t \in \mathscr{D}_t(s_t), \qquad\qquad \forall\, t = 1, \ldots, T; \tag{6.33}$$

$$s_{t+1} = f_t(s_t, d_t), \qquad\qquad \forall\, t = 1, \ldots, T - 1. \tag{6.34}$$

Objective function (6.32) minimizes the sum of the objective-contribution functions from the T stages. In a dynamic optimization problem the overall objective concerns the total objective-function value, which is obtained by adding the objective-contribution functions from each individual stage. The explicit constraints in each stage, which restrict what decisions are feasible in each stage, are given by (6.33). Equation (6.34) represents the dynamic of the problem, by specifying how the state variables evolve from one stage to the next.

The Investment Problem can be written in this form as:

$$\max_{d_1,d_2,d_3,s_2,s_3} \sum_{t=1}^{3} c_t(s_t, d_t)$$

$$\begin{aligned}
\text{s.t.} \quad & d_1 \in \{0, 1000, \dots\} \\
& d_1 \leq s_1 \\
& d_2 \in \{0, 1000, \dots\} \\
& d_2 \leq s_2 \\
& d_3 \in \{0, 1000, \dots\} \\
& d_3 \leq s_3 \\
& s_2 = s_1 - d_1 \\
& s_3 = s_2 - d_2,
\end{aligned}$$

where the objective function is written as a maximization, because the investor is concerned with maximizing return on investment.

By reorganizing the terms in objective function (6.32) and the constraints, we can rewrite generic optimization problem (6.32)–(6.34) as:

$$\begin{aligned}
\min_{d,s} \; c_1(s_1, d_1) \quad & + c_2(s_2, d_2) \quad && + \cdots + c_T(s_T, d_T) \\
\text{s.t.} \; d_1 \in \mathscr{D}_1(s_1) & \\
s_2 = f_1(s_1, d_1) & \\
& d_2 \in \mathscr{D}_2(s_2) \\
& s_3 = f_2(s_2, d_2) \\
& && \ddots \\
& && d_T \in \mathscr{D}_T(s_T).
\end{aligned}$$

All that is done here is breaking objective function (6.32) into the T components that are summed together. Moreover, the stage-t constraint and state-transition functions are stacked underneath the stage-t objective-contribution function. The Investment Problem can be written in this form as:

$$\max_{d,s} \ c_1(s_1, d_1) \quad\quad + c_2(s_2, d_2) \quad\quad + c_3(s_3, d_3)$$

$$\text{s.t.} \ \ d_1 \in \{0, 1000, \dots\}$$
$$d_1 \le s_1$$
$$s_2 = s_1 - d_1$$

$$d_2 \in \{0, 1000, \dots\}$$
$$d_2 \le s_2$$
$$s_3 = s_2 - d_2$$

$$d_3 \in \{0, 1000, \dots\}$$
$$d_3 \le s_3.$$

The key insight behind the Dynamic Programming Algorithm is that if the total sum of the T objective-contribution functions is minimized:

$$\min \ [c_1(s_1, d_1) + c_2(s_2, d_2) + \cdots + c_T(s_T, d_T)],$$

this is equivalent to minimizing from each stage forward:

$$\min [c_1(s_1, d_1) + \min [c_2(s_2, d_2) + \min [\cdots + \min [c_T(s_T, d_T)]]]].$$

Substituting this observation into the generic dynamic optimization problem gives:

$$\min_{d_1, s_2} c_1(s_1, d_1) + \cdots + \left[\begin{array}{l} \min_{d_{T-1}, s_T} \ c_{T-1}(s_{T-1}, d_{T-1}) + \left[\begin{array}{l} \min_{d_T} c_T(s_T, d_T) \\ \text{s.t.} \ \ d_T \in \mathscr{D}_T(s_T) \end{array} \right] \\ \quad \text{s.t.} \quad d_{T-1} \in \mathscr{D}_{T-1}(s_{T-1}) \\ \qquad\qquad s_T = f_{T-1}(s_{T-1}, d_{T-1}) \end{array} \right]$$
$$\text{s.t.} \ \ d_1 \in \mathscr{D}_1(s_1)$$
$$\qquad s_2 = f_1(s_1, d_1),$$

$$(6.35)$$

Note that the first 'min' operator has d_1 and s_2 (as opposed to s_1) listed underneath it. The reason for this is that in making our stage-1 decision, d_1, we are also implicitly determining our stage-2 state, s_2. This is because s_2 is impacted by d_1 through the stage-1 state-transition function. This same logic explains why d_{T-1} and s_T are listed underneath the second-to-last 'min' operator. No state variable is listed underneath the last 'min' operator, because after the final decision is made in stage T, there are no subsequent stages or state variables that are affected by d_T.

Substituting this same observation into the Investment Problem (and noting that the same result applies to maximizing the sum of the T objective-contribution functions) gives:

$$\max_{d_1,s_2} c_1(s_1, d_1) + \left[\begin{array}{l} \max_{d_2,s_3} c_2(s_2, d_2) + \left[\begin{array}{l} \max_{d_3} c_3(s_3, d_3) \\ \text{s.t.}\ \ d_3 \in \{0, 1000, \dots\} \\ \quad\ \ d_3 \leq s_3 \end{array} \right] \\ \text{s.t.}\ \ d_2 \in \{0, 1000, \dots\} \\ \quad\ \ d_2 \leq s_2 \\ \quad\ \ s_3 = s_2 - d_2 \end{array} \right]$$

$$\text{s.t.}\ \ d_1 \in \{0, 1000, \dots\}$$
$$\quad\ \ d_1 \leq s_1$$
$$\quad\ \ s_2 = s_1 - d_1. \tag{6.36}$$

Now, examine the term in the brackets in the upper right-hand corner of (6.35), which is:

$$\min_{d_T} c_T(s_T, d_T) \tag{6.37}$$

$$\text{s.t.}\ \ d_T \in \mathscr{D}_T(s_T). \tag{6.38}$$

The optimized objective-function value of this term is simply $J_T(s_T)$. That is because $J_T(s_T)$ is defined as the optimal objective-function value from stage-T forward, as a function of s_T. That is, it tells us what our total minimized objective-function value from stage T forward is if we enter stage T in stage s_T.

We can also see that $J_T(s_T)$ is the same as optimization problem (6.37)–(6.38) by examining Lines 4, 6, and 11 of the Dynamic Programming Algorithm: Backward Recursion, which is given in Section 6.4.2. In the Backward Recursion we compute $J_T(s_T)$ by first enumerating, in Line 4, all of the feasible decisions if the stage-T state is s_T. This enumeration corresponds to the $d_T \in \mathscr{D}_T(s_T)$ in constraint (6.38). We then compute the objective-contribution-function value for each feasible decision in Line 6 (which corresponds to the $c_T(s_T, d_T)$ in objective function (6.37)) and pick the best one in Line 11 (which corresponds to the 'min' operator in (6.37)).

Based on this observation, we can simplify (6.35) to:

$$\min_{d_1,s_2} c_1(s_1, d_1) + \cdots + \left[\begin{array}{l} \min_{d_{T-1},s_T} c_{T-1}(s_{T-1}, d_{T-1}) + J_T(s_T) \\ \text{s.t.}\ \ d_{T-1} \in \mathscr{D}_{T-1}(s_{T-1}) \\ \quad\ \ s_T = f_{T-1}(s_{T-1}, d_{T-1}) \end{array} \right]$$

$$\text{s.t.}\ \ d_1 \in \mathscr{D}_1(s_1)$$
$$\quad\ \ s_2 - f_1(s_1, d_1). \tag{6.39}$$

The optimization problem in (6.39) is the same as that in (6.35), except that the:

$$\min_{d_T} c_T(s_T, d_T)$$

$$\text{s.t.}\ \ d_T \in \mathscr{D}_T(s_T),$$

in the upper right-hand corner is replaced with $J_T(s_T)$.

We can apply this same sequence of arguments to the Investment Problem, by first examining the term in the brackets in the upper right-hand corner of (6.36), which is:

$$\max_{d_3} c_3(s_3, d_3) \tag{6.40}$$

$$\text{s.t. } d_3 \in \{0, 1000, \dots\} \tag{6.41}$$

$$d_3 \leq s_3. \tag{6.42}$$

What we are doing in (6.40)–(6.42) is examining each possible value that the stage-3 state variable can take. For each one we select the feasible decision that maximizes the stage-3 objective-contribution function. This is exactly how we examine stage 3 when solving the Investment Problem in Section 6.4.1 and how we find the decision-policy and value-function information that are recorded in Table 6.10. From this observation regarding (6.40)–(6.42) we know that the optimized value of objective function (6.40) is exactly equal to $J_3(s_3)$. Making this substitution into (6.36) yields:

$$\max_{d_1, s_1} c_1(s_1, d_1) + \begin{bmatrix} \max\limits_{d_2, s_3} c_2(s_2, d_2) + J_3(s_3) \\ \text{s.t. } d_2 \in \{0, 1000, \dots\} \\ d_2 \leq s_2 \\ s_3 = s_2 - d_2 \end{bmatrix} \tag{6.43}$$
$$\text{s.t. } d_1 \in \{0, 1000, \dots\}$$
$$d_1 \leq s_1$$
$$s_2 = s_1 - d_1.$$

Now, examine the term in the brackets in the upper right-hand corner of (6.39), which is:

$$\min_{d_{T-1}, s_T} c_{T-1}(s_{T-1}, d_{T-1}) + J_T(s_T)$$
$$\text{s.t. } d_{T-1} \in \mathscr{D}_{T-1}(s_{T-1})$$
$$s_T = f_{T-1}(s_{T-1}, d_{T-1}).$$

Substituting the equality constraint, $s_T = f_{T-1}(s_{T-1}, d_{T-1})$, into the objective function, this becomes:

$$\min_{d_{T-1}} c_{T-1}(s_{T-1}, d_{T-1}) + J_T(f_{T-1}(s_{T-1}, d_{T-1})) \tag{6.44}$$

$$\text{s.t. } d_{T-1} \in \mathscr{D}_{T-1}(s_{T-1}). \tag{6.45}$$

We can show that the optimized objective-function value of (6.44) is $J_{T-1}(s_{T-1})$, by following the same analysis that is used above to show that the optimized objective-function value of (6.37) is $J_T(s_T)$. Moreover, problem (6.44)–(6.45) exactly corresponds to the step in Line 11 of the Backward Recursion when examining stage $(T-1)$. Substituting this observation into (6.39) gives:

$$\min_{d_1, s_2} c_1(s_1, d_1) + \cdots + \begin{bmatrix} \min_{d_{T-2}, s_{T-1}} & c_{T-2}(s_{T-2}, d_{T-2}) + J_{T-1}(s_{T-1}) \\ \text{s.t.} & d_{T-2} \in \mathscr{D}_{T-2}(s_{T-2}) \\ & s_{T-1} = f_{T-2}(s_{T-2}, d_{T-2}) \end{bmatrix} \tag{6.46}$$
$$\text{s.t.} \quad d_1 \in \mathscr{D}_1(s_1)$$
$$s_2 = f_1(s_1, d_1).$$

We can also apply these same logical steps to examine (6.43) and conclude that the term in the brackets in the upper right-hand corner is equal to $J_2(s_2)$.

At this point, we can recursively apply the same set of arguments over and over again. Doing so shows that applying the steps of the Backward Recursion solves the generic dynamic optimization problem that is given by (6.32)–(6.34). However, the Dynamic Programming Algorithm solves the problem by looking at it one stage at a time.

The crux of the dynamic programming algorithm is as follows: suppose that the sequence of decisions, $d_1^*, d_2^*, \ldots, d_T^*$, is an optimal solution of a dynamic optimization problem. Suppose also that we follow this sequence of decisions up to some intermediate stage, t, where $1 < t < T$. We now, in stage t, ask the question, 'do we want to continue following the sequence of decisions, $d_t^*, d_{t+1}^*, \ldots, d_T^*$ for the remainder of the problem?' If we are looking from stage t forward, we do not care about anything that happens in stages 1 through $(t-1)$, only about the remaining stages of the problem. This is because once we are in stage t, the decisions that are made in the previous stages are fixed and they are no longer under our control [1, 2].

Suppose that there is a preferable set of decisions starting from stage t, which we denote $\hat{d}_t, \hat{d}_{t+1}, \ldots, \hat{d}_T$. Then, the original sequence of decisions, $d_1^*, d_2^*, \ldots, d_T^*$, cannot be optimal. That is because if we follow the decisions, $d_1^*, d_2^*, \ldots, d_{t-1}^*$, through stage $(t-1)$ and then, $\hat{d}_t, \hat{d}_{t+1}, \ldots, \hat{d}_T$, for the remaining stages, the overall objective-function value must be improved. The Dynamic Programming Algorithm finds a sequence of decisions in exactly this way. It works backwards through the stages of the problem and at each intermediate stage asks the question, 'what is the best sequence of decisions from this point forward?' By doing this, the resulting set of decisions is optimal over the full set of problem stages. This observation regarding the Dynamic Programming Algorithm is exactly Bellman's optimality principle [1, 2], which is the key concept underlying dynamic optimization.

To provide a more intuitive explanation of the optimality of the dynamic programming algorithm, consider the Production-Planning Problem, which is introduced in Section 6.1. Suppose that a sequence of production processes, $d_1^*, d_2^*, \ldots, d_5^*$, is optimal. Now suppose that we are in the third stage of the problem. If there is a better alternative than d_3^*, d_4^*, and d_5^* from stage 3 forward, then it cannot be the case that the sequence, $d_1^*, d_2^*, \ldots, d_5^*$, is optimal over the entire problem horizon.

This is, however, the exact way in which d_3^*, d_4^*, and d_5^* are found when applying the Dynamic Programming Algorithm to the Production-Planning Problem in Section 6.1.2. We first examine stage 5 and determine the best decision in stage 5 for each possible state-variable value. We next examine stage 4 and determine the best decision in stage 4 for each possible state-variable value, taking into account what subsequently occurs in stage 5. We do the same when we next examine stage 3.

6.4.4 Computational Complexity of the Dynamic Programming Algorithm

The computational benefits of using the Dynamic Programming Algorithm as opposed to brute force in the particular case of the Production-Planning Problem are discussed in Section 6.1. We close this section by generalizing this result to any dynamic optimization problem. More specifically, we estimate the number of calculations that it would take to solve a generic dynamic optimization problem using brute force and the Dynamic Programming Algorithm. For ease of analysis, consider a generic dynamic optimization problem with T stages. Moreover, suppose that at each stage there are $|S|$ possible state-variable values and $|D|$ feasible decision-variable values. Note, however, that for many problems, the number of states differs between stages. Similarly, the number of feasible decisions typically varies as a function of the state-variable value. These features can result in the specific number of calculations involved in solving a particular problem differing from the numbers that we present here. Nevertheless, we can draw insights into how much of a computational-cost savings the Dynamic Programming Algorithm provides.

We begin by determining how much computational effort must be expended to solve this generic problem by brute force. The brute-force approach explicitly enumerates all of the sequences of decisions and computes the total objective-function value for each. Because there are $|D|$ possible decisions at each stage, there are $|D|^T$ possible sequences of decisions. Moreover, to examine each one, the objective-contribution function must be evaluated T times (once for each stage). Thus, the total number of times that the objective-contribution function must be evaluated when using brute force is:

$$T \cdot |D|^T.$$

Now consider the Dynamic Programming Algorithm. In each stage there are $|S|$ possible state-variable values and $|D|$ feasible decision-variable values for each possible state. We must evaluate the objective-contribution function for each state-variable/decision-variable pair. Thus, we conduct $|S| \cdot |D|$ objective-contribution-function evaluations in each stage. Because there are T stages, we have a total of:

$$T \cdot |S| \cdot |D|,$$

objective-contribution-function evaluations.

To see the implications of this, consider the Energy-Storage Problem that is introduced in Section 6.3.1. There are 24 stages in this problem and at most 3 feasible decisions in each stage (depending on what the starting water level of the upper reservoir is in each stage, there may only be 2 feasible decisions in some stages after some sequence of actions). Thus, the brute-force method takes on the order of between:

$$24 \cdot 2^{24} \approx 4.03 \times 10^8,$$

and:

$$24 \cdot 3^{24} \approx 6.77 \times 10^{12},$$

objective-contribution-function evaluations. If the Dynamic Programming Algorithm is applied, there are five possible states (*i.e.*, the upper reservoir can have a starting water level in each stage between 0 and 4). Thus, the Dynamic Programming Algorithm takes on the order of:

$$24 \cdot 5 \cdot 3 = 360,$$

objective-contribution-function evaluations. It could take up to about 215 years to solve this problem by brute force if we could do 1000 objective-contribution-function evaluations per second. Conversely, the Dynamic Programming Algorithm would take less than one second to solve the problem if the objective-contribution function can be evaluated 1000 times per second.

Despite its efficiency compared to brute force, the Dynamic Programming Algorithm does run into tractability challenges once the problem size gets sufficiently large. In such cases, approximation techniques, which are beyond the scope of this book, must be implemented [3, 4, 7].

6.5 Final Remarks

Our analysis of dynamic optimization problems requires us to make a number of assumptions regarding the underlying problem structure. Some of the assumptions are required to apply the Dynamic Programming Algorithm, while others can be relaxed (with an associated increase in the complexity of the problem and in solving it).

The two assumptions that an optimization problem *must* satisfy to use dynamic optimization techniques are **overlapping subproblems** and **optimal substructure**. We do not formally define these properties. Rather, these assumptions are satisfied so long as we can formulate a problem to have the structure that is detailed in Section 6.2. Specifically, the assumption that the overall objective-function of the problem can be written as objective-contribution functions corresponding to each stage that are added to one another is crucial for application of the Dynamic Programming Algorithm. The other critically important feature is that the state-transition functions be written as single-stage transitions.

The other assumption underlying the Dynamic Programming Algorithm is that the problem can be decomposed into a finite number of stages, each of which has a finite number of possible state-variable values and feasible decision-variable values. If these assumptions are not satisfied, we cannot solve the problem using the enumeration technique that is used in the **for** loops in the Backward Recursion. These assumptions can be relaxed, but the analysis becomes considerably more difficult. Relaxing the assumption that the problem has a finite number of discrete decision stages requires the use of optimal control theory [6]. Such problems are considerably more difficult, because the state-transition function becomes a differential equation. Bertsekas [3, 4] discusses such techniques in detail.

Relaxing the assumption that there are a finite number of states or decisions also complicates the analysis. When doing so, approximation techniques must almost always be used. Solving a problem with continuous state or decision variables requires finding an explicit decision-policy and value function. This is difficult to do in all but the simplest problems. Otherwise, one can usually find an approximate solution to a problem with continuous state or decision variables by using a discrete approximation. For instance, suppose that a problem allows the state variable to take any value between 0 and 20. One may approximate the problem by restricting the state variable to take values of 0, 0.5, 1, . . . , 20 and apply the Dynamic Programming Algorithm to the approximated problem. The quality of the solution that is found is sensitive to how the finite state variable values are chosen. Interested readers are referred to more advanced texts that discuss these types of problems in further detail [2–5, 7, 8].

This chapter introduces one particular way to solve dynamic optimization problems. There are alternative techniques, for instance working forward through the problem stages, which can be more efficient for certain classes of problems. Interested readers are referred to other texts [3, 4, 7] for discussions of these methods.

6.6 Exercises

6.1 A company needs to staff a maintenance center over the next 10 days. On each day, the company can assign employees to work for either a single day, for two consecutive days, or for three consecutive days. Employees that are assigned to work for a single day are paid $100 for the day, while employees assigned to work for two consecutive days are paid $80 per day. Employees that are hired to work for three consecutive days are paid $70 per day. The company can assign all three types of employees on each day, except on the last two days. No one can be hired for three consecutive days on day nine. Only employees hired to work for one day can be hired on day 10. The company needs to have a minimum number of employees on hand on each day, which are given in Table 6.13. It can also have at most a total of 15 employees working at any given time. The company begins the first day with no employees carried over from the previous days. Formulate a dynamic optimization problem to determine the least-cost sequence of staffing decisions.

6.2 Claire needs to plan maintenance on her car over the next 12 months. At the beginning of each month, Claire must decide to either conduct maintenance or not. If she does, she incurs an immediate cost, which depends on how recently she has had the car maintained. After making her maintenance decision, Claire next incurs a monthly cost of operating her car. This operating cost depends on how recently the car has been maintained. Table 6.14 lists the monthly maintenance and operating costs, as a function of how recently the car has been maintained. If at any point the it has been four months since the most recent maintenance, Claire must have the car maintained immediately. As of today, Claire's car has most recently been maintained three months ago. Formulate a dynamic optimization problem to determine the least-cost sequence of maintenance decisions.

Table 6.13 Minimum number of employees that must be hired on each day in Exercise 6.1

Day	Minimum Employees
1	5
2	4
3	9
4	6
5	7
6	2
7	5
8	3
9	6
10	6

Table 6.14 Maintenance and operating costs in Exercise 6.2

Time Since Most Recent Maintenance as of the Beginning of the Month	Maintenance Cost	Operating Cost
0 Month	n/a	490
1 Month	40	500
2 Month	50	515
3 Month	55	535
4 Month	80	n/a

6.3 Use the Dynamic Programming Algorithm to solve the dynamic optimization problem that is formulated in Exercise 6.1.

6.4 Use the Dynamic Programming Algorithm to solve the dynamic optimization problem that is formulated in Exercise 6.2.

6.5 Consider the staffing problem, which is introduced in Exercise 6.1. Suppose that the company has the option of paying $95 to carry an employee over from the previous day to the first day. Would the company be willing to do this? Can you answer this question without re-solving the dynamic optimization problem that is formulated in Exercise 6.1 from scratch?

References

1. Bellman R (1954) The theory of dynamic programming. Bull Am Math Soc 60:503–515
2. Bellman R (2003) Dynamic programming, Reprint edn. Dover Publications, Mineola
3. Bertsekas DP (2012) Dynamic programming and optimal control, 4th edn. Athena Scientific, Belmont
4. Bertsekas DP (2012) Dynamic programming and optimal control, 4th edn. Athena Scientific, Belmont
5. Denardo E (2003) Dynamic programming: models and applications. Dover Publications, Mineola
6. Kirk DE (2004) Optimal control theory: an introduction. Dover Publications, Mineola
7. Powell WB (2011) Approximate dynamic programming: solving the curses of dimensionality, 2nd edn. Wiley-Interscience, Hoboken
8. Sniedovich M (2010) Dynamic programming: foundations and principles, 2nd edn. CRC Press, Boca Raton

Appendix A
Taylor Approximations and Definite Matrices

Taylor approximations provide an easy way to approximate a function as a polynomial, using the derivatives of the function. We know, from elementary calculus [1], that a single-variable function, $f(x)$, can be represented exactly by an infinite series. More specifically, suppose that the function $f(\cdot)$ is infinitely differentiable and that we know the value of the function and its derivatives at some point, x. We can represent the value of the function at $(x + a)$ as:

$$f(x + a) = \sum_{n=0}^{+\infty} \frac{1}{n!} f^{(n)}(x) a^n, \tag{A.1}$$

where $f^{(n)}(x)$ denotes the nth derivative of $f(\cdot)$ at x and, by convention, $f^{(0)}(x) = f(x)$ (*i.e.*, the zeroth derivative of the function is simply the function itself).

We also have that if we know the value of the function and its derivatives for some x, we can approximate the value of the function at $(x + a)$ using a finite series. These finite series are known as **Taylor approximations**. For instance, the first-order Taylor approximation of $f(x + a)$ is:

$$f(x + a) \approx f(x) + af'(x),$$

and the second-order Taylor approximation is:

$$f(x + a) \approx f(x) + af'(x) + \frac{1}{2} a^2 f''(x)$$

These Taylor approximations simply cutoff the infinite series in (A.1) after some finite number of terms. Of course including more terms in the Taylor approximation provides a better approximation. For instance, a second-order Taylor approximation tends to have a smaller error in approximating $f(x + a)$ than a first-order Taylor approximation does.

There is an exact analogue of the infinite series in (A.1) and the associated Taylor approximations for multivariable functions. For our purposes, we only require

© Springer International Publishing AG 2017
R. Sioshansi and A.J. Conejo, *Optimization in Engineering*,
Springer Optimization and Its Applications 120, DOI 10.1007/978-3-319-56769-3

first- and second-order Taylor approximations. Thus, we only introduce those two concepts here.

Suppose:

$$f(x) = f(x_1, x_2, \ldots, x_n),$$

is a multivariable function that depends on n variables. Suppose also that the function is once continuously differentiable and that we know the value of the function and its gradient at some point, x. The **first-order Taylor approximation** of $f(x + d)$ around the point, x, is:

$$f(x + d) \approx f(x) + d^\top \nabla f(x).$$

Suppose:

$$f(x) = f(x_1, x_2, \ldots, x_n),$$

is a multivariable function that depends on n variables. Suppose also that the function is twice continuously differentiable and that we know the value of the function, its gradient, and its Hessian at some point, x. The **second-order Taylor approximation** of $f(x + d)$ around the point, x, is:

$$f(x + d) \approx f(x) + d^\top \nabla f(x) + \frac{1}{2} d^\top \nabla^2 f(x) d,$$

Example A.1 Consider the function:

$$f(x) = e^{x_1 + x_2} + x_3^3.$$

Using the point:

$$x = \begin{pmatrix} 1 \\ 0 \\ 1 \end{pmatrix},$$

we compute the first- and second-order Taylor approximations of $f(x + d)$ by first finding the gradient of $f(x)$, which is:

$$\nabla f(x) = \begin{pmatrix} e^{x_1 + x_2} \\ e^{x_1 + x_2} \\ 3x_3^2 \end{pmatrix},$$

and the Hessian of $f(x)$, which is:

$$\nabla^2 f(x) = \begin{bmatrix} e^{x_1+x_2} & e^{x_1+x_2} & 0 \\ e^{x_1+x_2} & e^{x_1+x_2} & 0 \\ 0 & 0 & 6x_3 \end{bmatrix}.$$

Substituting in the given value of x gives:

$$f(x) = e + 1,$$

$$\nabla f(x) = \begin{pmatrix} e \\ e \\ 3 \end{pmatrix},$$

and:

$$\nabla^2 f(x) = \begin{bmatrix} e & e & 0 \\ e & e & 0 \\ 0 & 0 & 6 \end{bmatrix}.$$

The first-order Taylor approximation is:

$$f(x + d) \approx f(x) + d^\top \nabla f(x)$$

$$= e + 1 + \begin{pmatrix} d_1 & d_2 & d_3 \end{pmatrix} \begin{pmatrix} e \\ e \\ 3 \end{pmatrix}$$

$$= e + 1 + d_1 e + d_2 e + 3d_3$$

$$= e \cdot (1 + d_1 + d_2) + 3d_3 + 1,$$

and the second-order Taylor approximation is:

$$f(x + d) \approx f(x) + d^\top \nabla f(x) + \frac{1}{2} d^\top \nabla^2 f(x) d$$

$$= e \cdot (1 + d_1 + d_2) + 3d_3 + 1 + \frac{1}{2} \begin{pmatrix} d_1 & d_2 & d_3 \end{pmatrix} \begin{bmatrix} e & e & 0 \\ e & e & 0 \\ 0 & 0 & 6 \end{bmatrix} \begin{pmatrix} d_1 \\ d_2 \\ d_3 \end{pmatrix}$$

$$= e \cdot (1 + d_1 + d_2) + 3d_3 + 1 + \frac{1}{2} [e \cdot (d_1 + d_2)^2 + 6d_3^2]. \tag{A.2}$$

We can substitute different values for the vector, d, and obtain an approximation of $f(x + d)$ from these two expressions. The first-order Taylor approximation is a linear function of d while the second-order Taylor approximation is quadratic.

$\square$

A.1   Bounds on Quadratic Forms

The second-order Taylor approximation:

$$f(x + d) \approx f(x) + d^\top \nabla f(x) + \frac{1}{2} d^\top \nabla^2 f(x) d,$$

has what is referred to as a **quadratic form** as its third term. This is because when the third term is multiplied out, we obtain a quadratic terms involving d. This is seen, for instance, in the second-order Taylor approximation that is obtained in Example A.1 (specifically, the $[e \cdot (d_1 + d_2)^2 + 6d_3^2]$ term in (A.2)).

We are often concerned with placing bounds on such quadratic forms when analyzing nonlinear optimization problems. Before doing so, we first argue that whenever examining quadratic forms, we can assume that the matrix in the center of the product is symmetric. To understand why, suppose that we have the quadratic form:

$$d^\top A d,$$

where A is not a symmetric matrix. If we define:

$$\tilde{A} = \frac{1}{2} \left(A + A^\top \right),$$

we can first show that $\tilde{A}$ is symmetric. This is because the transpose operator distributes over a sum. Thus, we can write the transpose of $\tilde{A}$ as:

$$\tilde{A}^\top = \frac{1}{2} \left(A + A^\top \right)^\top = \frac{1}{2} \left(A^\top + A \right) = \frac{1}{2} \left(A + A^\top \right) = \tilde{A},$$

meaning that $\tilde{A}$ is symmetric. Next, we can show that for any vector d the quadratic forms, $d^\top A d$ and $d^\top \tilde{A} d$, are equal. To see this, we explicitly write out and simplify the second quadratic form as:

$$d^\top \tilde{A} d = d^\top \frac{1}{2} \left(A + A^\top \right) d$$

$$= \frac{1}{2} \left(d^\top A d + d^\top A^\top d \right). \tag{A.3}$$

Note, however, that $d^\top A^\top d$ is a scalar, and as such is symmetric. Thus, we have that:

$$d^\top A^\top d = \left(d^\top A^\top d \right)^\top = d^\top A d,$$

where the second equality comes from distributing the transpose across the product, $d^\top A^\top d$. Substituting $d^\top A^\top d = d^\top A d$ into (A.3) gives:

$$d^\top \tilde{A} d = \frac{1}{2} \left(d^\top A d + d^\top A^\top d \right) = \frac{1}{2} \left(d^\top A d + d^\top A d \right) = d^\top A d,$$

which shows that the quadratic forms involving A and $\tilde{A}$ are equal.

As a result of these two properties of quadratic forms, we know that if we are given a quadratic form:

$$d^\top A d,$$

where the A-matrix is not symmetric, we can replace this with the equivalent quadratic form:

$$\frac{1}{2} d^\top \left(A + A^\top \right) d,$$

which has a symmetric matrix in the middle. Thus, we henceforth always assume that we have quadratic forms with symmetric matrices in the middle.

The following result provides a bound on the magnitude of a quadratic form in terms of the eigenvalues of the matrix in the product.

Quadratic-Form Bound: Let A be an $n \times n$ square matrix. Let $\lambda_1, \lambda_2, \ldots, \lambda_n$ be the eigenvalues of A and suppose that they are labeled such that:

$$\lambda_1 \leq \lambda_2 \leq \cdots \leq \lambda_n.$$

For any vector, d, we have that:

$$\lambda_1 ||d|| \leq d^\top A d \leq \lambda_n ||d||.$$

Example A.2 Consider the function:

$$f(x) = e^{x_1 + x_2} + x_3^3,$$

which is introduced in Example A.1. We know from Example A.1 that if:

$$x = \begin{pmatrix} 1 \\ 0 \\ 1 \end{pmatrix},$$

then:

$$\nabla^2 f(x) = \begin{bmatrix} e & e & 0 \\ e & e & 0 \\ 0 & 0 & 6 \end{bmatrix}.$$

The eigenvalues of this Hessian matrix are $\lambda_1 = 0$, $\lambda_2 = 2e$, and $\lambda_3 = 6$. This means that for any vector, d, we have:

$$0||d|| \le d^\top \nabla^2 f(x)d \le 6||d||,$$

which simplifies to:

$$0 \le d^\top \nabla^2 f(x)d \le 6(d_1^2 + d_2^2 + d_3^2).$$

$\square$

A.2   Definite Matrices

The Quadratic-Form Bound, which is introduced in Section A.1, provides a way to bound a quadratic form in terms of the eigenvalues of the matrix in the product. This result motivates our discussion here of a special class of matrices, known as **definite matrices**. A definite matrix is one for which we can guarantee that any quadratic form involving the matrix has a certain sign. We focus on our attention on two special types of definite matrices—positive-definite and positive-semidefinite matrices. There are two analogous forms of definite matrices—negative-definite and negative-semidefinite matrices, which we do not discuss here. This is because negative-definite and negative-semidefinite matrices are not used in this text.

Let A be an $n \times n$ square matrix. The matrix, A, is said to be **positive definite** if for any nonzero vector, d, we have that:

$$d^\top A d > 0.$$

Let A be an $n \times n$ square matrix. The matrix, A, is said to be **positive semidefinite** if for any vector, d, we have that:

$$d^\top A d \ge 0.$$

One can determine whether a matrix is positive definite or positive semidefinite by using the definition directly. This is demonstrated in the following example.

Example A.3 Consider the matrix:

$$A = \begin{bmatrix} e & e & 0 \\ e & e & 0 \\ 0 & 0 & 6 \end{bmatrix}.$$

For any vector, d, we have that:

$$d^\top A d = e \cdot (d_1 + d_2)^2 + 6d_3^2.$$

Clearly, if we let d be any nonzero vector the quantity above is non-negative, because it is the sum of two squared terms that are multiplied by positive coefficients. Thus, we can conclude that A is positive semidefinite. However, A is not positive definite. To see why not, take the case of:

$$d = \begin{pmatrix} 1 \\ -1 \\ 0 \end{pmatrix}.$$

This vector is nonzero. However, we have that $d^\top A d = 0$ for this particular value of d. Because $d^\top A d$ is not strictly positive for all nonzero choices of d, A is not positive definite.

$\square$

While one can determine whether a matrix is definite using the definition directly, this is often cumbersome to do. For this reason, we typically rely on one of the following two tests, which can be used to determine if a matrix is definite.

A.2.1 Eigenvalue Test for Definite Matrices

The Eigenvalue Test for definite matrices follows directly from the Quadratic-Form Bound.

Eigenvalue Test for Positive-Definite Matrices: Let A be an $n \times n$ square matrix. The matrix, A, is positive definite if and only if all of its eigenvalues are strictly positive.

Eigenvalue Test for Positive-Semidefinite Matrices: Let A be an $n \times n$ square matrix. The matrix, A, is positive semidefinite if and only if all of its eigenvalues are non-negative.

We demonstrate the use of the Eigenvalue Test with the following example.

Example A.4 Consider the matrix:

$$A = \begin{bmatrix} e & e & 0 \\ e & e & 0 \\ 0 & 0 & 6 \end{bmatrix}.$$

The eigenvalues of this matrix are 0, $2e$, and 6. Because these eigenvalues are all non-negative, we know that A is positive semidefinite. However, because one of the eigenvalues is equal to zero (and is, thus, not strictly positive) the matrix is not positive definite. This is consistent with the analysis that is conducted in Example A.3, in which the definitions of definite matrices are directly used to show that A is positive semidefinite but not positive definite.

$\square$

A.2.2 *Principal-Minor Test for Definite Matrices*

The Principal-Minor Test is an alternative way to determine is a matrix is definite or not. The Principal-Minor Test can be more tedious than the Eigenvalue Test, because it requires more calculations. However, the Principal-Minor Test often has lower computational cost because calculating the eigenvalues of a matrix can require solving for the roots of a polynomial. The Principal-Minor Test can only be applied to symmetric matrices, whereas the Eigenvalue Test applies to non-symmetric square matrices. However, when examining quadratic forms, we always restrict our attention to symmetric matrices. Thus, this distinction in the applicability of the two tests is unimportant to us.

Before introducing the Principal-Minor Test, we first define what the principal minors of a matrix are. We also define a related concept, the leading principal minors of a matrix.

Let:

$$A = \begin{bmatrix} a_{1,1} & a_{1,2} & \cdots & a_{1,n} \\ a_{2,1} & a_{2,2} & \cdots & a_{2,n} \\ \vdots & \vdots & \ddots & \vdots \\ a_{n,1} & a_{n,2} & \cdots & a_{n,n} \end{bmatrix},$$

be an $n \times n$ square matrix. For any $k = 1, 2, \ldots, n$, the kth-order **principal minors** of A are $k \times k$ submatrices of A obtained by deleting $(n - k)$ rows of A and the corresponding $(n - k)$ columns.

Let:

$$A = \begin{bmatrix} a_{1,1} & a_{1,2} & \cdots & a_{1,n} \\ a_{2,1} & a_{2,2} & \cdots & a_{2,n} \\ \vdots & \vdots & \ddots & \vdots \\ a_{n,1} & a_{n,2} & \cdots & a_{n,n} \end{bmatrix},$$

be an $n \times n$ square matrix. For any $k = 1, 2, \ldots, n$, the kth-order **leading principal minor** of A is a $k \times k$ submatrix of A obtained by deleting the last $(n - k)$ rows and columns of A.

Example A.5 Consider the matrix:

$$A = \begin{bmatrix} e & e & 0 \\ e & e & 0 \\ 0 & 0 & 6 \end{bmatrix}.$$

This matrix has three first-order principal minors. The first is obtained by deleting the first and second columns and rows of A, giving:

$$A_1^1 = \begin{bmatrix} 6 \end{bmatrix},$$

the second:

$$A_1^2 = \begin{bmatrix} e \end{bmatrix},$$

is obtained by deleting the first and third columns and rows of A, and the third:

$$A_1^3 = \begin{bmatrix} e \end{bmatrix},$$

is obtained by deleting the second and third columns and rows of A. A also has three second-order principal minors. The first:

$$A_2^1 = \begin{bmatrix} e & 0 \\ 0 & 6 \end{bmatrix},$$

is obtained by deleting the first row and column of A, the second:

$$A_2^2 = \begin{bmatrix} e & 0 \\ 0 & 6 \end{bmatrix},$$

is obtained by deleting the second row and column of A, and the third:

$$A_2^3 = \begin{bmatrix} e & e \\ e & e \end{bmatrix},$$

is obtained by deleting the third column and row of A. A has a single third-order principal minor, which is the matrix A itself.

The first-order leading principal minor of A is:

$$A_1 = \begin{bmatrix} e \end{bmatrix},$$

which is obtained by deleting the second and third columns and rows of A. The second-order leading principal minor of A is:

$$A_2 = \begin{bmatrix} e & e \\ e & e \end{bmatrix},$$

which is obtained by deleting the third column and row of A. The third-order principal minor of A is A itself.

$\square$

Having the definition of principal minors and leading principal minors, we now state the Principal-Minor Test for determining if a matrix is definite.

Principal-Minor Test for Positive-Definite Matrices: Let A be an $n \times n$ square symmetric matrix. The matrix, A, is positive definite if and only if the determinants of all of its leading principal minors are strictly positive.

Principal-Minor Test for Positive-Semidefinite Matrices: Let A be an $n \times n$ square symmetric matrix. The matrix, A, is positive semidefinite if and only if the determinants of all of its principal minors are non-negative.

We demonstrate the use of Principal-Minor Test in the following example.

Example A.6 Consider the matrix:

$$A = \begin{bmatrix} e & e & 0 \\ e & e & 0 \\ 0 & 0 & 6 \end{bmatrix}.$$

The leading principal minors of this matrix are:

$$A_1 = \begin{bmatrix} e \end{bmatrix},$$

$$A_2 = \begin{bmatrix} e & e \\ e & e \end{bmatrix},$$

and:

$$A_3 = \begin{bmatrix} e & e & 0 \\ e & e & 0 \\ 0 & 0 & 6 \end{bmatrix},$$

which have determinants:

$$\det(A_1) = e,$$

$$\det(A_2) = 0,$$

and:

$$\det(A_3) = 0.$$

Because the determinants of the second- and third-order leading principal minors are zero, we can conclude by the Principal-Minor Test for Positive-Definite Matrices that A is not positive definite. To determine if A is positive semidefinite, we must check the determinants of all of its principal minors, which are:

$$\det(A_1^1) = 6,$$

$$\det(A_1^2) = e,$$

$$\det(A_1^3) = e,$$

$$\det(A_2^1) = 6e,$$

$$\det(A_2^2) = 6e,$$

$$\det(A_2^3) = 0,$$

and:

$$\det(A_3^1) = 0.$$

Because these determinants are all non-negative, we can conclude by the Principal-Minor Test for Positive-Semidefinite Matrices that A is positive semidefinite. This is consistent with our findings from directly applying the definition of a positive-semidefinite matrix in Example A.3 and from applying the Eigenvalue Test in Example A.4.

$\square$

Reference

1. Stewart J (2012) Calculus: early transcendentals, 7th edn. Brooks Cole, Pacific Grove

Appendix B
Convexity

Convexity is one of the most important topics in the study of optimization. This is because solving optimization problems that exhibit certain convexity properties is considerably easier than solving problems without such properties.

One of the difficulties in studying convexity is that there are two different (but slightly related) concepts of convexity. These are convex sets and convex functions. Indeed, a third concept of a convex optimization problem is introduced in Section 4.4. Thus, it is easy for the novice to get lost in these three distinct concepts of convexity. To avoid confusion, it is best to always be explicit and precise in referencing a convex set, a convex function, or a convex optimization problem.

We introduce the concepts of convex sets and convex functions and give some intuition behind their formal definitions. We then discuss some tests that can be used to determine if a function is convex. The definition of a convex optimization problem and tests to determine if an optimization problem is convex are discussed in Section 4.4.

B.1 Convex Sets

Before delving into the definition of a convex set, it is useful to explain what a set is. The most basic definition of a **set** is simply a collection of things (*e.g.*, a collection of functions, points, or intervals of points). In this text, however, we restrict our attention to sets of points. More specifically, we focus on the set of points that are feasible in the constraints of a given optimization problem, which is also known as the problem's feasible region or feasible set. Thus, when we speak about a set being convex, what we are ultimately interested in is whether the collection of points that is feasible in a given optimization problem is a convex set.

We now give the definition of a convex set and then provide some further intuition on this definition.

© Springer International Publishing AG 2017
R. Sioshansi and A.J. Conejo, *Optimization in Engineering*,
Springer Optimization and Its Applications 120, DOI 10.1007/978-3-319-56769-3

A set $X \subseteq \mathbb{R}^n$ is said to be a **convex set** if for any two points x^1 and $x^2 \in X$ and for any value of $\alpha \in [0, 1]$ we have that:

$$\alpha x^1 + (1 - \alpha) x^2 \in X.$$

Figure B.1 illustrates this definition of a convex set showing a set, X, (consisting of the shaded region and its boundary) and two arbitrary points, which are labeled x^1 and x^2, that are in the set. Next, we pick different values of $\alpha \in [0, 1]$ and determine what point:

$$\alpha x^1 + (1 - \alpha) x^2,$$

is. First, for $\alpha = 1$ we have that:

$$\alpha x^1 + (1 - \alpha) x^2 = x^1,$$

and for $\alpha = 0$ we have:

$$\alpha x^1 + (1 - \alpha) x^2 = x^2.$$

These two points are labeled in Figure B.1 as $\alpha = 1$ and $\alpha = 0$, respectively.

Next, for $\alpha = 1/2$, we have:

$$\alpha x^1 + (1 - \alpha) x^2 = \frac{1}{2}(x^1 + x^2),$$

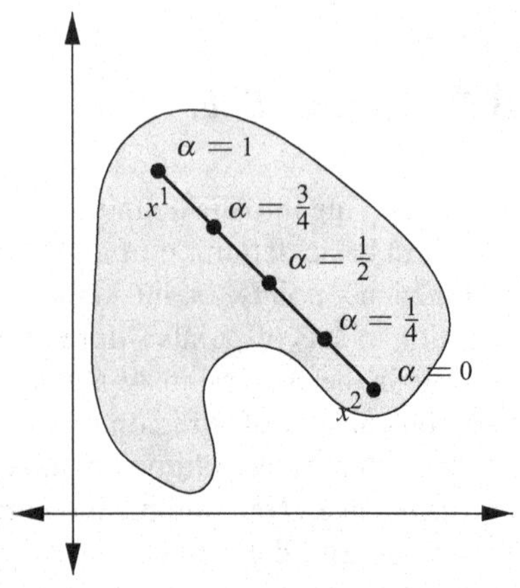

Fig. B.1 Illustration of definition of a convex set

which is the midpoint between x^1 and x^2, and is labeled in the figure. Finally, we examine the cases of $\alpha = 1/4$ and $\alpha = 3/4$, which give:

$$\alpha x^1 + (1 - \alpha)x^2 = \frac{1}{4}x^1 + \frac{3}{4}x^2,$$

and:

$$\alpha x^1 + (1 - \alpha)x^2 = \frac{3}{4}x^1 + \frac{1}{4}x^2,$$

respectively. These points are, respectively, the midpoint between x^2 and the $\alpha = 1/2$ point and the midpoint between x^1 and the $\alpha = 1/2$ point.

At this point, we observe a pattern. As we substitute different values of $\alpha \in [0, 1]$ into:

$$\alpha x^1 + (1 - \alpha)x^2,$$

we obtain different points on the line segment connecting x^1 and x^2. The definition says that a convex set must contain all of the points on this line segment. Indeed, the definition says that if we take *any* pair of points in X, then the line segment connecting those points must be contained in the set. Although the pair of points that is shown in Figure B.1 satisfies this definition, the set shown in the figure is not convex. Figure B.2 demonstrates this by showing the line segment connecting two other points, $\hat{x}^1$ and $\hat{x}^2$, that are in X. We see that some of points on the line segment connecting $\hat{x}^1$ and $\hat{x}^2$ are not in the set, meaning that X is not a convex set.

Fig. B.2 X is not a convex set

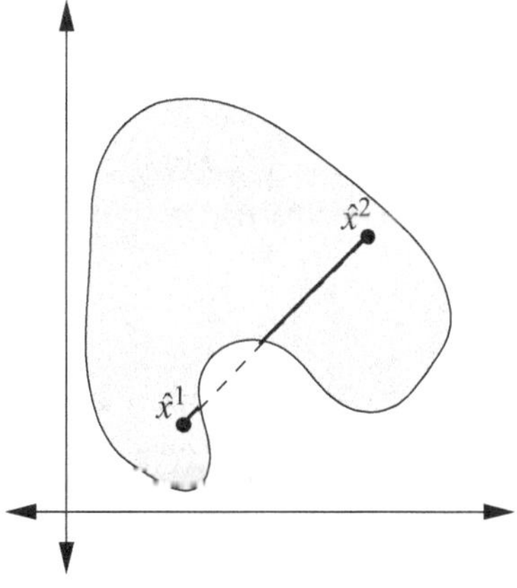

B.2 Convex Functions

With the definition of a convex set in hand, we can now define a convex function.

> Given a convex set, $X \subseteq \mathbb{R}^n$, a function defined on X is said to be a **convex function** on X if for any two points, x^1 and $x^2 \in X$, and for any value of $\alpha \in [0, 1]$ we have that:
>
> $$\alpha f(x^1) + (1 - \alpha) f(x^2) \geq f(\alpha x^1 + (1 - \alpha)x^2). \qquad \text{(B.1)}$$

Figure B.3 illustrates the definition of a convex function. We do this by first examining the right-hand side of inequality (B.1), which is:

$$f(\alpha x^1 + (1 - \alpha)x^2),$$

for different values of $\alpha \in [0, 1]$. If we fix $\alpha = 1$, $\alpha = 3/4$, $\alpha = 1/2$, $\alpha = 1/4$ and $\alpha = 0$ we have:

$$f(\alpha x^1 + (1 - \alpha)x^2) = f(x^1),$$

$$f(\alpha x^1 + (1 - \alpha)x^2) = f\left(\frac{3}{4}x^1 + \frac{1}{4}x^2\right),$$

$$f(\alpha x^1 + (1 - \alpha)x^2) = f\left(\frac{1}{2}(x^1 + x^2)\right),$$

$$f(\alpha x^1 + (1 - \alpha)x^2) = f\left(\frac{1}{4}x^1 + \frac{3}{4}x^2\right),$$

and:

$$f(\alpha x^1 + (1 - \alpha)x^2) = f(x^2).$$

Thus, the right-hand side of inequality (B.1) simply computes the value of the function, f, at different points between x^1 and x^2. The five points found for the values of $\alpha = 1$, $\alpha = 3/4$, $\alpha = 1/2$, $\alpha = 1/4$ and $\alpha = 0$ are labeled on the horizontal axis of Figure B.3.

Next, we examine the left-hand side of inequality (B.1), which is:

$$\alpha f(x^1) + (1 - \alpha) f(x^2),$$

for different values of $\alpha \in [0, 1]$. If we fix $\alpha = 1$ and $\alpha = 0$ we have:

$$\alpha f(x^1) + (1 - \alpha) f(x^2) = f(x^1),$$

Fig. B.3 Illustration of definition of a convex function

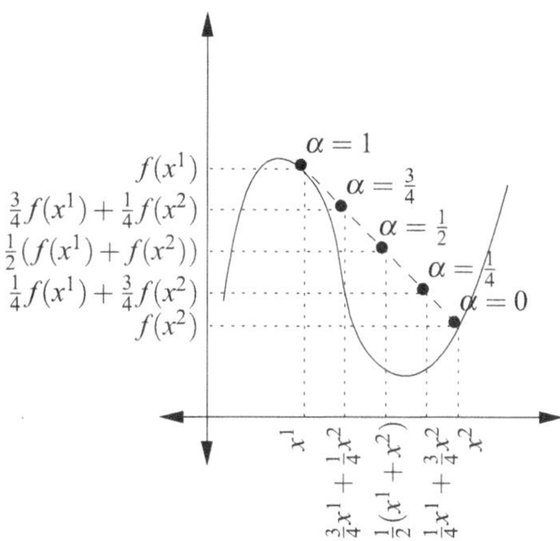

and:

$$\alpha f(x^1) + (1 - \alpha) f(x^2) = f(x^2),$$

respectively. These are the highest and lowest values labeled on the vertical axis of Figure B.3. Next, for $\alpha = 1/2$, we have:

$$\alpha f(x^1) + (1 - \alpha) f(x^2) = \frac{1}{2}(f(x^1) + f(x^2)),$$

which is midway between $f(x^1)$ and $f(x^2)$ and is the middle point labeled on the vertical axis of the figure. Finally, values of $\alpha = 3/4$ and $\alpha = 1/4$ give:

$$\alpha f(x^1) + (1 - \alpha) f(x^2) = \frac{3}{4} f(x^1) + \frac{1}{4} f(x^2),$$

and:

$$\alpha f(x^1) + (1 - \alpha) f(x^2) = \frac{1}{4} f(x^1) + \frac{3}{4} f(x^2),$$

respectively. These points are labeled as the midpoints between the $\alpha = 1$ and $\alpha = 1/2$ and $\alpha = 0$ and $\alpha = 1/2$ values, respectively, on the vertical axis of the figure. Again, we see a pattern that emerges here. As different values of $\alpha \in [0, 1]$ are substituted into the left-hand side of inequality (B.1), we get different values between $f(x^1)$ and $f(x^2)$. If we put the values obtained from the left-hand side of (B.1) above the corresponding points on the horizontal axis of the figure, we obtain the dashed line segment connecting $f(x^1)$ and $f(x^2)$. This line segment is known as the secant line of the function.

The definition of a convex function says that the secant line connecting x^1 and x^2 needs to be above the function, f. In fact, the definition is more stringent than that because *any* secant line that is connecting *any* two points needs to lie above the function. Although the secant line that is shown in Figure B.3 is above the function, the function that is shown in Figure B.3 is not convex. This is because we can find

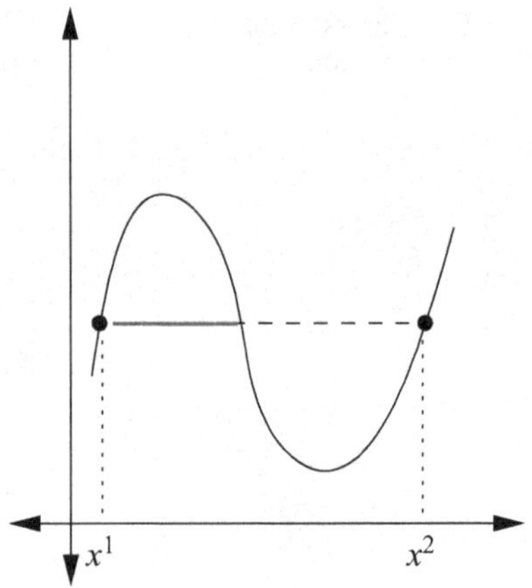

Fig. B.4 f is not a convex function

other points that give a secant line the goes below the function. Figure B.4 shows one example of a pair of points, for which the secant line connecting them goes below f.

We can oftentimes determine if a function is convex by directly using the definition. The following example shows how this can be done.

Example B.1 Consider the function:

$$f(x) = x_1^2 + x_2^2.$$

To show that this function is convex over all of $\mathbb{R}^2$, we begin by noting that if we pick two arbitrary points $\hat{x}$ and $\tilde{x}$ and fix a value of $\alpha \in [0, 1]$ we have that:

$$\begin{aligned}
f(\alpha\hat{x} + (1 - \alpha)\tilde{x}) &= (\alpha\hat{x}_1 + (1 - \alpha)\tilde{x}_1)^2 + (\alpha\hat{x}_2 + (1 - \alpha)\tilde{x}_2)^2 \\
&\leq (\alpha\hat{x}_1)^2 + ((1 - \alpha)\tilde{x}_1)^2 + (\alpha\hat{x}_2)^2 + ((1 - \alpha)\tilde{x}_2)^2 \\
&\leq \alpha\hat{x}_1^2 + (1 - \alpha)\tilde{x}_1^2 + \alpha\hat{x}_2^2 + (1 - \alpha)\tilde{x}_2^2,
\end{aligned}$$

where the first inequality follows from the triangle inequality and the second inequality is because we have that $\alpha, 1 - \alpha \in [0, 1]$ and as such $\alpha^2 \leq \alpha$ and $(1 - \alpha)^2 \leq 1 - \alpha$. Reorganizing terms, we have:

$$\begin{aligned}
\alpha\hat{x}_1^2 + (1 - \alpha)\tilde{x}_1^2 + \alpha\hat{x}_2^2 + (1 - \alpha)\tilde{x}_2^2 &= \alpha\hat{x}_1^2 + \alpha\hat{x}_2^2 + (1 - \alpha)\tilde{x}_1^2 + (1 - \alpha)\tilde{x}_2^2 \\
&= \alpha \cdot (\hat{x}_1^2 + \hat{x}_2^2) + (1 - \alpha)(\tilde{x}_1^2 + \tilde{x}_2^2) \\
&= \alpha f(\hat{x}) + (1 - \alpha)f(\tilde{x}).
\end{aligned}$$

Thus, we have that:

$$f(\alpha\hat{x} + (1 - \alpha)\tilde{x}) \leq \alpha f(\hat{x}) + (1 - \alpha)f(\tilde{x}),$$

for any choice of $\hat{x}$ and $\tilde{x}$, meaning that this function satisfies the definition of convexity.

$\square$

In addition to directly using the definition of a convex function, there are two tests that can be used to determine if a differentiable function is convex. In many cases, these tests can be easier to work with than the definition of a convex function.

> **Gradient Test for Convex Functions:** Suppose that X is a convex set and that the function, $f(x)$, is once continuously differentiable on X. $f(x)$ is convex on X if and only if for any $\hat{x} \in X$ we have that:
>
> $$f(x) \geq f(\hat{x}) + (x - \hat{x})^\top \nabla f(\hat{x}), \tag{B.2}$$
>
> for all $x \in X$.

To understand the gradient test, consider the case in which $f(x)$ is a single-variable function. In that case, inequality (B.2) becomes:

$$f(x) \geq f(\hat{x}) + (x - \hat{x})f'(\hat{x}).$$

The right-hand side of this inequality is the equation of the tangent line to $f(\cdot)$ at $\hat{x}$ (*i.e.*, it is a line that has a value of $f(\hat{x})$ at the point $\hat{x}$ and a slope equal to $f'(\hat{x})$). Thus, what inequality (B.2) says is that this tangent line must be below the function. In fact, the requirement is that *any* tangent line to $f(x)$ (*i.e.*, tangents at different points) be below the function.

This is illustrated in Figure B.5. The figure shows a convex function, which has all of its secants (the dashed lines) above it. The tangents (the dotted lines) are all below the function. The function shown in Figures B.3 and B.4 can be shown to be non-convex using this gradient property. This is because, for instance, the tangent to the function shown in those figures at x^1 is not below the function.

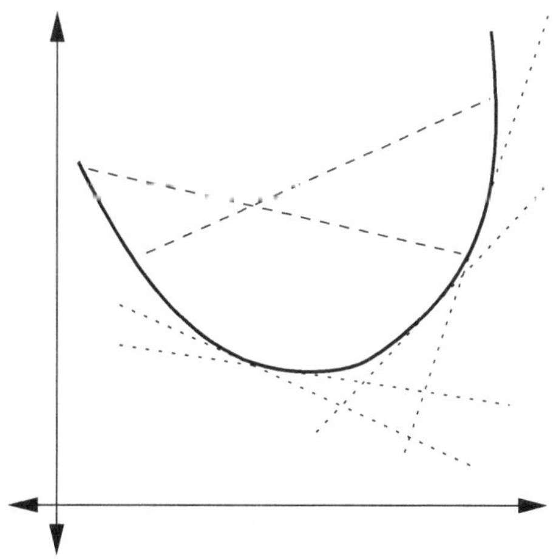

Fig. B.5 Illustration of Gradient Test for Convex Functions

Example B.2 Consider the function:

$$f(x) = x_1^2 + x_2^2,$$

from Example B.1. We have that:

$$\nabla f(\hat{x}) = \begin{pmatrix} 2\hat{x}_1 \\ 2\hat{x}_2 \end{pmatrix}.$$

Thus, the tangent to $f(x)$ at $\hat{x}$ is:

$$\begin{aligned} f(\hat{x}) + (x - \hat{x})^\top \nabla f(\hat{x}) &= \hat{x}_1^2 + \hat{x}_2^2 + \left(x_1 - \hat{x}_1 \;\; x_2 - \hat{x}_2 \right) \begin{pmatrix} 2\hat{x}_1 \\ 2\hat{x}_2 \end{pmatrix} \\ &= \hat{x}_1^2 + \hat{x}_2^2 + 2\hat{x}_1 \cdot (x_1 - \hat{x}_1) + 2\hat{x}_2 \cdot (x_2 - \hat{x}_2) \\ &= 2x_1\hat{x}_1 - \hat{x}_1^2 + 2x_2\hat{x}_2 - \hat{x}_2^2. \end{aligned}$$

Thus, we have that:

$$\begin{aligned} f(x) - [f(\hat{x}) + (x - \hat{x})^\top \nabla f(\hat{x})] &= x_1^2 - 2x_1\hat{x}_1 + \hat{x}_1^2 + x_2^2 - 2x_2\hat{x}_2 + \hat{x}_2^2 \\ &= (x_1 - \hat{x}_1)^2 + (x_2 - \hat{x}_2)^2 \\ &\geq 0, \end{aligned}$$

meaning that inequality (B.2) is satisfied for all x and $\hat{x}$. Thus, we conclude from the Gradient Test for Convex Functions that this function is convex, as shown in Example B.1. $\square$

We additionally have another condition that can be used to determine if a twice-differentiable function is convex.

> **Hessian Test for Convex Functions:** Suppose that X is a convex set and that the function, $f(x)$, is twice continuously differentiable on X. $f(x)$ is convex on X if and only if $\nabla^2 f(x)$ is positive semidefinite for any $x \in X$.

Example B.3 Consider the function:

$$f(x) = x_1^2 + x_2^2,$$

from Example B.1. We have that:

$$\nabla^2 f(x) = \begin{bmatrix} 2 & 0 \\ 0 & 2 \end{bmatrix},$$

which is positive definite (and, thus, positive semidefinite) for any choice of x. Thus, we conclude from the Hessian Test for Convex Functions that this function is convex, as also shown in Examples B.1 and B.2. $\square$

Index

C
Constraint, 1–3, 18, 129, 141, 347, 401
  binding, 111, 257
  logical, 129, 141
  non-binding, 111, 257
  relax, 152, 174, 306, 317, 325
Convex
  function, 225, 231, 404
  objective function, 231
  optimization problem, 219, 220, 232, 401
    optimality condition, 245, 250, 262
  set, 221, 402

D
Decision variable, 2, 3, 18, 343
  basic variable, 46, 50, 176
  binary, 123, 127, 139, 141
  feasible solution, 4
  index set, 22
  infeasible solution, 4
  integer, 123–125, 127, 138, 141
  non-basic variable, 46, 50, 176
Definite matrix, 394
  leading principal minor, 397
  positive definite, 394
  positive semidefinite, 394, 408
  principal minor, 396
Dynamic optimization problem, 11
  constraint, 347
  cost-to-go function, 366, 372
  decision policy, 366, 372
  decision variable, 343
  dynamic programming algorithm, 363
  objective-contribution function, 348
  stage, 343
  state variable, 343
    endogenous, 351, 355, 357, 358
    exogenous, 351, 352, 355
  state-transition function, 345, 368, 372
    state-dependent, 347
    state-independent, 346
    state-invariant, 346
  value function, 366, 372

F
Feasible region, 3, 19, 39, 125, 221, 401
  basic solution, 45
    artificial variable, 62, 67, 76
    basic feasible solution, 45
    basic infeasible solution, 46
    basic variable, 46, 50, 176
    basis, 51
    degenerate, 75
    non-basic variable, 46, 50, 176
    pivoting, 56
    tableau, 54
  bounded, 20, 41
  convex, 221
  extreme point, 21, 39, 174
  feasible solution, 4
  halfspace, 20
  hyperplane, 20, 21, 39, 222
  infeasible solution, 4
  polygon, 20
  polytope, 20, 39
  unbounded, 41
  vertex, 21, 39

H
Halfspace, 222
Hyperplane, 20, 21, 39, 222

© Springer International Publishing AG 2017
R. Sioshansi and A.J. Conejo, *Optimization in Engineering*,
Springer Optimization and Its Applications 120, DOI 10.1007/978-3-319-56769-3